Engineering Materials
Volume 1

Engineering Materials

Volume 1

Second Edition

R. L. Timings

An imprint of **Pearson Education**

Harlow, England · London · New York · Reading, Massachusetts · San Francisco
Toronto · Don Mills, Ontario · Sydney · Tokyo · Singapore · Hong Kong · Seoul
Taipei · Cape Town · Madrid · Mexico City · Amsterdam · Munich · Paris · Milan

Pearson Education Limited
Edinburgh Gate
Harlow
Essex CM20 2JE
England

and Associated Companies throughout the world

Visit us on the World Wide Web at:
www.pearsoned.co.uk

First published 1989
Second edition 1998
Reprinted 2001

British Library Cataloguing in Publication Data
A catalogue entry for this title is available from the British Library

ISBN-10: 0-582-31928-5
ISBN-13: 978-0-582-31928-8

7 6
09 08 07 06

Set by 32 in Times 9½/12 and Frutiger
Printed in Malaysia, PP

Contents

Preface

Engineering Materials, Volume 1 was originally written to provide a comprehensive text on engineering materials to satisfy the requirements of the Business and Technician Education Council (BTEC) standard units for students studying for an engineering qualification at the OND/ONC ('N' level).

Whilst still satisfying the original aims, this new edition has been rewritten to include additional material in order to give comprehensive coverage of the materials requirements for the Advanced GNVQ Engineering qualification. For example, there is now a chapter on material selection for engineered products. To make room for this additional material, the chapter on bearings and bearing materials has been moved to the new edition of *Engineering Materials*, Volume 2. At the same time, the opportunity has been taken to:

- Make the language less formal and more 'reader friendly'.
- Expand some of the explanations to improve their clarity.
- Correct some errors and omission.
- Modernise the general format of the book.
- Introduce self-assessment exercises at key points in the text.
- Make more use of bulleted summaries.

Each chapter is now prefaced by a summary of the topic areas to be covered and finishes with a selection of practice exercises.

This book leads naturally into *Engineering Materials*, Volume 2, intended for students studying at the HNC/HND level and beyond. The broad coverage of *Engineering Materials*, Volumes 1 and 2 ensures that they not only satisfy the requirements of technician engineers up to the highest level, but also provide an excellent technical background for undergraduates studying for a degree in Mechanical Engineering, Materials Engineering or Combined Engineering.

R. L. Timings
1998

Acknowledgements

The author and publishers are grateful to the following for permission to reproduce copyright material:

Hodder and Stoughton for our Fig. 2.14 (from *Engineering Metallurgy* by R. A. Higgins); The Castings Development Centre for our Figs 6.2, 6.3, 6.4, 6.6, 6.7 & 6.8; B & S Massey Ltd for our Figs 10.15 & 10.16; Davey-Loewy Ltd for our Fig. 10.18; Sir James Farmer Norton & Co. (International) Ltd for our Fig. 10.21; Samuel Denison Ltd (Denison Mays Group) for our Figs 11.1(a), 11.17, 11.26 & 11.27; TecQuipment Ltd for our Fig. 11.1(b).

Extracts from British Standards Institution publications (our Figs 6.9, 6.10, 6.11 & 14.6 and Table 6.4) are reproduced with the permission of BSI. Complete copies can be obtained by post from BSI Customer Services, 389 Chiswick High Road, London W4 4AL; Telephone: 0181 996 7000.

Introduction – How to use this book

There are many ways of using a text book. You can read through it from cover to cover and try to remember all that you have read. This is rarely successful unless you have a photographic memory, and few of us have. You may consider using it just as a reference book by looking up individual topics as and when you need that specific information. This can be useful as a reminder once you know the subject thoroughly but until then it can lead to misconceived ideas.

So, here are a few thoughts on how to maximise the benefit that you can get from this book.

This book consists of a number of chapters. Each chapter covers a major syllabus area. For example, **Chapter 4** covers plain carbon steels. It is divided up into sections. If you turn back to the Contents page you will find that:

- **Section 4.1** deals with ferrous metals.
- **Section 4.2** deals with the iron–carbon system.
- **Section 4.3** deals with critical change points, and so on.

Sometimes it is necessary to divide these sections up further. For example, Section 4.2 subdivides into:

- **Section 4.2.1**, which deals with cooling transformations for a steel with a eutectoid composition.
- **Section 4.2.2**, which deals with cooling transformations for a steel with hypo-eutectoid composition, and so on.

Don't let the technical jargon words put you off! To understand this chapter you need to have some prior knowledge of basic science, material properties, phase equilibrium diagrams, etc. So, where do you get this prior knowledge and how is the book organised for you to obtain the maximum benefit from it?

It may seem obvious, but start at the beginning. The first two chapters lay the foundations for all that is to follow. As with all the chapters, these two chapters are prefaced with lists of the main topic areas you are going to find in them.

As you work through the chapters, you will come across **self-assessment tasks** at key points. If you have understood what you have read previously, you should be able to complete these exercises *without looking back* at the text. If you have to look back, then you are not yet sure of your ground and there is no point in moving on. Try again and, if you are still not sure, have a chat with your tutor to clear up your difficulty.

At the end of each chapter there is a selection of more extended **exercises**. Your tutor will guide you as to their use and check your responses to ensure that you have the background knowledge required to understand the next topic area to be studied.

Having completed Chapters 1 and 2, you have to decide where to go next:

- If your interest lies in metals you will need Chapter 3 as well.
- For plain carbon steels and their heat treatment you will also require Chapters 4 and 5, but if you want cast irons you can go straight to Chapter 6.
- On the other hand, if you want non-ferrous metals and their alloys you can go to Chapter 7.
- If your interest lies in polymeric (plastic) materials you can turn to Chapters 8 and 9 after completing Chapters 1 and 2.

Whether you work through the book systematically from beginning to end, or through the foundation chapters and then move on to one of the specialist areas, you must complete the self-assessment tasks and end-of-chapter exercises as you come to them. This will ensure that you will understand the next stage of your journey through your study of engineering materials.

Although, as just indicated, this book is organised so that you can take short-cuts through the text to specific areas of interest, it is strongly recommended that, given the time, you should work through the whole book from beginning to end. This is because all engineers and materials scientists should have a general, background knowledge of materials before specialising in a particular field.

Finally, once you have qualified, you can still obtain benefit from this book. The numbering of the sections and subsections, the comprehensive list of contents and the extended index, all aid you to use this text as a quick reference book in your future career in engineering.

1 Introduction to materials and their properties

The topic areas covered in this chapter are:

- How material properties affect the selection of materials for any given application.
- The basic properties of materials and how these affect the performance of materials in service.
- The grouping of materials into their main categories.

1.1 Introduction

Since the earliest days of the evolution of mankind, the main distinguishing feature between human beings and other mammals has been the ability to use and develop materials to satisfy our human requirements. Initially this was to fashion simple weapons for hunting for food and for territorial defence. Then came tools and implements to cultivate the land and to manufacture the artefacts (such as cooking utensils) required for more civilised living. Woven cloths took the place of animal skins, and manufactured goods became increasingly more sophisticated. Nowadays we use many types of materials, fashioned in many different ways, to satisfy our requirements for housing, heating, furniture, clothes, transportation, entertainment, medical care, defence and all the other trappings of a modern, civilised society.

In the past, mankind looked upon the material resources of the planet Earth as being limitless and these resources were exploited with no thought for the future. However, the rapidly expanding world population and the mass production techniques required to satisfy its needs are using up what are now understood to be finite resources at an alarming rate. Therefore conservation of these resources is an ever-increasing and important issue that must be addressed. Wherever possible waste materials must be reclaimed and recycled, and materials from renewable sources must be used.

Engineers can play an important part in this conservation of material resources by a thorough understanding of the materials they use. This understanding is necessary in order to enable them select the most appropriate materials and to use them with the greatest efficiency, in minimum quantities whilst causing minimum pollution in their extraction, refinement and manufacture.

1.2 Selection of materials

Let's now start by looking at the basic requirements for selecting materials that are suitable for a particular application. This topic area will be returned to in the final chapter of this book in greater detail when all the requirements have been covered. Figure 1.1 shows three objects. The first is a connector joining electric cables. The plastic casing has been partly cut away to show the metal connector. Plastic is used for the outer casing because it is a good electrical insulator and prevents electric shock if a person touches it. It also prevents the conductors touching each other and causing a short circuit. As well as being a good insulator the plastic is cheap, tough, and easily moulded to shape. It has been selected for the casing because of these properties – that is, the properties of toughness, good electrical insulation, and ease of moulding to shape. It is also a relatively low cost material that is readily available.

The metal jointing piece and its clamping screws are made from brass. This metal has been chosen because of its special properties. These properties are: good electrical conductivity, ease of extruding to shape, ease of machining (cutting to length, drilling and tapping the screw threads), adequate strength and corrosion resistance. The precious metal silver is an even better conductor, but it would be far too expensive for this application and it would also be too weak and soft. It is also a relatively scarce material.

The second object in Fig. 1.1 is the connecting rod of a motor car engine. This is made from a special steel alloy. This alloy has been chosen because it combines the properties of strength and toughness with the ability to be readily forged to shape and finished by machining.

The third object shown in Fig. 1.1 is part of a machine tool. It is the tailstock casting for a small lathe. The metal used in this example is cast iron. This metal has been chosen because it combines the properties of adequate strength with ease of melting and casting to a complicated shape. It is also relatively easy to machine to its finished dimensions.

Fig. 1.1 *Material selection*

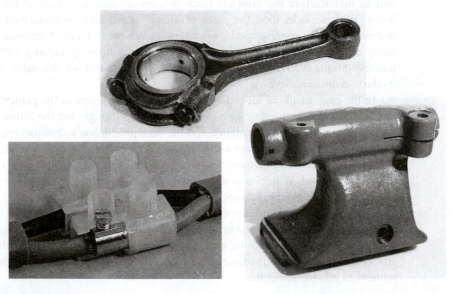

Thus the reasons for selecting the materials in the above example can be summarised as:

- Commercial factors such as:
  - (a) cost
  - (b) availability
  - (c) ease of manufacture
- Engineering properties of materials such as:
  - (a) Electrical conductivity
  - (b) strength
  - (c) toughness
  - (d) ease of forming by extrusion, forging and casting
  - (e) machinability
  - (f) corrosion resistance, etc.

1.3 Mechanical properties of materials

Let's now consider the principal properties of materials which are of importance to the engineer in selecting materials. These can be broadly divided into the *mechanical properties* which are concerned mainly with strength, and the *physical properties* which are concerned with such properties as melting temperature, density, electrical conductivity, thermal conductivity, etc. We will consider the mechanical properties first.

1.3.1 *Tensile strength*

This is the ability of a material to withstand tensile (stretching) loads without breaking. Figure 1.2 shows a heavy load being held up by a rod fastened to a beam. As the force of gravity acting on the load is trying to *stretch* the rod, the rod is said to be in *tension*. Therefore, the material from which the rod is made needs to have sufficient *tensile strength* to resist the pull of the load.

Fig. 1.2 *Tensile strength*

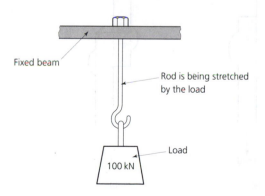

1.3.2 *Compressive strength*

This is the ability of a material to withstand compressive (squeezing) loads without being crushed or broken. Figure 1.3 shows a component being compressed by a heavy load. Therefore the component needs to be made from a material with adequate *compressive strength* to resist the load.

The applied *stress* is often used instead of the load when comparing the tensile and compressive strengths of materials. Stress is the applied force (load) divided by the cross-sectional area of the component measured at right angles to the direction of application of the force – that is, force per unit area. The units may be N/mm^2 or MN/m^2. The numerical value is the same in both cases; for example, $12\,N/mm^2$ equals $12\,MN/m^2$ (see also Section 11.3).

Fig. 1.3 *Compressive strength*

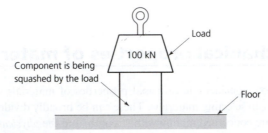

1.3.3 *Shear strength*

This is the ability of a material to withstand offset loads, or transverse cutting (shearing actions). Figure 1.4(a) shows a rivet joining two metal bars together. The forces acting on the two bars are trying to pull them apart. Because the loads are not exactly in line, they are

Fig. 1.4 *Shear strength*

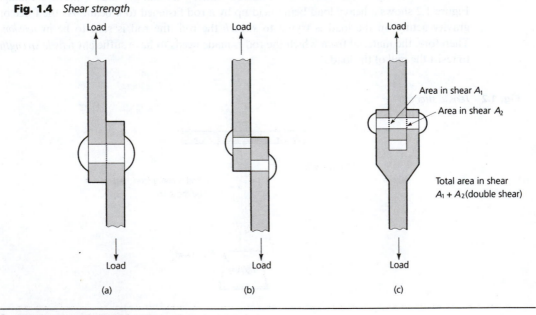

said to be *offset* and, therefore, the load on the rivet is called a *shearing load*; that is, the rivet is said to be in *shear*. If the rivet material does not have sufficient shear strength to resist the loads, the rivet will break (shear off) as shown in Fig. 1.4(b) and the bars and the loads acting on them will move apart. The same effect can be caused by loads pushing on the ends of the two metal bars joined by the rivet. Again we can use stress but this time it will be *shear stress* and again it is force per unit area. Only this time, it is the cross-sectional area of the rivet that is used. The units are the same as those given previously. The value for the shear stress for any given material is not the same as the value for tensile stress or for the compressive stress for that material. The rivet may be in single shear as shown in Fig. 1.4(a) and 1.4(b) or it may be in double shear as in Fig. 1.4(c). When in double shear the area in shear is effectively doubled for calculation purposes.

1.3.4 *Toughness (impact resistance)*

This is the ability of a material to withstand shatter. If a material shatters it is brittle (e.g. glass). Rubbers and most plastic materials do not shatter, therefore they are tough. Toughness should not be confused with strength. Figure 1.5 shows a metal rod in a vice being broken by impact loading. If the rod is made from a piece of high-carbon steel – for example, silver steel in the annealed (soft) condition (as normally supplied) – it will have only a moderate tensile strength, but under the impact of the hammer it will bend without breaking, therefore it is *tough*. If a similar specimen is made hard by making it red hot and quenching it (cooling it quickly in water), it will now have a very much higher tensile strength. However, although it is now stronger it will prove to be brittle and will break off easily when struck with a hammer. Therefore it now *lacks toughness*. A material is brittle and shatters because small cracks in its surface grow quickly under a tensile force. (As a material bends its outer layers tend to stretch and are in tension.) Therefore, any material in which the spread of surface cracks does not occur or only occurs to a small extent is said to be *tough*.

Fig. 1.5 *Impact strength*

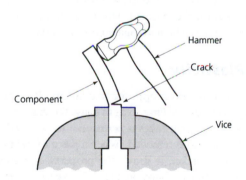

1.3.5 *Elasticity*

This is the ability of a material to deform under load and return to its original size and shape when the load is removed. Figure 1.6 shows a tensile test specimen. If it is made from an elastic material it will be the same length before and after the load is applied, despite the fact that it will be longer whilst the load is being applied. This is only true, for most

Fig. 1.6 *Elasticity*

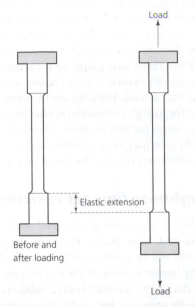

Load

Elastic extension

Before and
after loading

Load

materials, if the load is relatively small and within the elastic range of the material being tested. All materials possess elasticity to some degree and each has its own *elastic limit*. If stressed beyond this limit, permanent deformation (plastic deformation), and ultimately fracture, occurs. Materials are only stressed within the elastic range under normal service conditions.

At stress levels below the elastic limit the amount of deformation is directly proportional to the applied force which may be tensile or compressive. That is, if a tensile load on a component is doubled, the amount the component stretches will be doubled, providing, of course, that the stress produced by the load does not exceed the stress value at the elastic limit for the material. This relationship is known as 'Hooke's Law' after Robert Hooke (1635–1703) who investigated the property of elasticity in materials.

1.3.6 *Plasticity*

This property is the exact opposite to elasticity. It is the state of a material which has been loaded *beyond its elastic limit* so as to cause the material to *deform permanently*. Under such conditions the material takes a *permanent set* and will not return to its original size and shape when the load is removed. When a piece of mild steel strip is bent at right angles into the shape of a bracket, it shows the property of plasticity since it does not spring back straight again. This is shown in Fig. 1.7. *Ductility* and *malleability* are particular cases of the property of plasticity and they will now be considered separately.

Some metals such as lead have a good plastic range at room temperature and can be extensively worked. (In this context, the *working* of metal means squeezing, stretching or beating it to shape. See also Section 1.8.) This is an advantage for plumbers when beating lead flashings to shape on building sites. Other metals such as mild steel and copper can also be cold worked but to a more limited extent and only after annealing to soften them.

Fig. 1.7 *Plasticity*

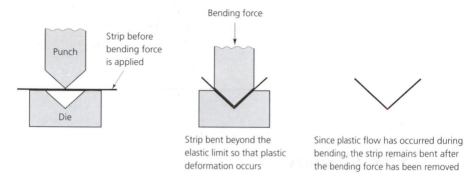

Strip before bending force is applied

Strip bent beyond the elastic limit so that plastic deformation occurs

Since plastic flow has occurred during bending, the strip remains bent after the bending force has been removed

They tend to harden whilst being formed to shape and may need to be annealed from time to time during processing. Annealing is a heat-treatment process and is described in Chapter 5 of this book.

1.3.7 *Ductility*

This is the term used when *plastic deformation* occurs as the result of applying a *tensile load*. A ductile material allows a useful amount of plastic deformation to occur under tensile loading before fracture occurs. Such a material is required for manipulation by such processes as wire drawing (Fig. 1.8), tube drawing, and cold pressing low-carbon steel sheets into motor car body panels.

Fig. 1.8 *Ductility*

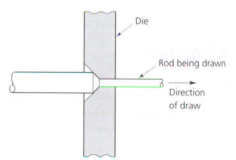

1.3.8 *Malleability*

This is the term used when *plastic deformation* occurs as the result of applying a *compressive load*. A malleable material allows a useful amount of plastic deformation to occur under compressive loading before fracture occurs. Such a material is required for manipulation by such processes as forging, rolling, and rivet heading, as shown in Fig. 1.9.

Fig. 1.9 *Malleability*

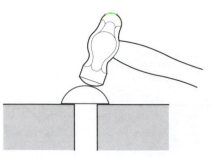

1.3.9 *Hardness*

This is defined as the ability of a material to withstand scratching (abrasion) or indentation by another hard body. It is an indication of the wear resistance of the material. Figure 1.10 shows a hardened steel ball being pressed first into a hard material and then into a soft material by the same load. The ball only makes a small indentation in the hard material, but it makes a very much deeper impression in the softer material. Hardness is often tested in this manner – see Section 11.11. Measurement of this property can only be relative to other materials and is given in the form of a *hardness number* with no units.

Fig. 1.10 *Hardness*

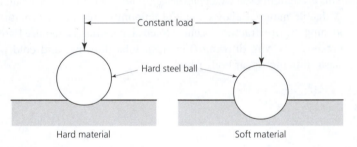

1.3.10 *Brittleness*

This is usually regarded as the opposite of ductility and malleability. It is the property of a material that shows little or no plastic deformation before fracture when a force is applied. For example, a steel rod can be bent but a grey cast iron rod snaps when you try to bend it. Therefore grey cast iron is a *brittle material*.

1.3.11 *Rigidity (stiffness)*

This is a measure of a material's ability not to deflect under an applied load. For example, although steel is very much stronger than cast iron the latter material is preferred for machine beds and frames because it is more rigid and less likely to deflect with consequent loss of alignment and accuracy. Consider Fig. 1.11(a): for a given load the cast iron bar deflects less than the steel bar because cast iron is a more rigid material. However, when the load is increased, as shown in Fig. 1.11(b), the cast iron bar will break, whilst the steel bar

Fig. 1.11 *Rigidity: (a) under a light load cast iron deflects less than steel since it is more rigid; (b) under a heavy load the cast iron breaks whilst the stronger but less rigid steel merely bends further*

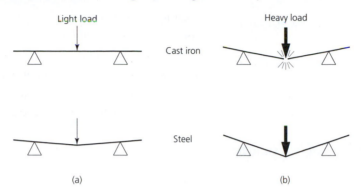

merely deflects a little further but does not break. Thus a material which is rigid is not necessarily strong. In fact, as has been shown, the opposite is more often true since rigid materials are often brittle.

SELF-ASSESSMENT TASK 1.1

1. State the main property required in each case by a material used in the manufacture of:

    (a) coins
    (b) copper wire
    (c) steel girders
    (d) the front wheel axle of a car
    (e) a cutting tool

1.4 Physical properties of materials

In addition to the mechanical properties just discussed, the design of an engineered product may require other properties such as electrical or thermal conductivity to be taken into account. Other properties of importance are: density, melting temperature, chemical stability (resistance to corrosion), magnetic properties, etc. These are referred to as the 'physical properties' and the more important of these will now be considered. Corrosion resistance will be considered in Chapter 13.

1.4.1 *Density*

Density is defined as mass per unit volume for a material. The derived unit usually used by engineers is the kg/m^3. Relative density is the density of a material compared with the density of pure water at $4\,°C$. The formulae for density and relative density are:

$$\text{density } (\rho) = \frac{\text{mass } (m)}{\text{volume } (V)}$$

$$\text{relative density } (d) = \frac{\text{density of the material}}{\text{density of pure water at } 4\,°C}$$

$$\text{or} \quad \text{relative density } (d) = \frac{\text{mass of a substance}}{\text{mass of an equal volume of pure water}}$$

Table 1.1 lists the densities and relative densities for a number of common substances. Note that relative densities do not have units since they are a ratio and are dimensionless.

Table 1.1 *Density*

Substance	Density (ρ) (kg/m^3)	(g/cm^3)	Relative density (d)
Aluminium	2 720	2.72	2.72
Brass	8 480	8.48	8.48
Cadmium	8 570	8.57	8.57
Chromium	7 030	7.03	7.03
Coal	1 440	1.44	1.44
Copper	8 790	8.79	8.79
Iron (cast)	7 200	7.20	7.20
Iron (wrought)	7 750	7.75	7.75
Lead	11 350	11.35	11.35
Nickel	8 730	8.73	8.73
Nylon	1 120	1.12	1.12
PVC	1 360	1.36	1.36
Rubber	960	0.96	0.96
Steel	7 820	7.82	7.82
Tin	7 280	7.28	7.28
Zinc	7 120	7.12	7.12
Alcohol	800	0.80	0.80
Mercury	13 590	13.59	13.59
Paraffin	800	0.80	0.80
Petrol	720	0.72	0.72
Water (pure)	1 000	1.00	1.00
Water (sea)	1 020	1.02	1.02
Acetylene	1.17	0.001 17	0.001 17
Air	1.30	0.001 3	0.001 3
Carbon dioxide	1.98	0.001 98	0.001 98
Carbon monoxide	1.26	0.001 26	0.001 26
Hydrogen	0.09	0.000 09	0.000 09
Nitrogen	1.25	0.001 25	0.001 25
Oxygen	1.43	0.001 43	0.001 43

Note: The figures for gases are at $0\,°C$ (273 K), 101,325 Pa (i.e. standard temperature and pressure).

1.4.2 *Melting temperature of materials*

Table 1.2 lists the melting temperatures and recrystallisation temperatures for some metals and alloys used in engineering. Note that only pure substances have a single melting temperature. Alloys of metals have a melting 'range' of temperatures and this will be considered later.

Table 1.2 *Melting and recrystallisation temperatures*

Material	Recrystallisation temperature (°C)	Melting point (°C)
Tungsten	1 200	3 410
Molybdenum	900	2 620
Nickel	600	1 458
Iron	450	1 535
Brasses	400	900–1 050
Bronzes	400	900–1 050
Copper	200	1 083
Silver	200	960
Aluminium	150	660
Magnesium	150	651
Zinc	70	419
Lead	20	327
Tin	20	232

1.4.3 *Electrical Conductivity*

Figure 1.12 shows a piece of electrical cable. In this example copper wire has been chosen for the conductor or core of the cable because copper has the property of very good electrical conductivity. That is, it offers very little resistance to the flow of electrons (electric current) through the wire. A plastic material such as polymerised vinyl chloride (PVC) has been chosen for the insulating sheathing surrounding the wire conductor. This material has been chosen because it is such a bad conductor very few electrons can pass through it. *Very bad conductors* such as PVC are called *insulators*. There is no such thing as a perfect insulator, only very bad conductors.

Care must be taken when comparing and interpreting tables of test data. For example, metallic conductors of electricity all increase in resistance as their temperatures rise. Pure

Fig. 1.12 *Electrical conductivity*

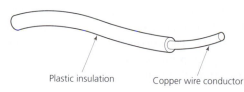

Plastic insulation   Copper wire conductor

metals show this effect more strongly than alloys. However, pure metals generally have a better conductivity than alloys at room temperature. The conductivity of metals and metal alloys improves as the temperature falls. Conversely, non-metallic materials used for insulators tend to offer a lower resistance to the passage of electrons, and so become poorer insulators, as their temperatures rise. Thus care must be taken if cables are bunched together without adequate ventilation. Glass, for example, is an excellent insulator at room temperature, but becomes a conductor if raised to red heat. Table 1.3 lists the properties of typical conductor materials, and Table 1.4 lists the properties of typical insulating materials. Reference should also be made to British Standard Specification BS 5714.

Resistivity is defined as the resistance offered by a unit cube of that material measured between opposite faces.

Resistivity (ρ) is usually given in units of ohm-metres (Ωm). That is the resistance between opposite faces of a metre cube of the material measured in ohms. However, in practice, it is often more convenient to work in micro-ohm-millimetres ($\mu\Omega$mm). For example, the resistivity of nickel is $0.000\,000\,136\,\Omega$m which is better written as $136\,\mu\Omega$mm. This latter unit is used in Table 1.3.

Table 1.3 *Properties of conductor materials*

Material	Resistivity ($\mu\Omega$mm)	Temperature coefficient α_0 (per °C)
Aluminium	28.0	42×10^{-4}
Carbon (graphitic)*	46.0×10^3	-5×10^{-4}
Copper (annealed)	17.2	43×10^{-4}
Mild steel	107.0	65×10^{-4}
Manganin alloy†	480.0	0
Nichrome alloy‡	1 090.0	53×10^{-4}
Nickel	136.0	56×10^{-4}
Silver (annealed)	15.8	41×10^{-4}

* The negative sign in the temperature coefficient column indicates a fall in resistance as temperature increases.
† Manganin is an alloy of copper, manganese and nickel and is used for wire wound resistors for use in measuring instruments where its zero value of temperature coefficient means that the ohmic value of such resistors are unaffected by temperature change.
‡ Nichrome is an alloy of nickel and chromium which resists oxidation at high temperatures. It is used for heating elements in electric radiators and furnaces.

1.4.4 *Semiconductors*

So far we have examined the conductivity of metals and the insulating properties of the non-metals (exception: carbon). In between conductors and insulators lies a range of materials known as *semiconductors*. These can be good or bad conductors depending upon their temperatures. The conductivity of semiconductor materials increases rapidly for relatively small temperature increases. This enables them to be used as temperature sensors in electronic thermometers.

Table 1.4 *Properties of insulating materials*

Material	Typical applications	Properties
Insulating oil	Switch gear, transformers, cable impregnation	Reduces arcing between switchgear contacts. Being a fluid insulator it can circulate by convection and cool the windings and cores of large transformers as well as insulating them. Used to impregnate the paper insulation in underground armoured mains cables.
Paper	Armoured mains cables	A good, relatively cheap, insulator for rigid cables. Must be oil impregnated and sealed against ingress of moisture which causes the oil to break down.
Rubber (natural)	Flexible cables	Flexible, high insulation resistance, reasonable mechanical properties when 'vulcanised' with 5% sulphur. Degrades rapidly (perishes) in strong sunlight and undue heat (max 55 °C). Sulphur content attacks copper conductors which, therefore, must be tinned.
Silicone rubber (synthetic)	Flexible cables	Similar to rubber, but suitable for applications up to 150 °C, very much more expensive.
Polyvinyl chloride (PVC)	Flexible cables	Although PVC has a much lower insulation resistance than rubber it is now more widely used because of its low cost and resistance to oils, petrol and chemical solvents. Max operating temperature 65 °C.
Mineral insulation (magnesium oxide)	Rigid metal sheathed (MIMS) cables sheathed heating elements	Can operate at temperatures up to dull-red heat. Limited to 660 volts. Terminations must be sealed as magnesium oxide powder is hygroscopic.
Thermosetting plastics	Moulded insulators (interior)	(a) Phenolic resins (Bakelite) used for moulded insulation blocks and switchgear components where strength is important and its dark colour acceptable. (b) Amino resins are used for domestic switchgear mouldings as it is available in white and light colours. Lower strength than phenolic resin.
Glass and ceramics	Moulded insulators (exterior)	Hard, highly glazed surface, prevents weathering. Weak in tension and shear, insulators must be designed to operate under compressive mechanical loads only. Used for high-voltage insulators for overhead transmission lines. High insulation resistance. Woven glass fibre (resin varnish impregnated) used for high-temperature insulation and sheathing in domestic cookers.

Semiconductor materials are capable of having their conduction properties changed during manufacture. Examples of semiconductor materials are silicon, germanium and certain metal oxides. They are used extensively in the electronics industry in the manufacture of solid-state devices such as diodes, thermistors, transistors and integrated circuits. These devices use specially treated (doped) crystals of semiconductor materials and their importance lies in their ability to control the current flow in a circuit and to provide signal amplification (see also Section 2.14). Semiconductor materials are dealt with in detail in *Engineering Materials*, Volume 2.

1.4.5 *Permittivity*

The branch of electricity called *electrostatics* studies electrical charges at rest. An electrostatic field accompanies a static charge and this phenomenon is exploited in devices known as capacitors. A *dielectric* (insulating) material is used to separate the charged surfaces of the electrodes (plates) of the capacitor. The ratio of the electric flux density in the dielectric to the electric field strength of the charge is called the *permittivity*. Permittivity values are dimensionless (they are ratios) and therefore carry no units. Some examples are listed in Table 1.5. The relevant British Standard Specification is BS 7663.

Table 1.5 *Typical permittivity values*

Material	Permittivity
Air	1
Polythene	2–3
Mica	3–7
Glass	5–10
Ceramics	6–1000

1.4.6 *Magnetic properties*

Reluctance

Just as some materials are good or bad conductors of electricity, some materials can be good or bad conductors of magnetism. The resistance of a *magnetic circuit* is referred to as *reluctance*. The good magnetic conductors have a low reluctance and examples are the ferromagnetic materials which get their name from the fact that they are made from iron, steel and associated alloying elements such as cobalt and nickel. All other materials are non-magnetic and offer a high reluctance to the magnetic flux field.

Permeability

This is a constant for any given material and, like all constants, has no units. It can be considered as the ease with which a material can be magnetised. Typical permeability curves for a nickel–iron alloy, a silicon–iron alloy and a soft iron are shown in Fig. 1.13. The absolute permeability (μ) for any material is the permeability for free space (μ_0) multiplied by the relative permeability (μ_r) for the material. You will notice that the permeability varies with the magnetising force. It reaches a peak and then falls again. This fall is the result of magnetic saturation and care must be taken not to 'overload' a magnetic circuit for a given material.

Fig. 1.13 *Permeability curves*

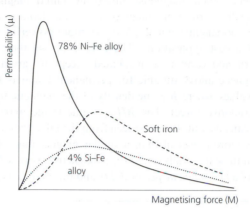

Hysteresis

Magnetic materials subjected to alternating fields as in, say, a transformer core are subjected to constantly changing conditions. They are magnetised in one direction, then demagnetised, then magnetised again in the opposite direction – this cycle being repeated many times per second depending upon the frequency of the magnetising current. Magnetic materials require time to adapt to these changes and tend to lag behind the current change. This lagging effect is referred to as *hysteresis* and Fig. 1.14 shows two typical hysteresis loops. Figure 1.14(a) is for a typical 'hard' magnetic material as used in a permanent magnet, whilst 1.14(b) is for a typical 'soft' magnetic material as used in transformer cores. Hysteresis causes energy loss which ends up as heat in the core of the transformer. The 'fatter' the curve the greater the losses. Therefore a material with a 'thin' hysteresis curve should be used for transformer cores and fluorescent light choke cores.

Fig. 1.14 *Magnetic hysteresis loops for (a) 'hard' and (b) 'soft' magnetic materials*

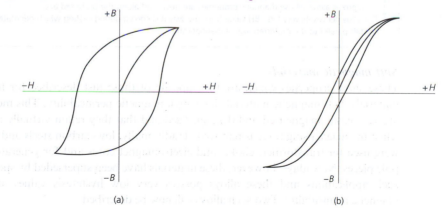

Classification

Magnetic materials can be classified as follows.

Hard magnetic materials

These retain their magnetism after the initial magnetising force has been removed. Traditionally, permanent magnets were made from quench-hardened high-carbon steels. Modern permanent magnet alloys are more powerful and are used for all but the most simple low-cost applications. These materials get their name from the fact that they are very hard and cannot be machined except by grinding. Table 1.6 lists some typical ferromagnetic alloys suitable for permanent magnets together with their corresponding BH_{max} values, where B is the density of the magnetic flux and H is the magnetising force (magnetomotive force). The BH_{max} value is the maximum magnetic energy which the magnet can give out. It can be seen from the table that an alloy such as 'Columax' is nearly 34 times more powerful than a similar magnet made from high-carbon steel. However, although these materials retain their magnetism very well once they have been magnetised, they have a *low magnetic permeability* and this makes them difficult to magnetise initially.

Table 1.6 *Permanent magnetic materials*

Name	Composition (%)*										BH_{max}
	C	Cr	W	Co	Al	Ni	Cu	Nb	Ti	Fe	
Quench-hardened high-carbon steel	1.0	—	—	—	—	—	—	—	—		1 560
35% cobalt steel	0.9	6.0	5.0	35.0	—	—	—	—	—		7 800
Alnico				12.0	9.5	17.0	5.0	—	—	Remainder	13 500
Alcomax III†				24.5	8.0	13.5	3.0	0.6	—		38 000
Hycomax III†				34.0	7.0	15.0	4.0	—	5.0		35 200
Columax‡				24.5	8.0	13.5	3.0	0.6	—		52 800

* C = carbon, Cr = chromium, W = tungsten, Co = cobalt, Al = aluminium, Ni = nickel, Cu = copper, Nb = niobium, Ti = titanium, Fe = iron.
† Anisopropic alloys whose magnetic properties are measured along the preferred axis.
‡ This alloy derives its very high BH value from the way it is cooled during casting which orientates its columnar crystals parallel to the preferred axis of magnetisation.

Soft magnetic materials

These show properties which are the opposite of those just described for hard magnetic materials. Soft magnetic materials have high magnetic permeability. This means that they are very easily magnetised and demagnetised and that they retain virtually no magnetism when the magnetising force is removed. Traditionally, low-carbon steels and wrought iron were used for transformer, choke, and electromagnet cores and for generator and motor pole pieces. Nowadays, however, these materials have been superseded by special alloys for such applications and these alloys possess very low hysteresis values and very high magnetic permeability. Two such alloys will now be described.

- *Mumetal* This is an alloy containing 74 per cent nickel, 5 per cent copper, 1 per cent manganese and 20 per cent iron. It is an expensive alloy, but is widely used as a shielding material in telecommunications equipment.
- *Silicon iron* This contains 4.0 per cent silicon, 0.3 per cent manganese, less than 0.05 per cent carbon, and the remainder iron. This alloy is very much cheaper than *mumetal* whilst still possessing very low hysteresis values and high magnetic permeability. It is widely used for lower frequency applications such as lamination stampings for the cores for transformers, chokes and motor and generator rotors and stators.

1.4.7 *Thermal properties*

Conductivity

This is the ability of a material to transmit heat energy by conduction. Figure 1.15 shows a soldering iron. The bit is made from copper which is a good conductor of heat and so will allow the heat energy stored in it to travel easily down to the tip and into the work being soldered. The wooden handle remains cool as it has a low thermal conductivity and resists the flow of heat energy.

Fig. 1.15 *Thermal conductivity*

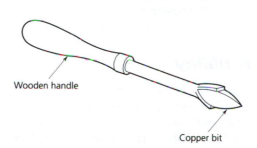

Wooden handle

Copper bit

Expansivity

This is the expansion of materials due to an increase in temperature. The linear expansivity or, more commonly, the *coefficient of linear expansion* of a material is a measure of the amount by which a unit length of the material expands when its temperature is raised by 1 °C.

Specific heat capacity

Different substances require different amounts of heat energy to produce the same rise in temperature. For example, water requires more heat energy to raise its temperature by any given amount than an equal mass of any other liquid. This is why water is such a good coolant. The heat capacity of unit mass of a substance is called its specific heat capacity. The derived unit is the joule per kilogram kelvin (J/kg K)

> *The specific heat capacity of a substance is defined as the heat energy required to raise the temperature of unit mass of that substance by one degree.*

Table 1.7 lists the thermal properties described above for a number of typical engineering materials.

Table 1.7 Thermal properties of materials

Material	Melting point	Conductivity
Aluminium	660 °C	Very good
Copper	1080 °C	Excellent
Iron	1535 °C	Good
Wood	No melting point (burns)	Poor (good insulator)
Polystyrene (expanded)	No defined melting point but softens at 100 °C	Very poor (excellent insulator)
Glass fibre	No defined melting point but softens at about 600–800 °C depending on composition	Very poor (very good insulator)
Fire bricks and clays for furnace linings	1595–1800 °C depending upon alumina content; softening and loss of strength is progressive and commences below these temperatures	Poor (good insulator)

1.4.8 *Fusibility*

This is the ease with which materials will melt. It can be seen from Fig. 1.16 that solder melts easily and so has the property of *high fusibility*. On the other hand, fire bricks used for furnace linings only melt at very high temperatures and so have the properties of *low fusibility*. Such materials which only melt at very high temperatures are called *refractory materials*. These must not be confused with materials which have a low thermal conductivity and are used as thermal insulators. Although expanded polystyrene is an excellent thermal insulator, it has a very low melting point (high fusibility) and in no way can it be considered a refractory material.

Fig. 1.16 *Fusibility*

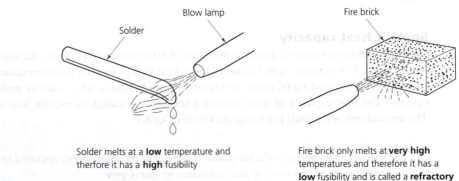

Solder melts at a **low** temperature and therfore it has a **high** fusibility

Fire brick only melts at **very high** temperatures and therefore it has a **low** fusibility and is called a **refractory**

1.4.9 *Temperature stability*

Substantial changes in temperature can have very significant effects on the structure and properties of materials and these will be considered later. However, there are two effects which changes in temperature can have on the dimensional stability of a component.

- *Thermal expansion* This has already been considered. All materials, to a greater or lesser extent, expand when heated and contract when cooled. This expansion or contraction is proportional to the change in temperature.
- *Creep* This is an important factor when considering polymeric (plastic) materials. It must also be considered when metals work continuously at high temperatures. For example gas-turbine blades. Creep is defined as the gradual extension of a material over a long period of time whilst the applied load is kept constant. The creep rate increases if the temperature is raised, but becomes less if the temperature is lowered. Creep will be considered in greater detail in Section 13.2 of this book.

SELF-ASSESSMENT TASK 1.2

1. Explain briefly the essential difference between the following terms as applied to the physical properties of metals:

   (a) resistance and reluctance
   (b) conductor and insulator
   (c) permittivity and permeability
   (d) specific heat capacity and thermal expansivity

2. List the material properties required for:

   (a) a 'hard' magnetic material
   (b) a 'soft' magnetic material

1.5 Engineering materials

Almost every substance known to man has found its way into the engineering workshop at some time or other. The most convenient way to study the properties and uses of engineering materials is to classify them into 'families' as shown in Fig. 1.17.

Fig. 1.17 *Classification of engineering materials*

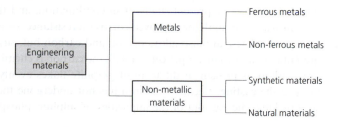

1.5.1 *Ferrous metals*

These are metals and alloys containing a high proportion of the element iron. They are the strongest materials available and are used for applications where high strength is required at relatively low cost and where weight is not of primary importance. For example: bridge building, the structure of large buildings, railway lines, locomotives and rolling stock and the bodies and highly stressed engine parts of road vehicles. However, lightweight materials such as aluminium alloys and polymers (plastics) are being increasingly used in railway and road vehicles to reduce their weight and make them more efficient in the use of energy. The ferrous metals themselves can also be classified into 'families', and these are shown in Fig. 1.18.

Fig. 1.18 *Classification of ferrous metals*

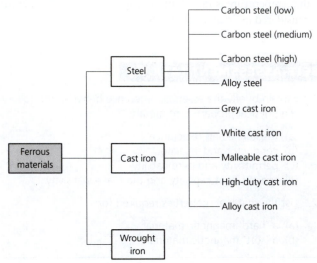

- *Plain carbon steels*   These are essentially alloys of iron and carbon in which the carbon is chemically combined with the iron. Although containing traces of other substances such as sulphur and phosphorus which are impurities carried over from the extraction process and a small amount of the metal manganese to offset the deleterious effects of the impurities. None of these substances is present in sufficient quantities to qualify as an *alloying element*, and these simple steels are referred to as *plain carbon steels*.
- *Alloy steels*   These are similar to low-carbon and medium-carbon steels with the addition of other metals such as manganese, nickel, chromium, molybdenum and vanadium, in sufficient quantities to materially alter and enhance the properties of the metal. They may be present singly or in combination, and the alloys so formed may have increased strength, or improved corrosion resistance, or improved heat resistance.
- *Cast irons*   These are also alloys of iron and carbon but, since the amount of carbon present is greater than 1.7 per cent, not all the carbon is chemically combined and some 'free carbon' is present in the form of graphite flakes evenly distributed through the mass of the casting. Because cast iron has not undergone the refinement processes of steel making, the amount of the impurities of sulphur, phosphorus and silicon will be

greater. Although tending to weaken the metal, they have the advantage of helping to reduce its melting temperature and increase its fluidity so that it can be cast in intricate moulds more easily. The properties of simple grey cast irons may be enhanced by such processes as *malleablising* (malleable cast irons) and by the addition of alloying elements (alloy cast irons).

- *Wrought iron*   This is the nearest commercial material to pure iron. It contains only 0.03 per cent carbon but up to 1.8 per cent impurities, mainly slag inclusions. Wrought iron bars can be readily forged to shape and joined by forge welding. The slag inclusions give it a fibrous structure which results in extreme toughness, malleability and ductility when cold. The slag fibres also improves the corrosion resistance of wrought iron compared with mild steel. It was the first malleable and ductile ferrous metal before steel making had been developed and was widely used in early engineering projects and equipment. Because of its relatively low strength it is now mainly used for decorative, architectural ironwork.

1.5.2 *Non-ferrous metals*

These materials refer to all the remaining metals known to mankind. The pure metals are rarely used as structural materials as they lack mechanical strength. However, they are used where their special properties such as corrosion resistance, electrical conductivity and thermal conductivity are required. Copper and aluminium are used as electrical conductors and, together with sheet zinc and sheet lead, are used as roofing materials. They are mainly used when alloyed with other metals to improve their strength. Even copper relies upon the inclusion of copper oxides (tough pitch copper) or traces of arsenic (arsenical copper) to increase its strength when used for chemical plant and for domestic plumbing. Of all the non-ferrous metals, copper is the easiest to join by soldering and brazing.

1.5.3 *Non-ferrous alloys*

- *Copper alloys*   Alloys such as bronze (copper and tin) are highly corrosion resistant, strong, easily machined and have relatively high melting points. Unfortunately they are heavier and much more costly than ferrous metals and alloys. Bronze alloys are widely used for steam and hydraulic valve components and for marine applications. The brass alloys (copper and zinc) are cheaper than the bronzes but are weaker and rather less corrosion resistant. Brass alloys are easily flow-formed to shape when hot and can be easily machined to a good finish. They are widely used in the manufacture of electrical components and domestic water fittings.
- *Aluminium alloys*   These materials are generally less strong than the ferrous- and copper-based alloys but have the advantage of being much lighter in weight. Like the copper alloys they are much more corrosion resistant than the ferrous metals and alloys, with the exception of the 'stainless steels'. Aluminium alloys are widely used in aircraft construction where low weight is of paramount importance. Unfortunately the strength of these alloys falls off rapidly as their temperature rises and for this reason supersonic aircraft are made from titanium alloy. Titanium is also a non-ferrous metal as light in weight as aluminium. However, it is as strong as steel and retains its strength at high temperatures. Unfortunately it is very costly and relatively difficult to shape. Non-ferrous metals and their alloys are considered in greater detail in Chapter 7.

Some widely used non-ferrous metals and alloys are classified in Fig. 1.19.

Fig. 1.19 *Classification of non-ferrous metals and alloys*

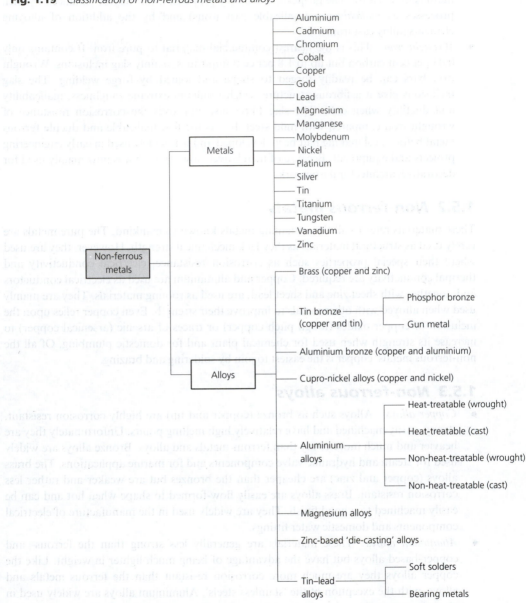

1.5.4 *Non-metals (synthetic)*

These are non-metallic materials that do not exist in nature, although they are manufactured from natural substances such as oil, coal and clay. Some typical examples are classified as shown in Fig. 1.20.

Fig. 1.20 *Classification of synthetic materials*

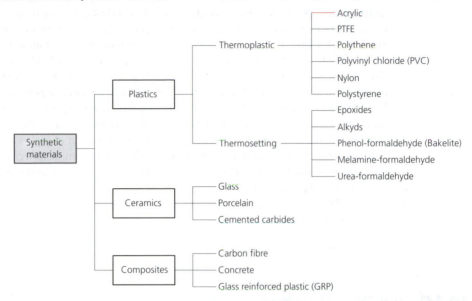

The so-called 'plastic' materials are used for an ever-increasing range of duties and, because they are synthetic, new materials are being created all the time. Generally they combine good corrosion resistance with ease of manufacture by moulding to shape and relatively low cost. Their properties can be matched to almost any design problem with the exception of high-temperature environments. Some of these materials can match the strength of alloy steels on a weight for weight basis, and are now replacing metals for many traditional applications. Synthetic adhesives are also being used for the joining of metallic components even in highly stressed applications. These materials are considered in greater detail in Chapters 2 and 8.

1.5.5 *Non-metals (natural)*

Such materials are so diverse that only a few can be listed here to give a basic introduction to some typical applications.

- *Wood* This is a naturally occurring fibrous composite material used for the manufacture of casting patterns.
- *Rubber* This is used for hydraulic and compressed air hoses and oil seals. Naturally occurring latex is too soft for most engineering uses but it is used widely for vehicle tyres when it is compounded with carbon black.
- *Glass* This is a hard wearing, abrasion-resistant material with excellent weathering properties. It is used for electrical insulators, laboratory equipment, optical components in measuring instruments and, in the form of fibres, is used to reinforce plastics. It is made by melting together the naturally occurring materials: silica (sand), limestone (calcium carbonate) and soda (sodium carbonate).
- *Emery* This is a widely used abrasive and is a naturally occurring aluminium oxide. Nowadays it is produced synthetically to maintain uniform quality and performance.

- *Ceramic* These are produced by baking naturally occurring clays at high temperatures after moulding to shape. Specially selected clays are used to make heat-resistant bricks for furnace linings. Kaolin 'china' clays are used to make porcelain mouldings for high-voltage insulators and high-temperature-resistant cutting tool tips. Commercially produced ceramics comprising metal oxides, carbides and nitrides are also widely used for high-performance tool tips because of their extreme hardness, resistance to abrasion and their ability to work at high temperatures without softening. Unfortunately, ceramic materials are all brittle and often require to be bonded with other materials to form *composites* to improve their strength properties.
- *Diamonds* These can be used for cutting tools for operation at high speeds for metal finishing where surface finish is of great importance. For example, internal combustion engine pistons and bearings. They are also used for dressing grinding wheels.
- *Oils* Used as bearing lubricants, cutting fluids and fuels.
- *Silicon* This is used as an alloying element and also for the manufacture of semiconductor devices.

These and other natural, non-metallic materials can be classified as shown in Fig. 1.21.

Fig. 1.21 *Classification of natural materials*

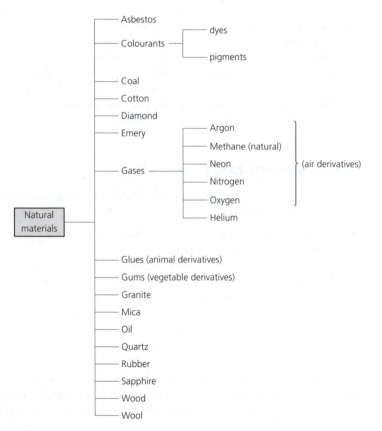

1.6 Composite materials (composites)

These are materials made up from, or *composed* of, a combination of different materials to take overall advantage of their different properties.

Naturally occurring composite materials include examples such as wood, bone and horn which are based upon naturally occurring fibres of cellulose, collagen and keratin respectively. In wood the tubular fibres of *cellulose* are bonded together in a matrix of *lignin*, as shown in Fig. 1.22

Fig. 1.22 *Wood as a natural composite*

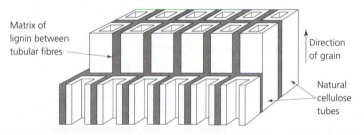

Matrix of lignin between tubular fibres

Direction of grain

Natural cellulose tubes

In man-made composites, the advantages of deliberately combining materials in order to obtain improved or modified properties was understood by ancient civilisations. An example of this was the reinforcement of air-dried bricks by mixing the clay with straw. This helped to reduce cracking caused by shrinkage stresses as the clay dried out. In more recent times, horse hair was used to reinforce the plaster used on the walls and ceilings of buildings. Again this was to reduce the onset of drying cracks.

Nowadays, especially with the growth of the plastics industry and the development of high-strength fibres, a vast range of combinations of materials is available for use in composites. For example, carbon fibre reinforced frames for tennis rackets and shafts for golf clubs have revolutionised these sports. Composites are dealt with in greater detail in Chapter 9. Some of the examples to be considered are:

- Reinforced concrete.
- Glass reinforced plastics.
- Sintered particulate materials including 'cermets'.

SELF-ASSESSMENT TASK 1.3

1. Sort the materials, iron, carbon, sulphur, low-carbon steel, bronze, polystyrene, brass, stainless steel, zinc, PVC, phosphorus, silicon, nickel–chrome alloy steel, duralumin, aluminium, titanium, melamine–formaldehyde and grey cast iron, into groups of ferrous metals, non-ferrous metals, non-metals.

2. Explain why most metallic materials used in engineering products are alloys.

1.7 Factors affecting material properties

The following are the more important factors which can influence the properties and performance of engineering materials.

1.7.1 *Heat treatment*

This is the controlled heating and cooling of metals to change their properties to improve their performance or to facilitate processing. Heat-treatment processes are considered in detail in Chapters 5 and 7. An example of heat treatment is the hardening of a piece of high-carbon steel rod. If it is heated to a dull red heat and plunged into cold water to cool it rapidly (quenching), it will become hard and brittle. If it is again heated to dull red heat but allowed to cool *very slowly* it will become softer and less brittle (more tough). In this condition it is said to be *annealed*. In this condition it will be too soft and the grain will be too coarse for it to machine to a good surface finish, but it will be in its best condition for *flow forming*.

During flow forming (working) the grains will be distorted and this will result in most metals becoming *work hardened* if flow formed at room temperature. To remove any locked in stresses resulting from the forming operations and to prepare the material for machining, the material has to be *normalised*. This consists of again heating the metal to dull red heat but cooling less slowly than for annealing. The hot metal is removed from the furnace and allowed to cool in still air away from draughts. This results in a finer grain than annealing and improves the metal's machining properties and strength. Thus by various heat-treatment processes a given material can be made to exhibit a range of properties.

1.7.2 *Processing*

Hot- and cold-working processes will be referred to many times throughout this book and it is essential to understand what is meant by the terms *hot* and *cold working* as applied to metals. Metals are said to be *worked* when they are squeezed or stretched or beaten into shape. Metals which have been shaped in this manner are said to be in the *wrought* condition. Thus metal which is *worked* into shape must possess the property of *plasticity* as described earlier in this chapter. Figure 1.23 shows examples of hot and cold working.

Metal is hot worked or cold worked depending upon the temperature at which it is flow formed to shape. These temperatures are not easy to define. For instance, lead hot works at room temperature and can be beaten into complex shapes without cracking, but steel does not hot work until it is red hot.

When metals are examined under the microscope it can be seen that they consist of very small grains. When most metals are bent or worked at room temperature (cold worked) these grains become distorted and the metal becomes hard and brittle. That is why metal which has been *cold worked* becomes *work hardened*. Care must be taken to avoid excessive cold working as this could cause the metal to crack. If considerable working is required to form a particular component, the metal must be softened from time to time during the processing by heat treatment (annealing).

Fig. 1.23 *Examples of (a) hot-working and (b) cold-working processes*

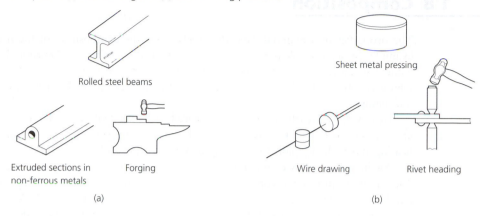

Rolled steel beams

Sheet metal pressing

Extruded sections in non-ferrous metals

Forging

Wire drawing

Rivet heading

(a)

(b)

When metals are hot worked the crystals are also distorted. However, they reform instantly into normal crystals because the process temperature is above the temperature of *recrystallisation* for the metal being used and work hardening does not occur. Thus *cold working* is the flow forming of metals *below* the temperature of recrystallisation, whilst *hot working* is the flow forming of metals *above* the temperature of recrystallisation. Hot and cold working together with recrystallisation will be considered in greater detail in Section 5.1.

1.7.3 *Environmental reactions*

The properties of materials can also be affected by reaction with the environment in which they are used. For example:

Dezincification of brass
Brass is an alloy of copper and zinc and when brass is exposed to a marine environment for a long time, the salts in the sea water spray react with the zinc content of the brass so as to remove it and leave behind a spongy, porous mass of copper. This obviously weakens the material which fails under normal working conditions.

Rusting of steel
Unless steel structures are regularly maintained by rust neutralisation and painting processes, rusting will occur. The rust will eat into the steel, reduce its thickness and, therefore, its strength. In extreme cases an entire structure made from steel may be eaten away.

Degradation of plastics
Many plastics degrade and become weak and brittle when exposed to the ultraviolet content of sunlight. Special dyestuffs have to be incorporated into the plastic to filter out these harmful rays. This will be considered further in Section 13.15.

1.8 Composition

The properties of a material depends largely upon the composition of the material. For example, consider the plain carbon steels. These are alloys of iron and carbon. If the carbon content is under 0.3 per cent, then the material is referred to as *low-carbon steel*. This is relatively soft and ductile with moderate strength and it cannot be hardened by heating and quenching.

However, if the carbon content is increased to 1.2 per cent, the material is referred to as *high-carbon steel*. This is much less ductile, but is stronger and it can be hardened by heating and cooling rapidly (quenching) to make cutting tools.

Again, a brass alloy containing 70 per cent copper and 30 per cent zinc has very great ductility and can be cold worked into complicated shapes without cracking. On the other hand, a brass alloy containing 60 per cent copper and 40 per cent zinc lacks ductility and cannot be readily cold worked. However, it hot works excellently and is widely used for hot stamping into such shapes as plumbing fittings, water taps, etc. It is widely used for hot extrusion into rods and sections.

The properties of plastic materials vary widely depending upon composition. A snooker ball is made from a hard plastic such as melamine–formaldehyde and has obviously different properties from the soft plastic insulation of a flexible electric cable. For this latter application a plastic material of very different composition is used such as polymerised vinyl chloride (PVC).

SELF-ASSESSMENT TASK 1.4

1. List the essential criteria for selecting a material for a specific application.

2. List the factors that can affect the properties of a material.

In this chapter a number of basic concepts have been introduced.

- The concept that material selection requires a knowledge of properties for a wide range of materials.
- The concept that properties can change with heat treatment, processing and environmental reactions.
- The concept that properties of materials depend upon their composition.

The following chapters will now examine these basic concepts in very much greater detail.

EXERCISES

1.1 Select suitable materials for each of the following applications and discuss the reason for your selection in each case in terms of cost and availability, suitability for the manufacturing processes required and suitability for the applications listed below.

(a) national electricity grid overhead cables

(b) telephone handset

(c) body casting for a steam valve

(d) filing cabinet

1.2 With the aid of diagrams explain the essential differences between tensile strength, shear strength and impact strength (toughness).

1.3 With reference to the physical properties of metals, choose a suitable material for:
(a) a wire-wound resistor that has to maintain a stable resistance at all normal working temperatures
(b) a motor car radiator
(c) the insulation of a refrigerator
(d) the lining of a furnace

1.4 List **three** factors which affect the mechanical properties of materials and, in each case, give an example of how the factor chosen affects their properties.

1.5 Briefly explain the essential differences between:
(a) plain carbon and alloy steels
(b) steels and cast irons
(c) brass and tin–bronze

1.6 State the main difference between thermoplastics and thermosetting plastics, and name two examples of each.

1.7 (a) Briefly explain what is meant by the 'heat treatment' of metals and why this is often necessary.
(b) Briefly explain the meaning of the following terms associated with heat treatment:
(i) annealing
(ii) quenching
(iii) normalising

1.8 State the essential difference between shaping metal by *machining* and *flow forming*.

1.9 (a) Explain the essential difference between the terms 'hot working' and 'cold working' as applied to metals.
(b) Explain what is meant by the term 'work hardening' as applied to metals.

1.10 (a) Name one corrosion-resistant ferrous metal and one corrosion-resistant non-ferrous metal found in most homes.
(b) State the main cause of plastic degradation and how it can be overcome.
(c) Find out what is meant by the term 'biodegradable' as applied to plastics. Discuss the main advantage and the main disadvantage of biodegradable plastics.

2 Basic science of materials

The topic areas covered in this chapter are:

- Atoms, molecules and lattices.
- Elements, compounds and mixtures.
- Solids liquids and gases.
- Allotropy.
- Crystals, crystal growth, grain structure, solidification and solidification defects.
- Macro- and microscopical examination.
- Semiconductor materials.
- Polymeric materials.
- Polymer 'building blocks'.
- Crystallinity in polymers and orientation
- Melting (T_m) temperatures in polymers, glass transition (T_g) temperatures in polymers and the effect of operating temperature on polymer applications.
- Memory effects.

2.1 Introduction

Figure 1.1 showed a number of components made from metal. Although the properties of the metals used varied widely they all had one thing in common. No matter what their composition, no matter what changes they had gone through during extraction from the ore, refinement and processing, they were all crystalline. However, before studying the crystalline structure of metals and alloys it is advisable to revise some basic concepts.

2.2 Atoms

Atoms can be considered as the smallest particles of a substance which can exhibit all the properties of that substance. An atom is electrically neutral because it has an equal number of negatively charged electrons and positively charged protons. Figure 2.1(a) shows, diagrammatically, an atom of the gas hydrogen (the simplest atom) and Fig. 2.1(b) shows an atom of the metal copper. It can be seen that both atoms consist of a nucleus around which are orbiting one or more electrons. Figures 2.1(a) and (b) are called *Bohr diagrams* after the name of their originator. Although it is convenient to represent the electrons spending their time in 'shells' as shown in Fig. 2.1(b), it is now assumed that the electrons are free to form a 'cloud' around the nucleus. The probability of an electron occupying an orbit at a particular radius from the nucleus can be shown graphically as in Fig. 2.1(c). This results in areas of greater and lesser electron density forming in the cloud, rather than the rigid rings of the Bohr atomic model. We will consider this in greater detail in *Engineering Materials*, Volume 2.

Fig. 2.1   *Typical atomic structures: (a) hydrogen; (b) copper; (c) movement of electrons round the nucleus*

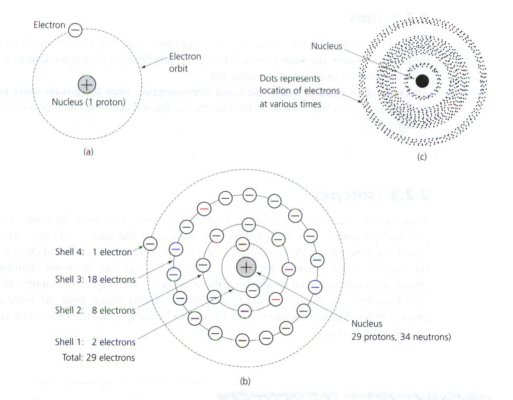

2.2.1 *Basic structure of the atom*

- *Nucleus*   This is the basic core of the atom and consists of *protons* and *neutrons*.
- *Protons*   These are *positively* charged particles of very much greater mass than the electrons.

- *Neutrons* These particles have the same mass as protons but carry *no electrical charge*. The mass of the atom is generally considered to be the sum of the mass of the protons and neutrons present since the mass of the electrons is negligible by comparison.
- *Electrons* These particles are *negatively* charged and orbit the nucleus like planets around the sun. Although electrons are very small and have only 1/1836 the mass of a proton or a neutron, they are supremely important when considering how atoms bond together to form molecules and lattice structures. The chemical properties of an atom – that is, how it combines with other atoms – are determined by the number of electrons it has. Electrons also determine the electrical and magnetic properties of a material.

2.2.2 *Ions*

Ions are atoms which have gained or lost one or more electrons. Loss of an electron makes the atom *electropositive* since there will be a positively charged proton without its balancing electron. Such an ion is called a *positive ion*.

Gaining an electron makes the atom *electronegative* since there is no spare positively charged proton in the nucleus to balance the additional electron. Such an ion is called a *negative ion*.

2.2.3 *Isotopes*

Since the electron is so small compared with the proton, the mass of the atom can – for all practical purposes – be considered as concentrated in the nucleus. As has already been stated, the neutron has the same mass as the proton but with no electrical charge. Thus, if the number of neutrons in the nucleus changes, the mass of the atom changes but its chemical properties do not change since there has been no change in the number of protons and, therefore, the number of electrons present. Atoms which have the same chemical properties but different atomic masses are said to be *isotopic* and are referred to as *isotopes* of a given substance.

SELF-ASSESSMENT TASK 2.1

1. (a) Name the particles that make up an atom and state what electrical charge they carry, if any.
   (b) State which of the particles are found in the nucleus of an atom.

2. (a) Briefly explain the difference between an atom and an ion.
   (b) Briefly explain what is meant by the term isotope.

2.3 Molecules and lattices

So far, we have considered the atom as a single free particle. However, apart from the noble gases such as neon (used in electric discharge tubes for advertising) and argon (used as a gas shield for some welding processes), atoms rarely occur as single particles but are generally associated with other atoms in small or large groups. Let's now consider how atoms bond together.

2.3.1 Covalent bond

Figure 2.2 shows how two hydrogen atoms bond together. It can be seen that they form a bond by sharing their electrons, and that the shared electrons are in the outer or *valency* shell of the atom. Since the bond is formed by sharing electrons it is referred to as a *covalent bond*. When atoms bond together covalently, a particle called a *molecule* is formed. Some molecules contain thousands of individual atoms – particularly those found in polymeric materials. These are called *macromolecules*.

Fig. 2.2 *Covalent bonding*

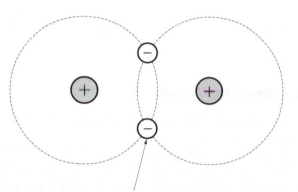

In the hydrogen molecule the electrons are shared to give a covalent bond

2.3.2 Ionic bond

All ionic solids are compounds. No elements have the properties of ionic solids. Ionic solids are formed by the complete transference of an electron from one atom to another. Common table salt (sodium chloride), for example, has a molecule in which there is an ionic bond between a sodium atom and a chlorine atom. One isolated sodium atom has only one electron in its outer (valency) shell, but eight electrons in its next, inner shell. One isolated chlorine atom has seven electrons in its outer (valency) shell. In becoming a stable compound both atoms need to have 'complete' outer shells of eight electrons. To achieve this, the sodium atom gives up the sole electron in its outer shell and becomes electropositive. The chlorine atom gains this electron and becomes electronegative. That is, both atoms become ions. Since particles with unlike charges attract each other, the sodium ion is attracted to the chlorine atom and an *ionic bond* is formed. At the same time both atoms attain full (eight electron) valency shells. This is shown diagrammatically in Fig. 2.3.

Fig. 2.3 *Ionic bonding: (a) isolated atoms of sodium and chlorine; (b) ionic bond formed by transfer of electron*

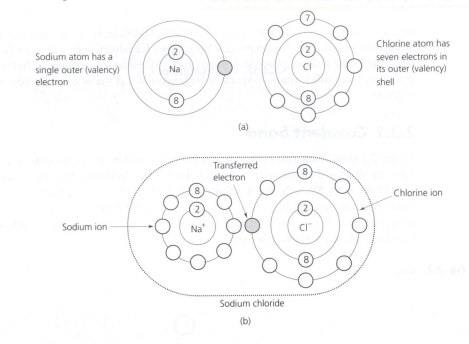

Sodium atom has a single outer (valency) electron

Chlorine atom has seven electrons in its outer (valency) shell

(a)

Transferred electron

Chlorine ion

Sodium ion

Sodium chloride

(b)

2.3.3 *Metallic bond*

This is also a form of ionic bond but instead of pairs of atoms bonding, electrons are shared among all the atoms in the lattice. As its name implies, a metallic bond is only formed between metallic atoms of the same element. The atoms lose electrons and become ionised. The metal ions do not form molecules but they form large, three-dimensional, geometrical *lattice structures*. It is this form of ionic bonding together with the lattice structure that gives metals their special properties:

- Metals can be bent and formed into complex shapes in the solid condition.
- Metals readily conduct heat and electricity.
- Metals have shiny surfaces when cut.

Let's look at this in more detail. Figure 2.4(a) shows how the three-dimensional lattice would appear if we could see it under a microscope that is much more powerful than any existing instrument. Figure 2.4(b) shows a portion of the bonding in greater detail. The metal atoms give up one or more electrons from their valency shells to become positively charged metal ions. The electrons that have escaped from the atoms are free to move about between the metal ions.

Now let's consider Fig. 2.5. For simplicity this shows the forces acting between two metal ions and a free electron.

- The positive ions (denoted by circles containing plus signs) have like charges so they repel each other and keep apart.

Fig. 2.4 *Metallic bonding: (a) metal lattice structure; (b) free electrons form a 'cloud' within the lattice of positive metal ions*

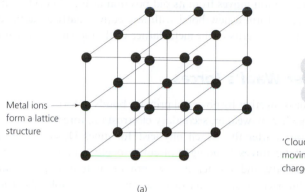

Metal ions form a lattice structure

Positively charged metal ions

'Cloud' of negatively charged free electrons moving at random between the positively charged metal ions

(a)

(b)

Fig. 2.5 *Forces in a metallic bond*

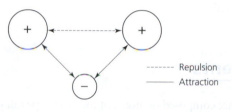

------- Repulsion
———— Attraction

- The positive ions are attracted to the free electron (denoted by the circle with the minus sign). This free electron is *delocalised*, which means it is free to move about. Remember: like charges repel each other; unlike charges attract each other.
- The attractive forces between the positive ions and the electrons are greater than the repulsive forces between the ions themselves. This is because the ions are nearer to the electrons than they are to each other.
- The forces of attraction between the positively charged ions and the negatively charged electrons holds the ions in position in the lattice.
- Remember: the lattice is three dimensional.

Let's now see how this arrangement of the ions and electrons enables metals to be bent and formed in the solid state. Figure 2.6 shows that the metal ions are free to form bonds with any adjacent electrons in the electron 'cloud' that exists between the ions.

Fig. 2.6 *The effect of an external mechanical force: (a) no force applied; (b) external force applied: ions take up new positions but retain their bonds*

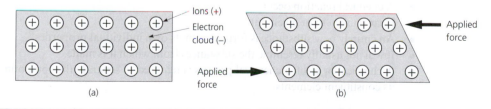

Ions (+)

Electron cloud (–)

Applied force

Applied force

(a)

(b)

Therefore, when an external mechanical force is applied to the metal as shown, the 'layers' of ions in the lattice are free to slide over each other. Sliding can occur without the lattice being broken. When an ion moves from its old position in Fig. 2.6(a) to a new position in Fig. 2.6(b), it is free to form a new bond with any conveniently adjacent electrons. This maintains the structural integrity of the metal lattice while allowing it to change shape.

2.3.4 *Van der Waal's forces*

Just as atoms are held together by powerful primary bonds to form molecules, so molecules can be bonded together by weaker, secondary electrostatic forces. These forces are called *Van der Waal's* forces after the Dutch physicist Johannes Diderik Van der Waal. The secondary electrostatic forces bonding molecules together influence such properties as melting point, solubility and the tensile strength of materials (particularly polymers). Adhesives of the 'impact' type rely on these secondary intermolecular forces to form a bond. Although the attractive force between each pair of molecules is relatively weak, there are many, many molecules in even a small piece of material and, collectively, they show great strength.

2.4 Elements

A substance composed of atoms of the same type (all with the same number of electrons) is an *element*. Elements are pure substances incapable of further division and consisting of particles formed entirely from one type of atom. For example: iron, carbon, sodium, chlorine and copper. Steel is not an element since it contains both iron atoms and carbon atoms. Table salt is not an element because it contains both sodium atoms and chlorine atoms. There are 103 elements known at the present time.

2.5 Compounds

It has just been stated that table salt (sodium chloride) is not an element since its molecules consist of different types of atom (sodium atoms and chlorine atoms). When two or more atoms of different elements combine together to form a single molecule a compound is said to have been formed. Thus atoms of the elements sodium and chlorine combine together to form the compound molecule sodium chloride.

When compounds are formed:

- A chemical reaction occurs.
- Heat is taken in or given out.
- An entirely new substance is formed with its own individual properties.
- It will not usually resemble the substances from which it is made.
- A chemical or electro-chemical process is required to separate the compound back into its constituent elements.

Generally a compound molecule will have totally different properties to those of the constituent atoms. Further, a chemical process is required to separate a compound back into its constituent elements. For example, the highly reactive metal element sodium reacts violently with the poisonous gaseous element chlorine to form the compound sodium chloride which is a stable and edible salt used in cooking. Again, the gases hydrogen and oxygen combine together to form the liquid called water. Different metals sometimes combine together to form *intermetallic compounds*, but these are rare. Different metals normally combine together to form alloys. Generally, these are intimate mixtures of two or more metals which form liquid solutions (when molten) or *solid solutions* at room temperature (when cooled).

2.6 Mixtures

Mixtures are formed when two or more substances are in close association (mixed together) without any chemical reaction taking place, without any new substance (compound) being formed, and with the individual substances being separable by physical means. For example, iron filings and sand can be mixed together. No new substance is formed and the iron filings can be removed with a magnet. Again, a mixture of sand and salt can be separated by dissolving the salt in water, filtering off the sand and recovering the salt by boiling the water so that evaporation occurs.

2.7 Solids, liquids and gases

Most substances can exist as solids, liquids or gases depending upon their temperature. A notable exception is iodine which sublimes directly from the solid state into the gaseous state when heated without becoming a liquid. Most substances behave like water. Below its freezing point water is a *solid* (ice). Above its freezing point and below its boiling point it is in the *liquid* state. If its temperature is increased still further it boils and becomes a *vapour* (steam) before turning into a *gas* with further heating. The fact that a substance can exist in the solid state, the liquid state and the gaseous state is due to the fact that the atoms and molecules of a substance are in a permanent state of vibration, providing the temperature is above *absolute zero* ($-273\,°C$), at which temperature all atomic and molecular movement stops.

- When the temperature is sufficiently low for a given substance to be in its solid state, the vibration is of small amplitude and the atoms and molecules only move to a small extent about a fixed point.
- When the temperature of a solid is raised to above its melting point, the atoms and molecules of the substance vibrate more violently. They no longer move about a fixed position but are free to move about within the constraints of the container holding the liquid.
- Finally, if the temperature is raised still further until it is above the temperature of vaporisation for the substance, the atoms and molecules move so freely that they can disperse until they completely fill the vessel containing them and, if they escape, they continue to disperse throughout the atmosphere indefinitely.

The changes of state of a substance do not proceed evenly as the temperature rises, but in a series of steps as shown in Fig. 2.7. This is due to a change in state being accompanied by the taking in or the giving out of *latent heat* – that is, the heat energy associated with a change of state without an accompanying change of temperature. Heat energy which causes a change of temperature without a change of state is referred to as *sensible heat*.

Fig. 2.7 *Cooling curve for water*

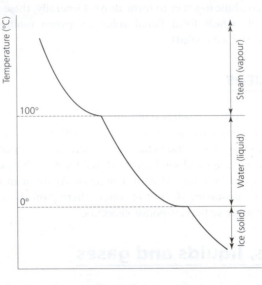

SELF-ASSESSMENT TASK 2.2

1. Briefly describe the essential differences between an element, a compound and a mixture.

2. Name the three states in which matter can exist.

3. Considering the fundamental structures of materials, name the essential differences between:

   (a) a covalent bond
   (b) an ionic bond
   (c) a metallic bond

2.8 Crystals

Many substances, including metals, have a crystalline structure in the solid state. Metal crystals form when the molten metal cools and solidifies, whereas crystals of other substances, for example copper sulphate and sodium chloride (common salt), form when a saturated solution of the compound evaporates causing the solid to crystallise out. A closer examination of crystals shows that their basic particles are arranged in definite three-dimensional patterns of rigid geometrical form. This pattern is repeated many times, provided the crystal is allowed to grow freely. Figure 2.8 shows some copper sulphate crystals and their regular geometric shape is clearly apparent.

Fig. 2.8 *Copper sulphate crystals growing from solution*

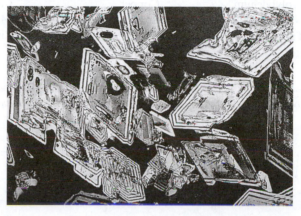

Non-crystalline solids such as pitch, glass and many polymeric (plastic) materials do not have their basic particles arranged in a geometric pattern. Their particles have a random formation, and as a result, such substances are said to be *amorphous* (without shape).

The structure of crystals is better understood if their constituent atoms are considered to be spherical in shape. Figure 2.9(a) shows a simple cubic crystal built up from eight spherical particles. The dotted lines joining the centres of the spheres represents the *unit cell* of this simple crystal. The unit cell is the geometric figure which illustrates the fundamental grouping of the particles in the solid. To form the crystal this unit cell is repeated many times to form the *space lattice* as shown in Fig. 2.9(b).

Fig. 2.9 *The crystal structure: (a) unit cell for sodium chloride (common salt) crystal; (b) part of the space lattice for sodium chloride (eight unit cells shown)*

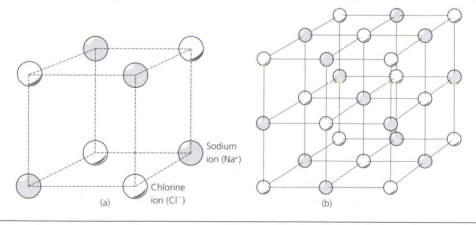

All crystal structures can be analysed into fourteen basic space lattices called the *Bravais space lattices*. For simplicity only the unit cells of each lattice is shown in Fig. 2.10. Of these fourteen possible space lattice formations, only six are met with in metal crystals. Of these, the most common are:

Body-centred-cubic	*Face-centred-cubic*	*Close-packed-hexagonal*
Chromium	Aluminium	Beryllium
Molybdenum	Copper	Cadmium
Niobium	Lead	Magnesium
Tungsten	Nickel	Zinc
Iron	Iron	

Fig. 2.10 *Bravais space lattices: (a) P-type (primitive space lattices); (b) C-type (base centred on 'ab' face); (c) I-type (body centred); (d) F-type (face centred); (e) R-type; (f) H-type*

| Triclinic ($\alpha \neq \beta \neq \gamma$) | Monoclinic ($\angle = 90°$ except β) | Orthorhombic ($\angle = 90°$) | Cubic ($\angle = 90°$) | Tetragonal ($\angle = 90°$) |

(a)

| Base-centred orthorhombic | Base-centred monoclinic | Body-centred orthorhombic | Body-centred tetragonal | Body-centred cubic |

(b)   (c)

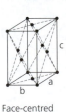
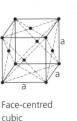

| Face-centred orthorhombic | Face-centred cubic | Rhombohedral ($\alpha \neq 90°$) | Close-packed hexagonal |

(d)   (e)   (f)

Crystals of the metal iron can have either a body-centred-cubic form or a face-centred-cubic form depending upon the temperature of the metal. At room temperature, the atoms are arranged in a body-centred-cubic form as shown in Fig. 2.11(a). When the metal is heated and reaches a temperature of 910 °C, the atoms rearrange themselves into face-centred-cubic crystals as shown in Fig. 2.11(b). If the metal is heated to the still higher temperature of 1400 °C the atoms again rearrange themselves, this time back into a body-centred-cubic form.

Note that when the atoms change from a body-centred-cubic arrangement to a face-centred-cubic arrangement they are more closely packed together. This results in a volumetric shrinkage of the metal. Therefore when making calculations for the thermal expansion or contraction of iron and steel, you must be careful that the temperature range involved does not include a change of crystal structure. If it does, the linear relationship between temperature and expansion will not hold true.

Other metals such as zinc, magnesium and cadmium are arranged in a *close-packed hexagonal* pattern as shown in Fig. 2.11(c).

Fig. 2.11 *Typical unit cells: (a) body-centred cubic crystal; (b) face-centred cubic crystal; (c) close-packed hexagonal crystal*

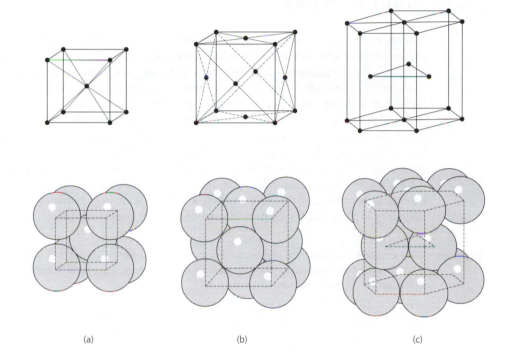

(a)   (b)   (c)

2.9 Allotropy

Allotropy is the ability of a substance to exist in more than one physical form. The non-metal carbon is said to be allotropic since it can exist both as diamond (the hardest substance known to man) and as graphite (the soft 'lead' of a pencil). Both these substances consist solely of carbon atoms, but it is the crystal structure and the way in which the carbon atoms are bonded together where the difference lies. The metal iron is another allotropic material, which is why it is shown in two columns at the top of page 40.

- Below 910 °C, iron has a body-centred cubic space lattice and is referred to as alpha (α) iron.
- Between 910 °C and 1400 °C, iron has a face-centred cubic space lattice and is referred to as gamma (γ) iron.
- Above 1400 °C, iron again has a body-centred cubic space lattice and is referred to as delta (δ) iron.

The allotropy of solids which relies solely on differences in their crystal structure (space lattice) is referred to as *polymorphism*. Many metals are allotropic.

2.10 Grain structure

Although metals are crystalline solids, this is not immediately apparent when they are examined under a microscope. Figure 2.12(a) shows the appearance under a microscope of a typical metal specimen which has been polished and etched. Although obviously granular, it is difficult to identify the geometric regularity expected of crystals. This is because crystals can only achieve geometric regularity when they are free to grow without restraint or interference. In a metal, many crystals commence to grow at the same time and eventually collide with each other so that growth becomes restricted and their boundaries become distorted. The term grain is used to describe crystals whose geometric shape has been distorted by contact with adjacent crystals so that their growth is impeded. Figure 2.12(b) shows, diagrammatically, how the atoms within a grain can have the regular geometric space lattice expected of a crystal, and how that pattern breaks down at the grain boundary to make way for the geometric space lattice in an adjacent grain.

Fig. 2.12 *Grain structure: (a) appearance of granular structure of metal under the microscope after etching; (b) despite the irregular appearance of the grain structure due to boundary interference, the crystal lattice within the grain is correctly ordered*

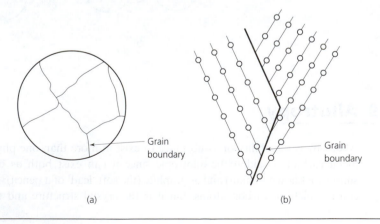

Grain boundary

Grain boundary

(a)

(b)

2.11 Crystal growth

When metals are in the liquid state there is no orderly arrangement of the atoms which are free to move about with respect to each other (Section 2.7) thus, in the liquid state, metals possess *mobility*. As the temperature of a molten metal falls, a point is reached where the metal starts to solidify. At this point the atoms change from a disordered or amorphous state to an ordered or *crystalline* state.

Like all pure crystalline substances, pure metals solidify at a fixed temperature as shown in Figure 2.13(a). However, under industrial conditions, such purity will not be obtained and the crystal nucleus will form around an impurity particle such as a particle of slag. If the purity is of a high order, some under-cooling may occur before *nucleation* (the formation of the nucleus of the crystal) commences as shown in Figure 2.13(b). Amorphous (non-crystalline) solids such as glass, pitch and some polymers (plastics) exhibit no such change point and there is a gradual change from the liquid state to the solid state, as shown in Figure 2.13(c).

Fig. 2.13 *Cooling curves: (a) pure metal: no undercooling; (b) pure metal: some undercooling; (c) amorphous solid: no single freezing temperature*

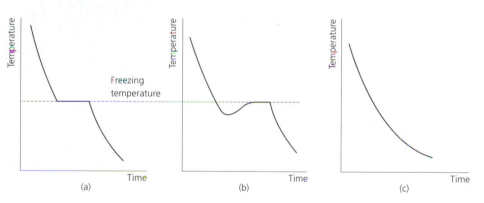

Once the nucleus of the crystal forms, it provides a solid/liquid interface where crystallisation will proceed. For metals, the nuclei will generally be made up of face-centred cubic, body-centred cubic or close-packed hexagonal unit cells. As the crystals grow they tend to develop spikes and change into 'tree-like' shapes called *dendrites* (Greek: *dendron* = a tree). Figure 2.14 shows a typical metal dendrite.

The dendritic crystal grows until the spaces between the branches fill up. Growth of the dendrite ceases when the branches of one dendrite meets those of an adjacent dendrite. This process continues until eventually the entire liquid solidifies. At this point there is little trace of the dendritic structure left, and it is only possible to see the grains into which the dendrites have grown. The steps in the growth pattern of a crystal from nucleus to grain are shown in Fig. 2.15.

Fig. 2.14 *Metallic dendrite growth (after: R. A. Higgins)*

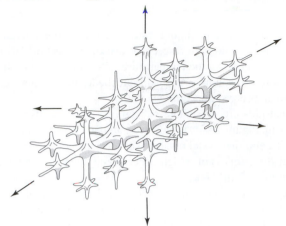

Fig. 2.15 *Crystal growth*

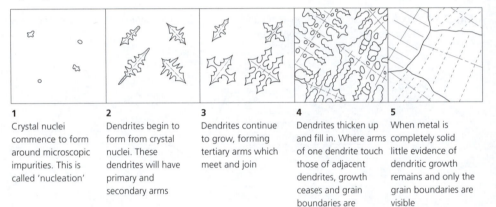

1	2	3	4	5
Crystal nuclei commence to form around microscopic impurities. This is called 'nucleation'	Dendrites begin to form from crystal nuclei. These dendrites will have primary and secondary arms	Dendrites continue to grow, forming tertiary arms which meet and join	Dendrites thicken up and fill in. Where arms of one dendrite touch those of adjacent dendrites, growth ceases and grain boundaries are established	When metal is completely solid little evidence of dendritic growth remains and only the grain boundaries are visible

The reason for dendritic growth is as follows. When a solid metal is heated, its atoms vibrate about fixed points called lattice points, each atom being held in place by electrostatic forces of attraction. As the temperature of the metal increases the energy of each atom increases. At a certain temperature (depending upon the metal) the atoms become so agitated that they can escape from their fixed positions. Thus the structure begins to lose rigidity. That is, it begins to melt.

In moving from their fixed positions the atoms and molecules work against their binding forces and energy is used up. This energy is replaced by the heat source of the furnace. The replacement heat energy producing fusion (melting), rather than causing a rise in temperature, is called the *latent heat of fusion*.

When a molten metal at its fusion point (melting point) solidifies it gives up its latent heat energy. That is, it gives up the heat energy originally required to cause melting. Thus, during solidification, the metal/liquid interface is warmed up by the release of latent heat energy. This slows down or stops further solidification occurring in that direction, and solidification recommences in some other direction where the temperature is sufficiently low. The result of this action is for spikes of solid metal to occur in regions where the liquid is coolest. As these new spikes warm up, forward growth is again retarded and secondary and even tertiary spikes are formed to produce the typical dendrite.

Although it is hard to relate a dendrite to the well-ordered crystal structures previously considered, you must remember that the unit cells and space lattices are very, very small even when compared in size with a dendritic spike. Thus, during solidification and crystallisation, the ordered pattern of the space lattice is still being built up, but the rate of growth is not uniform in all directions.

Another factor that influences the size and shape of the grains as a metal cools is the rate of cooling. In a casting this rate of cooling varies from the surface to the centre of a casting. Figure 2.16 shows a section through the corner of a large casting. You can see that the grain structure is not uniform but varies with the shape of the mould and the rate of cooling. Fine chill, grains occur at the surface of the casting where the molten metal first comes into contact with the cold surface of the mould. At the centre of the casting, where the rate of cooling is relatively slow, large equi-axed grains are formed. At an intermediate position, long columnar grains are formed at right angles to the surface of the casting. These result in a 'plane of weakness' at each corner of the casting which can be overcome, to some extent, if the corners are given a large radius.

Fig. 2.16 *Effect of cooling rate on grain growth*

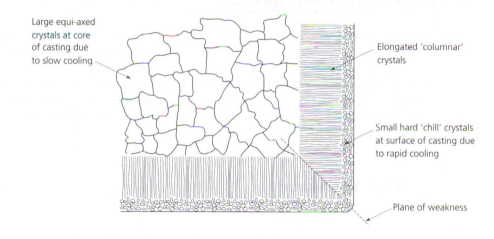

Large equi-axed crystals at core of casting due to slow cooling

Elongated 'columnar' crystals

Small hard 'chill' crystals at surface of casting due to rapid cooling

Plane of weakness

Sand and metal oxides (slag) adhering to the surface of the casting, together with fine, hard chill crystals just below the surface can cause premature wear and even serious damage to cutting tools. It is essential, therefore, for the first cut to be sufficiently deep for the tip of the cutting tool to operate below this surface zone as shown in Fig. 2.17. To ensure that the machined faces of the casting will 'clean up' during machining the pattern maker must allow sufficient additional metal (machining allowance) to be left on such faces when the casting pattern is manufactured. The pattern maker must also allow for the shrinkage of the metal, as the casting cools, by making the pattern oversize by a controlled amount. To do this the pattern maker uses a *contraction rule* to suit the metal being cast. This rule is marked with normal dimensions but they are all oversize by the required amount.

Fig. 2.17 *Machining allowance*

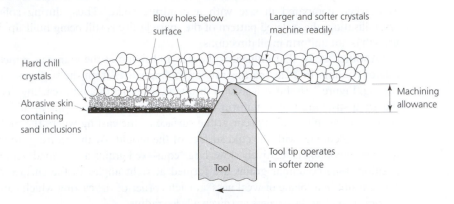

2.12 Solidification defects

The solidification processes described above assume pure or virtually pure metals solidifying under ideal conditions. Under practical industrial conditions impurities will inevitably be present. Also the dendrites will be growing in highly competitive conditions within the confines of the mould and each branch of each dendrite will be 'fighting' for space to develop. This overcrowding causes the dendritic branches or spikes to press against each other causing deformation. Let's now consider the more important solidification defects which can occur.

2.12.1 *Shrinkage*

This can occur on two scales:

- Very fine shrinkage leading to porosity between the dendrite branches.
- Large-scale shrinkage cavities resulting from poor mould design which prevents the feeding of molten metal to compensate for the normal volumetric shrinkage that occurs whenever metal solidifies and cools.

Shrinkage cavities occurring between the dendritic branches and spikes is known as *interdendritic porosity* and is caused by over-rapid cooling of the cast metal. As the molten metal between the dendrite branches and spikes cools and solidifies, it shrinks. Under normal cooling conditions additional molten metal has time to flow into the cavities so formed and no discontinuity occurs. If, however, cooling is too rapid there is not sufficient time for the shrinkage cavities to fill and porosity occurs along the dendrite branches. These fine cavities should not be confused with those due to gas porosity.

When a casting cools and solidifies there is nearly always considerable volumetric shrinkage. (A notable exception is cast iron, which *expands* as it solidifies due to structural changes in the crystal lattice. It then contracts normally as cooling continues in the solid state.) In a well-designed mould there are sufficient runners and risers (see Section 10.1) to feed additional molten metal back into the mould as shrinkage takes place. If the feeding of metal is inadequate during solidification, *drawing* may occur. This is where the solidifying and shrinking metal in the smaller sections of the casting draws molten metal from adjacent, thicker sections of the casting instead of from the runners and risers. This leaves large cavities which weaken the casting. It also leaves unsightly sunken surfaces which may not clean up during machining.

2.12.2 *Misorientation*

It has already been stated that dendritic growth forms a highly competitive environment and that the outermost branches and spikes of the dendrites are distorted as they interfere with each other at the grain boundaries. This results in the strict crystalline geometry of the space lattice breaking down at the grain boundary and producing an amorphous layer some two or three atoms thick, as shown in Fig. 2.18. This amorphous layer behaves like a highly viscous liquid and allows slight movement to occur between the grains. This accounts for the creep which occurs in metals stressed over long periods of time, and why creep is greater at elevated temperatures when the viscosity of the amorphous layer is reduced. This also accounts for the reason why metals tend to fracture by transverse cracking of the crystals at low temperatures when the amorphous layers are more viscous and resistant to intercrystalline movement. At high temperatures failure is more likely to occur by intercrystalline cracking than by transverse cracking since the amorphous layer is less viscous, weaker and more likely to allow intercrystalline movement.

Fig. 2.18 *Misorientation*

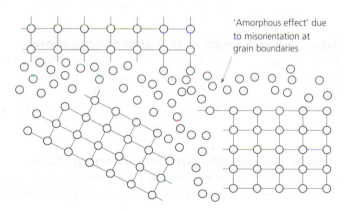

'Amorphous effect' due to misorientation at grain boundaries

2.12.3 *Segregation and inclusions*

Once solidification is complete, there will be no evidence of dendritic growth providing the metal is absolutely pure and no shrinkage cavities or inclusions are present. However, such conditions never occur commercially and it is necessary to examine the effects of such inclusions. Various types of segregation and inclusions may be present and these will now be considered:

- Dissolved inclusions with minor segregations.
- Dissolved inclusions with major segregations.
- Undissolved inclusions.

Dissolved inclusions

These tend to remain in the molten metal as long as possible so that the inclusions finally solidify in the spaces between the branches and spikes of dendrites along with the residual host metal. The presence of these dissolved inclusions often causes discoloration of the host metal outlining the shape of the original dendritic formation when examined under the microscope. Thus the dissolved inclusions are segregated from the host metal and are referred to as minor segregations, as shown in Fig. 2.19. This figure also shows interdendritic porosity.

Fig. 2.19 *Minor segregation and interdendritic porosity*

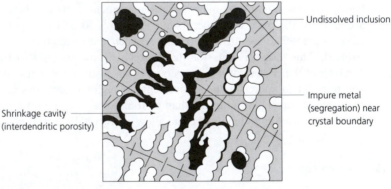

Undissolved inclusion

Impure metal (segregation) near crystal boundary

Shrinkage cavity (interdendritic porosity)

Dissolved inclusions with minor segregations

These are the most harmful since they lead to brittleness and cracking in the casting and may result, in the case of an ingot, to cracking and crumbling during subsequent forging and rolling. This is referred to as 'shortness' in the metal. Metal which is *hot short* crumbles when hot worked, whilst metal referred to as *cold short* crumbles and cracks when cold worked.

Dissolved inclusions with major segregations

These are only likely to be found in large castings such as ingots where large columnar crystals are present. As these move inwards they push the residual molten metal and dissolved inclusions ahead of them. Thus all the inclusions tend to become concentrated in the central 'pipe' where final solidification takes place. The appearance of these segregated inclusions in a sectioned and etched ingot casting is shown in Fig. 2.20.

Fig. 2.20 *Major segregations*

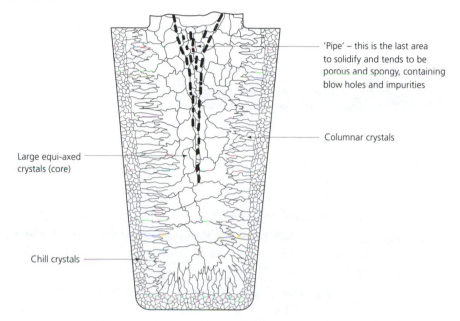

'Pipe' – this is the last area to solidify and tends to be porous and spongy, containing blow holes and impurities

Columnar crystals

Large equi-axed crystals (core)

Chill crystals

Undissolved inclusions

These are substances such as sand particles washed from the sides of the mould by the molten metal and also metal oxide particles (scale). Generally such inclusions are less dense than the molten metal so they tend to float to the surface of the runners and risers where they can be discarded. However, they sometimes become trapped in the casting where they form discontinuities from which fatigue cracks can originate. Such inclusions also adversely affect the strength and machining properties of the casting. A cutting tool hitting a hard inclusion will have its cutting edge blunted or chipped.

2.12.4 *Gas porosity*

Gases are frequently dissolved in the hot, molten metal but are expelled as the metal cools and solidifies. The sources of these gases may be from the furnace atmosphere, or from chemical reactions which take place during the melt. Most aluminium alloys and some copper alloys are susceptible to gassing. These metals tend to absorb hydrogen gas from the furnace atmosphere or from the moisture in the foundry atmosphere. The hydrogen is driven off by the addition of suitable chemicals immediately prior to pouring. This is known as degassing. Adequate ventilation is required during this operation as the gas driven off is often in the form of hydrogen chloride which turns into hydrochloric acid on contact with atmospheric moisture.

When ferrous metals such as steel and cast iron are cast, carbon monoxide gas may be present as the carbon in these metals tends to combine with the oxygen in the air. Any gases generated will bubble out as the metal cools and become trapped between the branches of the dendrites to form small random cavities.

The moisture in green-sand moulds may boil to steam which will be trapped at or just below the surface of the casting to form larger blow-holes if adequate venting of the mould is not provided. When exposed during machining, blow holes appear as spherical or oval cavities with shiny surfaces.

SELF-ASSESSMENT TASK 2.3

1. Briefly explain what is meant by the terms:

   (a) allotropy
   (b) polymorphism
   (c) crystalline
   (d) amorphous

2. Name the **three** most common crystalline structures found in metals and, for each structure, name **two** metals that possess that structure at room temperature.

3. Briefly explain the essential difference between crystals and grains when referring to the structure of a metal.

4. Name any **three** solidification defects and write brief notes on the cause and effect of each one.

2.13 Macro- and microscopical examination

The grain structure of metals and any inclusions and discontinuities which may be present can be studied by macro- or by microscopical examination (microscopy).

2.13.1 *Macro-examination*

Macro-examination implies the use of the unaided eye or the use of a low power magnifying glass. The sample component is sectioned and the surface is ground smooth. Since grinding tends to 'drag' the surface slightly, it is usual to hand finish the surface using a grade 0 or 00 abrasive paper. To reveal the grain structure it is necessary to etch the specimen. This is done by dripping a suitable etchant onto the surface of the specimen. The etchant will eat away the grain boundaries so that the individual grains stand out in relief. The specimen is then washed and examined. Details can frequently be seen more clearly whilst the surface is still wet. Table 2.1 lists a number of suitable etchants for macro-examination, whilst Fig. 2.21 shows a typical example of the appearance of a component prepared for macro-examination. A slag inclusion in the weld is just visible.

Table 2.1 *Etchants for macro-examination*

Material	Composition	Application
Steel	50% hydrochloric acid (conc.) 50% water	Specimen boiled in etchant for 5–15 minutes. For revealing flow lines, structure of fusion welds, cracks, porosity, case depth.
	25% nitric acid (conc.) 75% water	As above, but can be applied by cold swabbing for large specimens.
	Stead's reagent	Reveals dendritic structure in steel castings, and phosphorous segregation.
Aluminium and aluminium alloys	20% hydrofluoric acid (conc.) 80% water	Reveals flow lines and general grain structure and impurities (undissolved inclusions).
	45% hydrochloric acid (conc.) 15% hydrofluoric acid (conc.) 15% nitric acid (conc.) 25% water	As above, but more reactive reagent. Avoid contact with skin.
Copper and copper alloys	25 g ferric chloride in 100 ml 25% hydrochloric acid (conc.) 75% water	Reveals dendritic structure of α phase solid solutions.
	33% ammonium hydroxide (0.880) 33% ammonium persulphate (5%) 34% water	Reveals β phase structure.

Fig. 2.21 *Specimen as it appears for macro-examination*

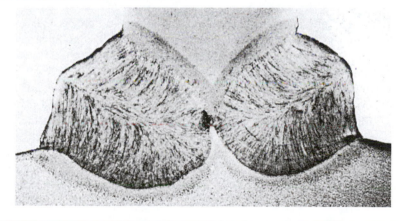

2.13.2 *Microscopy*

Microscopical examination (microscopy) requires much more careful preparation of the specimen. Since this has to be mounted on a microscope slide, it is usual to cut a small specimen from the component to be examined. The specimen is then often mounted in a low melting point alloy or a thermosetting plastic matrix for ease of handling as shown in Fig. 2.22. Initial grinding must proceed with the utmost care to avoid overheating the specimen and altering its micro-structure. Intermediate finishing is then carried out using progressively finer grades of abrasive paper. The paper is placed flat on a piece of plate glass and the specimen is moved back and forth so that the abrasive marks are a series of straight lines. At each change to a finer grade of abrasive the specimen is worked so that the new abrasive marks are at right-angles to the previous ones. Treatment continues until the previous abrasive marks are no longer visible. Finally, the specimen is polished on a rotary metallurgical polishing machine. A suspension of jeweller's rouge or a suspension of diamantine is dripped onto the rotating pad of 'selvyt' cloth. Absolute cleanliness is essential to avoid scratching the polished surface and all traces of polishing agent must be removed before etching. Polishing continues until no marks can be seen on the unetched specimen under the microscope.

Fig. 2.22 *Mounted specimen*

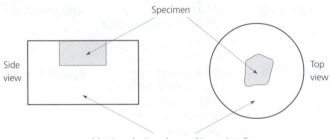

The etchant used and the method of application depends upon the metal being examined and the particular characteristics of the grain structure to be exposed. Microscopical metallography is a detailed study in its own right and only the very basic principles can be considered in this section. Table 2.2 lists a few of the more widely used etchants and their applications for microscopical examination. Since metallurgical specimens are opaque, metallurgical microscopes use reflected light. The light source is adjustable for intensity and colour filters are provided so that the light can be tinted to give the optimum contrast and clarity. Usually a turret of objective lenses is provided so that the magnification can be quickly and easily changed. Figure 2.23 shows the appearance of a typical etched specimen as seen under a microscope. The slag inclusion first identified in Fig. 2.21 is more easily visible under the microscope, as is the grain structure of the metal.

Table 2.2 _Etchants for microscopical examination_

Material	Composition	Application
Steel	Nital 2–5% nitric acid (conc.) 95–98% ethyl or methyl alcohol	General purpose etching reagent for cast irons and cast and wrought steels. Reveals pearlite, martensite, troostite and ferrite boundaries.
	Picral 4% picric acid (conc.) 96% ethyl or methyl alcohol	Reveals details of pearlitic and spheroidised structures – does not attack ferrite boundaries. Excellent for most cast irons except alloy and ferritic cast irons.
Aluminium and aluminium alloys	0.5% hydrofluoric acid (conc.) 99.5% water	General purpose cold-swabbing etchant.
	Keller's reagent 1.0% hydrofluoric acid (conc.) 1.5% hydrochloric acid (conc.) 2.5% nitric acid (conc.) 95% water	Immersion etching of wrought, heat-treatable alloys such as 'duralumin'.
Copper and copper alloys	10 g ammonium persulphate 20% ammonium hydroxide 80% water	General purpose etchant which reveals the grain boundaries of brasses, tin bronzes and cupro-nickel alloys. Must be freshly made before use.
	10 g ferric chloride 35% hydrochloric acid (conc.) 65% water	Suitable for $\alpha\beta$ brasses, bronzes, aluminium bronzes and cupro-nickel alloys. Darkens the β phase and improves the contrast.

Fig. 2.23   _Specimen under microscopical examination_

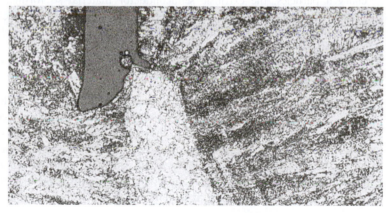

2.14 Semiconductor materials

Further discussion of these materials has been included here because they are based on *crystalline* materials. All the semiconductor materials have a valency of four, so they are *quadravalent*; that is, they have four electrons in their outer shells. Let's consider one of the most widely used, namely *silicon*.

A single 'super' crystal of silicon of extremely high purity is grown under very closely controlled conditions into a long rod of up to 75 mm diameter. It is then sliced into thin 'wafers'. In this pure or 'intrinsic' form of silicon the conduction of electricity is achieved by means of *thermally released* free electrons. Therefore the conduction of this material is *temperature controlled*.

For example, a piece of pure silicon is used in the temperature sensor of your car engine. It is wired in series with a current-measuring device, such as a milliammeter on the instrument cluster. As the temperature of the engine increases so does the magnitude of the current conducted by the sensor and this, in turn, increases the reading of the milliammeter on the instrument cluster. The dial of this instrument is marked with a temperature scale. The higher the temperature the better the conductivity of the silicon in the sensor, the higher the current it passes and the higher the reading on the instrument readout. The lower the temperature the poorer the conductivity of the silicon in the sensor, the lower the current it passes and the lower the reading on the instrument readout.

In order to make semiconductor devices such as transistors and diodes, the composition and structure of the pure (intrinsic) semiconductor material is changed by adding small amounts of other substances or 'impurities'. This is called 'doping' the semiconductor material and it has a dramatic effect on its conducting properties.

2.14.1 *n-type material*

Figure 2.24 shows a piece of pure (intrinsic) silicon. It consists solely of atoms of silicon with all their valency bonds taken up by covalent bonding. In Fig. 2.25(a) a *pentavalent* material has been added. In this example it is phosphorus. Pentavalent means that there are five electrons in the outer (valency) shell of the atom. Only one atom has been shown for simplicity but you can see that there is now a spare electron which can break free very easily with the thermal energy available at room temperature. Since electrons are negatively charged particles, the extrinsic semiconductor material so formed is called an *n-type* material and the flow of electric current due to the movement of free electrons through the material is called *extrinsic current flow*. A doping element that provides free electrons is called a *donor* element.

2.14.2 *p-type material*

Alternatively a *trivalent* material can be used as the dopant and, in the example shown in Fig. 2.25(b), it is boron. Trivalent means there are only three electrons in the outer (valency) shell. This leaves a 'hole' in the system. Since an electron is negatively charged, holes are considered to be *positively* charged. Current flow occurs by electrons moving from hole to hole. Elements that leave holes in the system are called *acceptor* elements since they can accept free electrons. Again, p-type materials are called extrinsic semiconductor materials and any current flow through them is called extrinsic current flow.

Fig. 2.24 *The silicon atom and its covalent bond: (a) single silicon atom with four electrons in the outer (valency) shell; (b) covalent bonding between adjacent atoms results in the valency shells being 'filled' with 8 electrons; in practice this occurs three dimensionally*

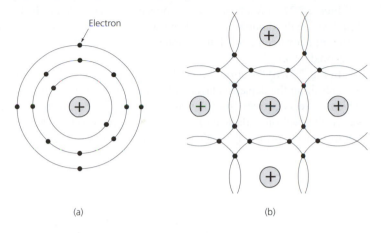

(a)   (b)

Fig. 2.25 *Extrinsic semiconductor material: (a) n-type; (b) p-type*

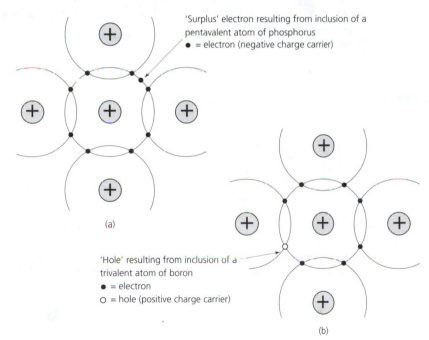

'Surplus' electron resulting from inclusion of a
pentavalent atom of phosphorus
● = electron (negative charge carrier)

(a)

'Hole' resulting from inclusion of a
trivalent atom of boron
● = electron
○ = hole (positive charge carrier)

(b)

The simplest device we can make using n-type and p-type semiconductor materials is a *junction diode*. This consists of a piece of semiconductor material that is doped to give it p-type characteristics for half its thickness and n-type characteristics for the other half of its thickness. Where these two regions meet is called the *junction*, hence the name 'junction diode'. Such a diode is shown diagrammatically in Fig. 2.26. Where the positive and negative charge carriers face each other across the junction, they tend to neutralise each other to produce a *depletion layer*.

Figure 2.27 shows two ways of connecting such a diode. In Fig. 2.27(a) the lamp *will light* because the diode has not only been biased in the *forward* direction, but the bias is also sufficient to overcome the 'barrier potential' of the depletion layer (some 0.6 to 0.7 volts for a silicon diode).

In Fig. 2.27(b) the lamp *will not light* because *reverse* bias has been applied to the diode. This results in the depletion layer widening so as to provide an insulating gap in the circuit which prevents current from flowing. Thus our diode can be considered as an electronic switch that will only allow current to flow in one direction. Semiconductor materials and semiconductor devices are dealt with in detail in *Engineering Materials*, Volume 2.

Fig. 2.26 *Junction diode*

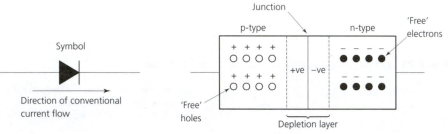

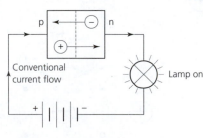

Fig. 2.27 *Diode operation: (a) forward bias – lamp on; (b) reverse bias – lamp off*

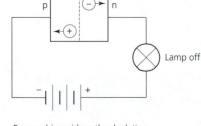

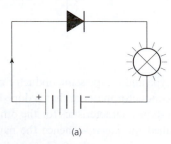

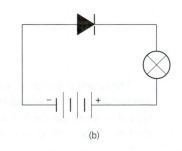

(a)　　　　　　　　　　　　　　　　　(b)

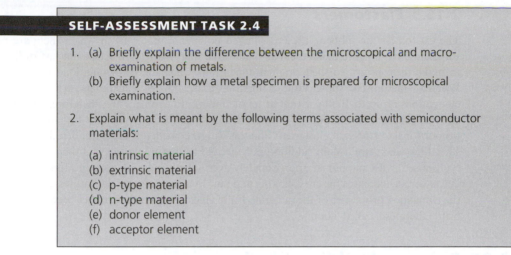

2.15 Polymeric materials

There is an ever-increasing number of synthetic, polymeric materials available under the popular name of *plastics*. This is a misnomer since polymeric materials rarely show plastic properties in their finished condition. In fact many show elastic properties. The name 'plastics' comes from the fact that during the moulding process by which they are shaped, they are reduced to a plastic condition by heating them to just above the temperature of boiling water. There are three main groups of polymeric or 'plastic' materials.

2.15.1 *Thermosetting plastics (thermosets)*

This group of polymeric materials undergoes chemical change during the moulding process and can never again be softened by reheating. These materials are generally hard, rigid and rather brittle. A typical example is melamine formaldehyde used for making such articles as snooker and billiards balls, table-wear and domestic electrical fittings. The strength of thermosetting plastics can be greatly increased by reinforcing them with fibrous materials (see Chapter 8).

2.15.2 *Thermoplastics*

These become soft and can be remoulded each time they are reheated. They are not so rigid as thermosetting plastics but tend to be tougher. For example, rigid polymerised vinyl chloride (PVC) is used for rain water guttering and down-piping on buildings. Thermoplastics can also be soft and pliable. For example, the non-rigid PVC is used for the insulation of flexible cables (see Chapter 8).

2.15.3 *Elastomers*

The elastomers, or rubbers, are cross-linked polymeric materials in which there are not sufficient cross-links to make them as rigid as the thermosetting plastics, but just sufficient to make them return to their original dimensions when the deforming load is removed. Whereas thermosets show little elongation under stress, elastomers are capable of elongations of up to 1000 per cent at tensile failure. Elastomers are, therefore, capable of extreme elastic deformation at low levels of stress. Unlike metals, the strain is not proportional to stress for elastomers and this will be considered in greater details in Section 11.7. Elastomers are usually addition polymerised as thermoplastics and then cross-linked (vulcanised) with sulphur at approximately every five-hundredth carbon atom. Increased vulcanisation increases the cross-linking and this, in turn, increases the stiffness and reduces the elongation properties of the material. Fully vulcanised, natural rubber becomes a rigid, brittle thermoset called 'ebonite'.

2.16 Polymer building blocks

The polymeric (plastic) materials introduced in Section 2.15 are all built from carbon atoms in association with other elements such as oxygen, hydrogen, nitrogen, chlorine and fluorine. Carbon atoms have four chemical bonds or, as chemists would say, a valency of four. Hydrogen atoms have a valency of one, so if hydrogen and carbon are combined in the simplest way to give a molecule of methane (natural) gas, the molecule would appear as:

$$\begin{array}{c} \text{H} \\ | \\ \text{H—C—H} \\ | \\ \text{H} \end{array}$$

Thus four hydrogen atoms combine with one carbon atom to make one molecule of methane gas. This molecule is given the chemical formula CH_4 and, because it consists solely of hydrogen and carbon, it is referred to as a *hydrocarbon*. The hydrocarbons are found in crude oil, coal and natural gas. They can be classified into four main groups as follows.

2.16.1 *Alkanes*

The *alkane series* was previously known as the *paraffin series*. These are the simplest of the four groups of hydrocarbons. They have a general formula of C_nH_{2n+2}. For example, in the methane molecule just considered there is only one carbon atom so $n = 1$, and the number of hydrogen atoms is $2(1) + 2 = 4$. This agrees with the formula already stated as CH_4. One way of recognising alkanes is the fact that their names always end in 'ane' (as in methane, propane, octane, etc.).

Alkanes are *saturated* hydrocarbons and, as such, they contain the maximum number of hydrogen atoms in each case, as shown in Fig. 2.28, and this makes them rather inactive chemically. The alkanes are the most common group of hydrocarbons appearing in crude oil.

Fig. 2.28 *Common alkanes: H = hydrogen, C = carbon*

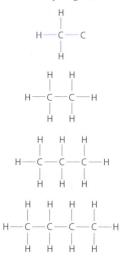

Methane (CH_4) – natural gas

Ethane (C_2H_6) – converted into plastics

Propane (C_3H_8) – heating fuel

Butane (C_4H_{10}) – heating fuel or converted into synthetic rubbers

2.16.2 *Olefins*

These are *unsaturated* hydrocarbons or *alkenes*. That is, additional hydrogen atoms have to be added to olefins in order to saturate them. This unsaturated condition makes them chemically reactive and olefins form the basis of many thermoplastic and elastomer materials. When their general formula is C_nH_{2n} they are called mono-olefins and are given names ending in 'ylene' (as in ethylene, propylene, etc.). There are more complex forms of the olefins but they are beyond the scope of this book. Olefins are usually produced in the course of oil refinery operations, but they are not abundant in crude oil. They are used as a feed stock for the polymer (plastics) industry where they are known as chemical intermediates. Two typical examples are shown in Fig. 2.29.

Fig. 2.29 *Common olefins: H = hydrogen, C = carbon*

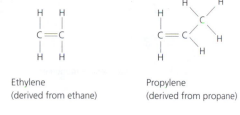

Ethylene
(derived from ethane)

Propylene
(derived from propane)

2.16.3 *Naphthenes and aromatics*

These both have ring-shaped molecules, as shown in Fig. 2.30. Materials made from a ring-shaped molecule have improved mechanical properties, for example the high tensile strength of nylon. Naphthenes have saturated molecules and names beginning with 'cyclo' (as in cyclohexane). Aromatics, on the other hand, are unsaturated and are chemically

Fig. 2.30 *Common naphthenes and aromatics: H = hydrogen, C = carbon*

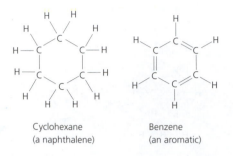

Cyclohexane
(a naphthalene)

Benzene
(an aromatic)

highly reactive, being used in solvents and explosives. Aromatics are rare in crude oils (except those found in California), but occur in coal. They form the basis of the styrene group of plastics.

2.17 Polymers

Consider the manufacture of simple polymeric (plastic) material such as polyethylene. The alkane *ethane* is first converted into its corresponding olefin (hence the term 'chemical intermediate'). A single molecule of the olefin *ethylene* is referred to as a *monomer*, and the next stage of the process is to combine several monomers together to form a much larger molecule called a *polymer* ('poly' means many). In this example it is *polyethylene*.

In the form of a polymer the olefin takes the characteristics of a plastic material. Two examples are shown in Fig. 2.31. There are some simple basic rules which govern the number of monomers which can be brought together to form a polymer. For example, at room temperature ethylene, which is made up of single molecules (monomers), is a gas.

Fig. 2.31 *Simple polymers*

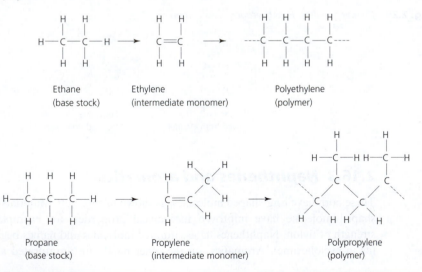

Ethane
(base stock)

Ethylene
(intermediate monomer)

Polyethylene
(polymer)

Propane
(base stock)

Propylene
(intermediate monomer)

Polypropylene
(polymer)

A polymer of 6 monomers of ethylene is a liquid; a polymer of 36 polymers is a grease; a polymer of 140 monomers is a wax; and a polymer of 500 or more monomers is a solid plastic material. The upper limit is about 2000 monomers. At this point there is little further increase in strength, but a considerable increase in hardness and brittleness. This rule applies to most plastic materials.

Polymeric materials containing only carbon and hydrogen are highly flammable. To render these materials less flammable (and the rules governing building applications insist on this) at least one of the hydrogen atoms in each monomer has to be replaced by a chlorine atom, as shown in Fig. 2.32. The resulting polyvinyl chloride (PVC) is a non-flammable plastic suitable for extruding into rain guttering. Fluorine may be added instead of chlorine to produce a more expensive material with superior mechanical and fire-resistant properties. It is also more resistant to sunlight.

Fig. 2.32 *Chlorinated plastics: H = hydrogen, C = carbon, Cl = chlorine*

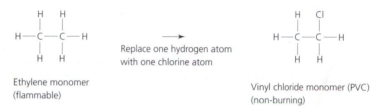

Ethylene monomer
(flammable)

Replace one hydrogen atom
with one chlorine atom

Vinyl chloride monomer (PVC)
(non-burning)

Thus it becomes obvious that all polymeric materials have two things in common.

- They are all made up of long chains of individual unit molecules called monomers. When large numbers of these monomers are repeated over and over again to form a long chain molecule they are referred to as polymers. Hence such materials are known as polymeric materials. This is a much more accurate description of these materials than the popular word 'plastic'.
- They are all based on a chain of monomers which builds up a giant molecule. It is the shape of this chain as well as its composition that determines the properties of polymeric materials.

Figure 2.33 shows some of the shapes which the molecular chain of a polymeric material may take. The linear chain shown in Fig. 2.33(a), and the linear chain with side branches shown in Fig. 2.33(b), are typical of thermoplastic materials. The simple linear chains with no side branches can easily move past each other. This results in a non-rigid thermoplastic material which can be flexed and stretched. Such materials melt at low temperatures and easily return to their original state when they cool down. Polyethylene is an example of such a material.

Since it is more difficult for branched linear chains to move past each other, materials with monomers in this configuration are more rigid, harder and stronger. Also they are less dense since the side branches prevent the chains being packed so closely together. Heat energy is required to break down the side branches and this raises the melting temperature above that for materials with a simple linear chain. An example of a thermoplastic material with a branched linear chain is polypropylene.

Fig. 2.33 *Typical polymer chains: (a) linear polymer chain; (b) branched polymer chain; (c) cross-linked polymer chain*

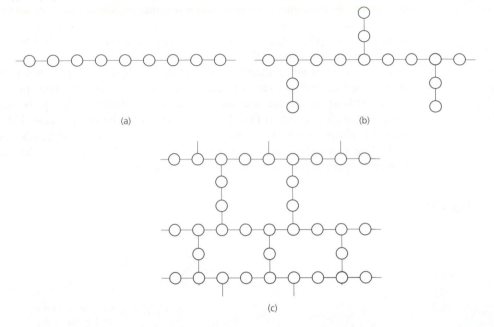

(a)

(b)

(c)

The cross-linked molecular chain shown in Fig. 2.33(c) is typical of the thermosetting plastics. These are rigid and tend to be brittle once the links have been formed by curing the material during the moulding process. Once curing (a chemical reaction) has occurred the material cannot be softened by reheating as the process is not reversible. If heated sufficiently they char or burn and are destroyed.

Thermosetting plastics differ from thermoplastic materials in the way in which polymerisation (curing) occurs. In thermoplastics, polymerisation occurs through the addition of monomers at the time of manufacture and no further curing occurs during the moulding process. In thermosetting plastics, polymerisation usually occurs through condensation. In this latter process the plastic moulding material reacts within itself, or with some other chemical (the hardener), when heated to a critical temperature. At this temperature the moulding material releases or 'condenses' out small molecules such as water and polymerisation becomes complete. This loss of water results in volumetric shrinkage which has to be allowed for in the moulding process and, also, the moulds have to be designed with vents to allow the steam generated during polymerisation (curing) to escape. The principle of polymerisation by condensation is shown in Fig. 2.34.

2.18 Crystallinity in polymers

Crystals have already been described as having their particles arranged in recurring well-ordered geometric patterns. Materials which do not have this ordered arrangement of geometric patterns as their basic structure have been described as amorphous (without shape). For example, the polymeric material PTFE (used for coating non-stick cooking

Fig. 2.34 *Curing of thermosetting plastics*

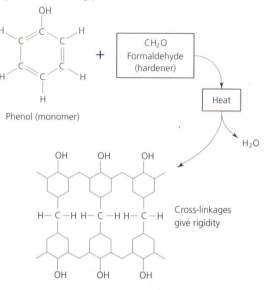

Phenol (monomer)

CH₂O
Formaldehyde
(hardener)

Heat

H₂O

(Water condenses out as the polymer is formed – this causes shrinkage, and moulds have to be vented as water is driven off as steam)

Cross-linkages
give rigidity

Phenol-formaldehyde polymer (Bakelite)

utensils) has a carbon chain which is helical with 14 atoms per turn of the helix and to which are attached side chains. It is hardly surprising that a polymer with the shape of a coil spring with side chains of attached methyl groups (CH₃), or with aromatic rings, has little chance of taking up the ordered patterns of a crystalline material. Hence most polymeric materials are amorphous or even glass-like (supercooled high-viscosity liquids).

Such amorphous arrangements of polymer chains are often represented by a tangle of lines as shown in Fig. 2.35(a). Each line represents an individual molecular chain, but does not show the individual atoms for the practical reason that there would be too many and they would be too small to draw. However, simple linear chains without side branches or cross-links may show some degree of ordering on a submicroscopic scale, Such ordered regions are called *crystallites* and there may be several such regions along a single molecular chain. Such an arrangement is shown in Fig. 2.35(b) where it can be seen that the individual molecular chains extend through several crystalline and non-crystalline regions.

Fig. 2.35 *Crystallinity in polymeric materials: (a) linear amorphous polymer chains; (b) crystallites amongst amorphous chains*

Crystallite

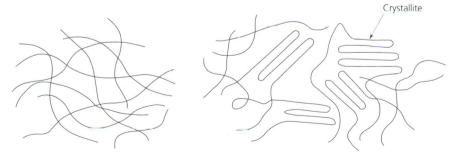

The *crystallinity* of a polymeric material is defined as the ratio between the mass of the crystallites and the total mass of the material being considered. For example, a material having 80 per cent crystallinity will consist of 80 per cent crystallite structure and 20 per cent non-crystallite (amorphous) structure. Since the monomers making up the polymer chain are packed more closely together in crystallites, it follows that materials with a high crystallinity will be more dense than materials with a low crystallinity. For example, low-density polyethylene with a crystallinity of only 50 to 70 per cent will have a density of about 920 kg/m³ and a melting point of 115 °C; whereas high-density polyethylene with 75 to 95 per cent crystallinity will have a density of about 950 kg/m³ and a melting point of 135 °C. The crystallinity of a polymeric material has a marked effect upon its properties. For example, increasing the crystallinity of a material:

- Increases its melting point and, instead of softening gradually with increased temperature, it will exhibit a sharper melting point which is similar to that of fully crystalline materials.
- Increases the resistance of the material to the absorption of water and to solvent attack since it is more difficult for the water and solvent to penetrate the high-density crystallites than it is to penetrate the more open amorphous structure.
- Prevents the penetration of plasticisers and this reduces the ultimate elongation of the material.
- Makes the material more impervious to gases and this may be useful in food packaging and protective coatings. However, this high level of impermeability is a disadvantage in polymer fibres which must be coloured by dyeing.

The effect of crystallinity on the ultimate tensile strength and percentage elongation of a typical polymeric material is shown in Fig. 2.36. The relative crystallinity can be modified by heat treatment. For example, a crystallisable polythene can be given a crystallinity of 80 per cent by slow cooling or a crystallinity of only 65 per cent by rapid cooling (quenching).

Fig. 2.36 *Effect of crystallinity on the properties of polyethylene*

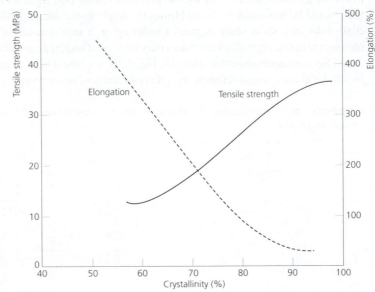

2.19 Orientation

The intermediate condition of orientation lies between the amorphous and the crystalline state. The processing of the plastic material and the effects produced are similar to those involved in the cold working of metals. The mechanical working of amorphous and crystalline polymers has the following effects upon their morphology and properties.

2.19.1 *Uniaxial orientation*

If a polymer is drawn into fibres through a die (similarly to metal wire drawing), the molecules and the crystallites of the polymer are aligned parallel to the direction of drawing – that is, the structure has become oriented in the direction of drawing. This oriented condition greatly increases the tensile strength and the impact strength levels compared with the same polymer in the bulk condition.

To understand how this improvement in properties comes about, consider the typical stress–strain graph for a crystalline polymer as shown in Fig. 2.37. From O to A the material suffers elastic deformation and on removal of the stress the displaced atoms return to their original position. From the point A, plastic deformation occurs. Providing the strain rate (the speed at which the material is deformed) is low enough, two things will happen. Firstly, the polymer chains will unfold and straighten out and, secondly, they will slip over each other. Thus the polymer chains will end up aligned in the direction of drawing. They will also be packed together in a very orderly manner. This results in the increased tensile strength and toughness mentioned earlier.

Fig. 2.37 *Typical stress–strain curve for a crystalline polymer*

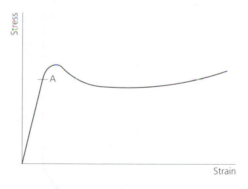

Unfortunately, materials which have been subjected to uniaxial working (working in one direction only) only benefit from improved properties in the direction of working (drawing). They still remain relatively weak and tend to split when forces are applied at an angle to the direction of orientation. For fibres this is relatively unimportant since they are nearly always loaded in the direction of polymer chain orientation; however, for films (sheet) this can be a serious defect.

2.19.2 *Biaxial orientation*

Films and sheets are produced so that the polymer chains are biaxially oriented. That is, the material is stretched longitudinally and transversely during production. This ensures uniform strength in whichever direction the film or sheet is stressed.

Film

This is the term used for flat material which is less than 0.25 mm in thickness. The process of film blowing is shown in Fig. 2.38. The plastic material is extruded vertically as a tube from an annular die. The wall thickness of the tube so produced is usually in the order of 0.4 mm to 0.6 mm. The tube is closed by pinch rolls high above the point of extrusion, and air is blown into the tube through the centre of the die mandrel to inflate the tube into a thin-walled bubble. The wall thickness of the bubble is the final film thickness and stretching is uniform in all directions. Usually the bubble diameter is 1.5 to 3 times the die diameter. The tube is finally slit and opened out to make a flat film.

Fig. 2.38 *Film blowing*

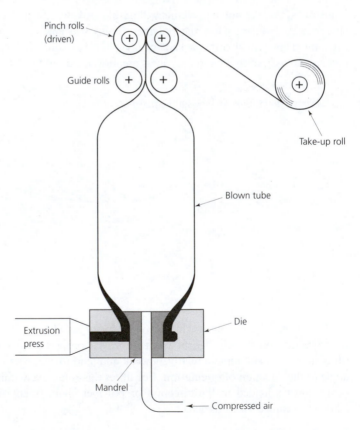

Sheet

Sheet is produced by extruding the plastic material through a sheeting die as shown in Fig. 2.39(a). The extruded sheet is then calendered between rolls, as shown in Fig. 2.39(b), to orientate the polymer chains, reduce the sheet to its final thickness (over 0.25 mm for sheet, under 0.25 mm for film) and impart the required surface finish.

The optical properties of polyolefins are greatly improved by biaxial orientation. The best gloss and clarity is given by rolling, although the clarity of blown film improves as the blow ratio (ratio of bubble diameter to die annulus diameter) is increased.

Fig. 2.39 *Plastic sheet production: (a) sheeting die; (b) calendering*

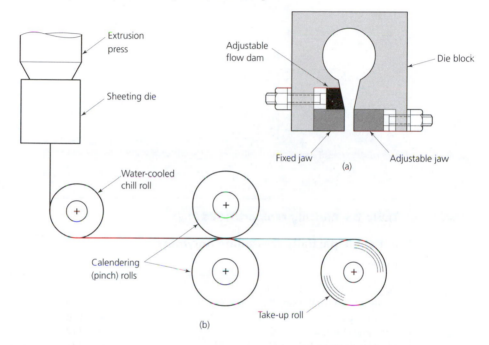

2.20 Melting points of crystalline polymers (T_m)

Crystalline materials such as metals show well-defined melting points when heated to sufficiently high temperatures. However, amorphous solids do not show a clearly defined melting point when heated. They merely become progressively less rigid until they eventually become liquid. Amorphous thermoplastic materials behave in this way. At room temperature they are so viscous that they behave as solids, but as the heat rises the material becomes progressively less viscous until it becomes liquid without showing any clearly defined melting point.

Other thermoplastic materials show some crystallinity, and for these it is possible to determine a melting temperature (T_m). The melting temperature is determined by plotting the specific volume of the material against temperature rise as shown in Fig. 2.40. Initially the smaller, less perfect crystallites become amorphous and this is indicated by the portion of the curve marked AB. At the point B, a rapid increase in specific volume with temperature occurs. This point is regarded as the melting temperature (T_m), and is defined as the point where the material loses its crystallinity completely and becomes amorphous. The greater the crystallinity of any thermoplastic material, the higher will be its melting temperature. Table 2.3 lists the melting points for some partially crystalline polymers.

Fig. 2.40 *Melting temperature (T_m) of a partially crystalline polymer*

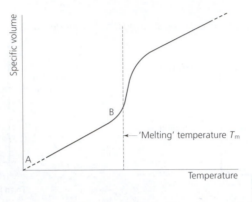

Table 2.3 *Melting temperatures (T_m)*	
For some partially crystalline polymers	
Material	*T_m (°C)*
Polyethylene (low density)	120
Polyethylene (high density)	135
Polypropylene	180
Polyvinyl chloride	212
Polytetrafluoroethylene (PTFE)	327
Natural rubber	30

2.21 Glass transition temperature (T_g)

Polymethyl methacrylate (Perspex) is a rigid, glass-like plastic material with excellent optical properties at room temperature. At just above the temperature of boiling water it becomes soft (but not molten) and can be moulded into streamline shapes for aircraft cockpit canopies.

Polythene is a flexible material at room temperature and is widely used for mouldings and in sheet form. However, if it is cooled to about −120 °C it becomes a hard, brittle material. A rubber ball which can be bounced indefinitely at room temperature would shatter into fragments if it were dropped immediately after cooling in liquid nitrogen. The temperature at which a polymeric material changes from being rigid and brittle to being flexible and rubbery is called the glass transition temperature (T_g).

The glass transition temperature (T_g) is less well defined than the melting temperature (T_m) and is difficult to determine. However, at the glass transition temperature, the tensile modulus (see Section 11.7) undergoes an abrupt change, as shown in Fig. 2.41, and this can be used to determine the glass transition temperature. Below the glass transition temperature polymeric materials show a relatively high tensile modulus, with little extension and a high level of rigidity. Above the glass transition temperature, the tensile modulus is lower, the level of rigidity is lower and the extension is very considerably increased as shown in Fig. 2.42. The glass transition temperature varies widely from one polymeric material to another, as can be seen from Table 2.4.

Fig. 2.41 *Glass transition temperature (T_g)*

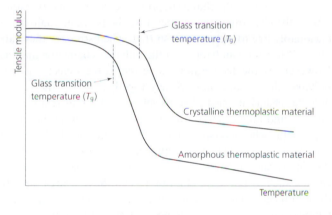

Fig. 2.42 *Effect of the glass transition temperature (T_g) on the mechanical properties of a typical thermoplastic material*

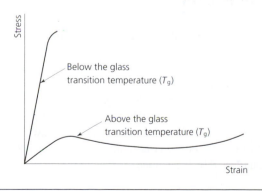

Table 2.4 *Glass transition temperatures (T_g)*

For some typical thermoplastic materials

Material	T_g (°C)
Polyethylene	−120
Natural rubber	−70
Polypropylene	−30
Polyvinylidene chloride	−17
Polymethyl methacrylate (Perspex)	0
Polyvinyl acetate	27
Polyethylene terephthalate (Terylene)	70
Polyvinyl chloride	80
Polystyrene	100
Cellulose acetate	120
Polytetrafluoroethylene (PTFE)	130

The reason for the change in properties at the glass transition temperature is as follows. Above the glass transition temperature the polymer chains for thermoplastic materials are reasonably free to move about so that, when stressed, they can uncoil and slide over each other. When this can happen to the polymer chains the material is soft, elastic and tough. However, as the temperature is lowered, the density of the material increases as the polymer chains pack more closely together. Below the glass transition temperature, the polymer chains are packed so closely together that relative movement becomes extremely difficult and little extension occurs, thus the material becomes hard and does not extend appreciably when stressed – that is, it becomes brittle. Polymers with linear molecules with no side branches, such as polyethylene, have a very low glass transition temperature since the chains can slide over each other relatively easily. However, polymers having branched chains, such as polyvinyl chloride (PVC), have a high glass transition temperature since their branched chains interfere with each other as they start to pack together when the temperature falls. This interference makes relative movement difficult.

2.22 The effect of temperature on polymer applications

Since polymeric materials with a high crystallinity have a well-defined melting temperature (T_m), they can be hot formed (moulded) above this temperature, and cold formed between their T_g and T_m temperatures when they will be solid but soft and flexible. For example, polyethylene with a 95 per cent crystallinity has a T_g of −120 °C and a T_m of +138 °C. Thus it is soft and flexible over a wide range of temperatures. The maximum service temperature is usually taken as approximately 85 per cent of the melting temperature, which in this case is 120 °C. Compare this with a polythene with only 60 per cent crystallinity where the T_m is reduced to 115 °C and a service temperature of only 98 °C. The glass transition temperature (T_g) is unaffected by the change in crystallinity.

Amorphous polymeric materials are usually moulded or formed above their glass transition temperature (T_g) where they are soft and pliable, but used below this temperature where they are rigid. For example, rigid polyvinyl chloride (PVC) is an amorphous polymer with a T_g of 87 °C. It is normally softened by hot air blast or radiant heat before manipulating to shape. Since amorphous plastics do not have a well-defined melting temperature, the service temperature is taken as 85 per cent of the glass transition temperature which, for rigid PVC, is 70 °C.

2.23 Memory effects

Polymeric materials with uniaxial and biaxial polymer chain orientation can have their crystallinity restored by heat treatment. This heat treatment consists of heating the material between its T_g and T_m temperatures so that the polymer chains lose their orientation and recover their original amorphous or their original semicrystalline structure. This ability to return to the prestretched, disoriented state is referred to as the *memory effect* of the material.

Use is made of the memory effect in the packaging industry by heat shrinking protective foil around prepacked food and other commodities. The commodity to be packed is wrapped loosely in the foil which is then heated to above its glass transition temperature but well below its melting temperature. This causes restoration of the original disoriented polymer structure and shrinkage occurs resulting in the commodity becoming tightly packed.

SELF-ASSESSMENT TASK 2.5

1. State the essential differences between:

    (a) thermoplastics
    (b) thermosetting plastics
    (c) elastomers

2. (a) An alkane molecule contains **four** carbon atoms. State its chemical formula and show how you arrived at your answer.
    (b) An olefin molecule contains **two** carbon atoms. State its chemical formula and show how you arrived at your answer.
    (c) Explain the essential differences between a **saturated** and an **unsaturated** hydrocarbon molecule and state how this affects their reactivity.

3. (a) Explain the essential differences between a monomer and a polymer.
    (b) Explain the essential difference between a wax polymer molecule and a solid polymer molecule.

4. Explain what is meant by the term 'crystallinity'.

2.1 Sketch a typical representation of an atom and describe the particles which are to be found in the nucleus and orbiting around the nucleus.

2.2 Explain in detail the mechanism of crystal formation and growth as a metal cools from the liquid state to the solid state.

2.3 Explain with the aid of sketches:
(a) the difference between unit cell and space lattice
(b) how dendritic growth occurs
(c) the difference between the crystal structure of a metal and its irregularly shaped grains

2.4 Explain what is meant by the 'latent heat of fusion' and why the temperature of the metal remains constant whilst fusion occurs.

2.5 Explain the following terms associated with the solidification of a molten metal:
(a) interdendritic porosity
(b) drawing
(c) segregations
(d) undissolved inclusions
(e) gas porosity

2.6 Explain what is meant by the term 'misorientation' and state how this can affect the 'creep' resistance of a metal.

2.7 Describe the differences between the following groups of substances and explain how they affect the polymeric materials made from them:
(a) paraffins
(b) olefins
(c) naphthenes
(d) aromatics

2.8 With the aid of diagrams, show what is meant by the following terms related to polymer chains and how they affect the properties of the polymer:
(a) linear
(b) branching
(c) cross-linked

2.9 With reference to polymeric materials, explain what is meant by:
(a) uniaxial orientation
(b) biaxial orientation
(c) memory effect

2.10 With reference to polymeric materials, explain what is meant by:
(a) glass transition temperature (T_g)
(b) melting temperature (T_m)

3 Alloying of metals

The topic areas covered in this chapter are:

- Alloys and alloying elements.
- Solubility and solid solutions.
- Intermetallic compounds.
- Cooling curves.
- Phase and alloy types.
- Phase equilibrium diagrams.
- Coring.

3.1 Alloys

Pure metal objects are used where good electrical conductivity, good thermal conductivity, good corrosion resistance or a combination of these properties are required. However, pure metals usually lack the strength required for structural materials. Therefore alloys are mainly used for structural materials since they can be formulated to give superior mechanical properties. For example, properties such as tensile strength, yield strength and hardness are improved by alloying. However, ductility is reduced. Since alloys can be designed to give specific properties they can be 'tailored', to suit a particular application.

An alloy is an intimate association of two or more component materials which form a single metallic liquid or solid. The component materials may be metal elements, or they may be metal elements and non-metal elements, or they may be metal elements and chemical compounds. It is important to distinguish between alloying elements and impurities. *Alloying elements* (see Section 3.2) are deliberately added in controlled quantities to modify the properties of a material to match a particular specification. Impurities are undesirable and are usually carried over from some previous process such as smelting or casting. Since they impair the properties of the material, steps are usually taken to remove the impurities or reduce them to a level where their effects become insignificant. Impurities must not be confused with alloying elements.

Useful alloys can only be produced from component materials which are soluble in each other in the molten state. That is, they must be completely *miscible*. It would be useless to try to form an alloy from zinc and lead. The molten zinc would float on top of the molten lead and, upon cooling, they would form separate layers in the solid state with only tenuous bonding at the interface. Alloys are formed in three ways.

- If the alloying components in the molten solution have similar chemical properties, and their atoms are of similar size, they will not react together but will form a *solid solution* on cooling.
- If the alloying components in the molten solution have different chemical properties they may attract each other and form chemical compounds. Where the alloying components are both metals, these compounds are referred to as *intermetallic compounds*. Upon cooling the crystals will consist of a mixture of such compounds.
- In a situation where atoms with different chemical properties attract each other less than those with similar chemical properties, then both intermetallic compounds and solid solutions will be present at the same time. Upon cooling they will tend to separate out at the grain boundaries to form a *heterogeneous mixture*.

In any alloy, the metal which is present in the larger proportion is referred to as the *parent metal* or *solvent*, whilst the metal (or non-metal) present in the smaller proportion is known as the *alloying component* or *solute*. Commercial alloys often contain more than one alloying element. For example, 'gun-metal' bronze alloy contains both tin and zinc in addition to the parent metal copper, whilst phosphor-bronze contains both tin and phosphorus in addition to the parent metal copper.

3.2 Alloying elements

Alloy steels are widely used in engineering, so let's now look at some of the more important alloying elements and the effects we can expect them to have. These alloying elements may be used individually or in combination.

Aluminium
The presence of up to 1 per cent aluminium in alloy steels enables them to be given a hard, wear-resistant skin by a heat treatment process called *nitriding*.

Chromium
The presence of small amounts of chromium improves the ability of steels to respond to hardening by heat treatment by forming hard carbide particles in the alloy. Unfortunately the presence of chromium also promotes grain growth. Grain growth reduces the strength and toughness of all metals. For this reason chromium is only used in small amounts when it is the only alloying element present. More often it is associated with nickel in alloy steels (nickel–chrome alloys), this is because nickel tends to refine the grain structure. The presence of large amounts of chromium improves the corrosion resistance and heat resistance of steels (stainless steels).

Cobalt
The presence of cobalt induces sluggishness into the heat-treatment transformations and improves the ability of tool steels to operate at high temperatures without softening. It is an important alloying element in 'super' high-speed steels.

Copper

The presence of 0.5 per cent copper helps to improve the corrosion resistance of alloy steels.

Lead

The presence of up to 0.2 per cent lead improves the machinability of steels. Unfortunately it also reduces the strength of the steel to which it is added. Leaded steels are widely used for the mass production of lightly loaded turned parts produced on automatic and CNC lathes.

Manganese

This element is always present in steels, both plain carbon steels and alloy steels. It combines with any residual sulphur which is an impurity carried over from the smelting (extraction) process. The presence of sulphur, in the form of iron sulphide, causes brittleness and reduces the strength and toughness of the steel. In larger quantities (up to 12.5 per cent) manganese improves the wear resistance of steels by spontaneously forming a hard skin when subjected to abrasion. Manganese alloy steels can also have high strength and toughness and are being increasingly used in place of the more expensive nickel–chrome alloys in stressed components for the motor industry.

Molybdenum

The presence of molybdenum in alloy steels raises their high-temperature creep strength; stabilises their carbides; improves the 'red-hardness' of high-speed steel-cutting tools and allows them to operate at continuously higher temperatures, and reduces 'temper brittleness' and 'weld-decay' in nickel–chromium steels.

Nickel

The presence of nickel in alloy steels results in increased strength by grain refinement, it also improves the corrosion resistance of steels. Unfortunately nickel is a powerful graphetiser and it reduces the presence of any carbides present. For this reason it is only associated with low-carbon steels when it is the sole alloying element. Usually nickel and chromium are used together. By stabilising the carbon, promoting the formation of carbides and offsetting the graphetising effects of the nickel, the chromium offsets the disavantages of nickel as an alloying element. At the same time, the grain refinement properties of the nickel offsets the tendency of the chromium to cause grain growth. Thus chromium improves the ability of the alloy to respond to heat treatment, whilst the nickel increases the strength of the alloy both by grain refinement and directly because of its own strength (see also 'Chromium').

Phosphorus

This is a residual element from the smelting (extraction) process. It causes weakness in the steel and, usually, considerable care is taken to reduce its presence below 0.05 per cent. However, a trace of phosphorus can improve machinability and, in larger quantities, can improve the fluidity of casting steels when maximum strength and toughness are not of prime importance. It is also included in cast irons where high fluidity is required during the casting process.

Silicon

The presence of up to 0.3 per cent silicon also improves the fluidity of casting steels without the reduction of strength associated with phosphorus. Up to 1 per cent silicon improves the heat resistance of steels and also improves the magnetic properties by increasing the permittivity of the alloy (see Section 1.4, 'Magnetic properties'). Unfortunately silicon, like nickel, is a powerful graphetiser and is never added in larger quantities to medium- and high-carbon steels.

Sulphur

This is a residual element from the smelting (extraction) process. Since the presence of iron sulphide reduces the strength and toughness of the steel, every effort is made to eliminate it. Fortunately sulphur has a greater affinity for manganese that it has for iron, and manganese sulphide does not impair the mechanical properties of steels. However, sulphur is sometimes added to low-carbon steels to improve their machinability when the reduction in component strength can be tolerated.

Tungsten

The presence of tungsten in alloy steels promotes the formation of very hard carbides, and induces sluggishness into the heat-treatment transformations during hardening. This enables steels to retain their hardness at high temperatures. Tungsten is the main alloying element in high-speed steels used for cutting tools. It is also used in high-duty die steels.

Vanadium

This element enhances the performance of the other alloying elements and is never used alone. Its effects in alloy steels are many and various:

- Promotes the formation of carbides.
- Stabilises the martensite and improves hardenability.
- Reduces grain growth.
- Enhances the 'hot hardness' of tool steels and die steels.
- Improves the fatigue resistance of steels (particularly spring steels).
- Improves the life of valve steels used in internal combustion engines.

3.3 Solubility

In order to understand the formation of alloys, it is first necessary to understand the principles of solubility in the liquid and solid states. Sodium chloride (common table salt) dissolves readily in cold water. At room temperature, approximately 35 g of sodium chloride will dissolve in 100 g of water. The exact amount will depend upon the temperature of the water. If more sodium chloride is added to the solution it will not dissolve because the solution has already taken up all the salt it can dissolve and is said to be *saturated*. The excess salt will remain as a residue. The solubility of sodium chloride increases only slightly as the temperature of the water increases. In this example:

- The water is the *solvent*.
- The sodium chloride is the *solute*.
- The resulting liquid is the *solution*.

Figure 3.1 shows the difference between complete and partial solubility. In the following example the salt used is *copper sulphate* dissolved in *water*. Unlike sodium chloride, the solubility of copper sulphate increases substantially as the temperature of the water increases. This is shown in Fig. 3.2, where point A represents 50 g of copper sulphate being dissolved in 100 g of water.

Fig. 3.1 *Solubility: (a) complete; (b) partial*

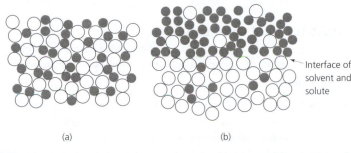

(a)   (b)

Fig. 3.2 *Solubility curve for copper sulphate*

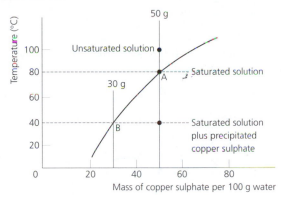

- Above 80 °C the water is capable of dissolving more than 50 g of copper sulphate, so the solution is said to be *unsaturated*.
- At 80 °C the water will dissolve a maximum of 50 g of copper sulphate so the solution is said to be *saturated*.
- Below 80 °C the water dissolves less than 50 g of copper sulphate. For example, at 40 °C (point B) only 30 g of copper sulphate can be dissolved in 100 g of water (can be 'held in solution') and the balance of 20 g of copper sulphate will be precipitated out of solution as a solid residue.

Substances which will not dissolve in a solvent are said to be *insoluble*. However, a substance which is insoluble in one solvent may be soluble in a different solvent.

3.4 Solid solutions

Most metals are completely and mutually soluble (they are miscible) in the liquid state – that is, when they are molten. Some, such as copper and nickel, not only form solutions in the molten or liquid state but remain in solution upon cooling and solidifying to become *solid solutions*. There are two sorts of solid solutions:

- *Substitutional* solid solutions.
- *Interstitial* solid solutions.

Substitutional solid solutions

The copper–nickel alloy mentioned previously is an example of a substitutional solid solution. The more important factors governing the formation of a substitutional solid solution are:

- *Atomic size*   The atoms of the solute and the solvent must be approximately the same size. If the atom diameters vary by more than 15 per cent the formation of a substitutional solid solution is highly unlikely.
- *Electrochemical series*   If there is only a small difference in charge between the alloying components then they will probably form a solid solution. Conversely, if their charges are very dissimilar they are more likely to form intermetallic compounds.

Fig. 3.3 *Substitutional solid solution: (a) face-centred cubic crystal of copper; (b) face-centred cubic crystal of nickel; (c) substitutional solid solution of copper and nickel*

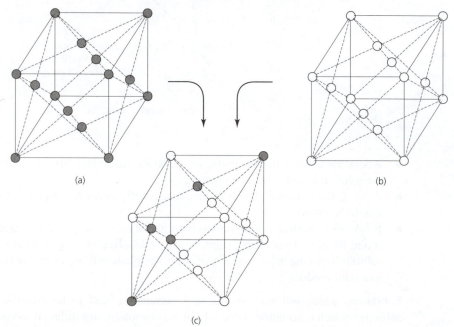

(a)

(b)

(c)

- *Valency*   A metal of lower valency is more likely to dissolve one of higher valency than the other way round, assuming the conditions stated above are also favourable. This holds good particularly for monovalent metals such as copper, silver and gold.

Figure 3.3 shows that both copper and nickel form face-centred-cubic crystals. When these two metals are in solid solution they form a single face-centred-cubic lattice with atoms of nickel replacing atoms of copper in the lattice. Hence the term *substitutional solid solution*. The substitution can be ordered, with the solute atoms taking up regular fixed positions of geometric symmetry in the lattice. However, most solid solutions are disordered, with the solute atoms appearing virtually at random throughout the solvent lattice.

Interstitial solid solutions

These are formed when the solute atoms are small enough to lie between the solvent atoms as shown in Fig. 3.4. For example, carbon atoms can form an interstitial solid solution with face-centred-cubic crystals of iron.

Fig. 3.4   *Interstitial solid solution*

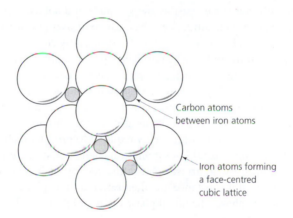

Carbon atoms
between iron atoms

Iron atoms forming
a face-centred
cubic lattice

3.5 Intermetallic compounds

It has already been stated that where the components of the alloy are sufficiently different chemically, they will tend to form compounds rather than solid solutions. In general, intermetallic compounds tend to be hard and brittle and are thus less useful for engineering alloys than the tough and ductile solid solutions. Intermetallic compounds are most widely found in bearing alloys where they form hard, wear-resistant pads with a low coefficient of friction, set in a matrix of tough, ductile solid solution.

3.6 Cooling curves

Most substances can exist as gases, liquids and solids, depending upon their temperature. Water is one such substance, which can exist as a gas or vapour (steam) if it is sufficiently hot, as a liquid, and as a solid (ice) if sufficiently cold. If water is raised to its boiling point and allowed to cool slowly, the change in temperature with time can be plotted as a graph, as shown in Fig. 3.5. Such a graph is called a cooling curve. It can be seen from the graph that where a change of state occurs (such as liquid water to solid ice) there is a short pause in the cooling process. This pause is referred to as an *arrest point* and is the result of the water giving up latent heat energy as it changes into ice.

Latent heat is the heat energy required to produce a change of state in a substance at a constant temperature. Thus a physical change of state during cooling or heating is always accompanied by an arrest point in the cooling or heating curve. The gaseous, liquid and solid states of a substance are often referred to as phases, and substances are said to be in the gaseous phase, the liquid phase or the solid phase. The term 'phase' will be dealt with more fully in Section 3.7.

The cooling curve shown in Fig. 3.5 is typical of all pure substances and applies equally well to pure metals. Alloys, however, consist of two or more components and, to understand their behaviour on cooling, the above explanation must now be extended to encompass a solution. A suitable solution is that of domestic table salt (sodium chloride) in water. Figure 3.6 shows the cooling curve for a sodium chloride–water solution compared with the cooling curve for pure water. It can be seen that the salt water solution has two arrest points and that both of these are below the freezing point of pure water. Only the solid and liquid states are considered in this figure.

A salt–water solution has a lower freezing point than pure water and at 0 °C no change of state occurs. However, as cooling continues, droplets of pure water separate out from solution and immediately turn into ice particles. This occurs at the upper arrest point, which is usually not too well defined. The process of separation continues as the temperature of the remaining solution is further reduced. Thus as the temperature continues to fall, more and more water separates out and freezes, causing the concentration of the remaining salt water to increase. When the lower arrest point is reached, even the concentrated salt–water solution freezes and no liquid phase is left. The solid so formed consists of a mixture of fine crystals of pure water (ice) and fine crystals of salt (sodium chloride).

Fig. 3.5 *Cooling curve for water*

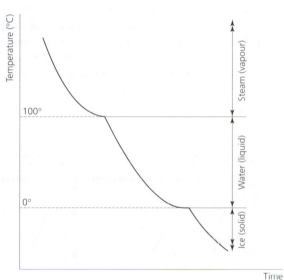

Fig. 3.6 *Cooling curve for a salt–water solution*

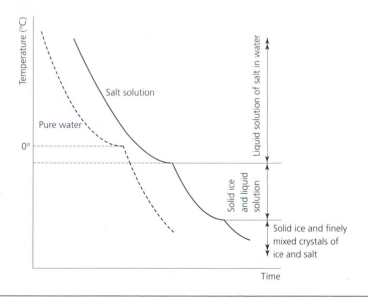

If the experiment is repeated several times using stronger and weaker salt–water solutions, a 'family' of cooling curves can be plotted on the same axes as shown in Fig. 3.7. Reference to this figure shows that:

- The temperature of the lower arrest point remains constant.
- The temperature of the upper arrest point falls as the concentration of the solutions increases until a point is reached where the temperatures of the upper and lower arrest points coincide.
- The ratio of solid to liquid, where the temperatures of the arrest points coincide, is referred to as a *eutectic* composition. Solutions with a lower concentration of solid to liquid are referred to as *hypo-eutectic* solutions. Solutions with higher concentrations of solid to liquid are referred to as *hyper-eutectic* solutions.
- When the concentration of the solution increases beyond that of the eutectic composition the temperature of the upper arrest point rises once more.

Fig. 3.7 *'Family' of cooling curves*

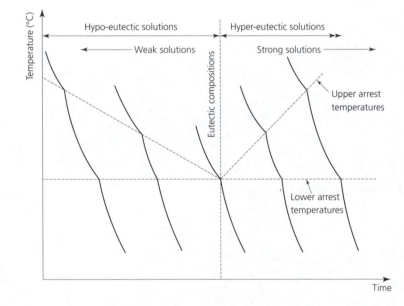

Since the amount of salt which can be held in solution with water varies with temperature, water separates out as ice crystals between the arrest points of hypo-eutectic solutions as the temperature falls, and salt crystals separate out between the arrest points of hyper-eutectic solutions as the temperature falls, therefore the remaining solution is always of a constant concentration and has a constant arrest point temperature. This concentration is the eutectic composition. The fact that excess water or salt is rejected from solution so that a eutectic 'balance' or 'equilibrium' is always ultimately achieved, results in the diagram formed from the cooling curves (Fig. 3.7) being referred to as a *phase equilibrium diagram*. (Also referred to as a 'thermal equilibrium diagram' in some older texts.)

3.7 Phase

The term 'phase' has already been introduced in the previous section. Let us now consider it in more detail. A phase may be defined as:

A portion of a system which is of uniform composition and texture throughout, and which is separated from the other phases by clearly defined surfaces.

Thus for the salt–water solution just considered there are four possible phases:

- Water vapour (steam).
- Liquid salt solution (sodium chloride in water).
- Crystals of water (ice).
- Crystals of salt (sodium chloride).

Each of these four phases is of uniform composition and is separated from adjacent phases by definite boundaries. It is necessary to distinguish between crystals (grains) and phases, since each phase is homogeneous but not necessarily continuous. Thus, the ice phase may appear as separate lumps with each lump containing many single-phase water (ice) crystals and the single-phase crystalline sodium chloride may appear as lumps of salt containing many separate crystals.

Extending this argument to metal alloys, when a liquid solution of two metals (a binary alloy) solidifies one of the following conditions will occur.

- Metals which are soluble in the liquid state may become totally insoluble in the solid state and separate out as grains of two pure metals. Thus there will be two phases present, with each phase consisting of many grains of the same composition.
- Metals which are soluble in the liquid state may remain totally soluble in the solid state resulting in a 'solid solution'. Thus a single-phase solid solution will be present consisting of many grains of the same composition.
- The two metals may react together chemically to form an 'intermetallic compound'. Again a single phase consisting of many grains of the same composition will be present.

Therefore a binary alloy may be built up in a number of different ways and may consist of:

- Two pure metals existing entirely separately in the structure. In practice this is extremely rare since there is usually some solubility of one metal in another.
- A single solid solution of one metal dissolved in another.
- A mixture of two solid solutions if the metals are only partially soluble in each other.
- An intermetallic compound and a solid solution mixed together.

The individual grains found in any of these phases may vary considerably in size. Some are large enough to see with the unaided eye, whilst others are so small that a high-powered microscope is required. Note that although the phases found in alloys, as described above, are formed from two metals, they may equally well be formed between a metal and a non-metal. For example, *austenite* is a solid solution of carbon in iron, whilst *cementite* is the compound iron carbide.

3.8 Alloy types

As has already been stated, alloys consisting only of two component metals are referred to as binary alloys. Even when more than two components are present, a lot of useful information can be obtained from a study of the binary diagram of the two principal components present. The constituent components of most commercially available binary alloys are completely soluble in each other in the liquid (molten) state and, in general, do not form intermetallic compounds. (The exceptions being some bearing metals.) However, upon cooling into the solid state, binary alloys can be classified into the following types.

- *Simple eutectic type*   The two components are soluble in each other in the liquid state, but are completely insoluble in each other in the solid state.
- *Solid solution type*   The two components are completely soluble in each other both in the liquid state and in the solid state.
- *Combination type*   The two components are completely soluble in the liquid state, but are only partially soluble in each other in the solid state. Thus this type of alloy combines some of the characteristics of both the previous types, hence the name 'combination type' phase equilibrium diagram.

Let's now consider these three types of binary alloy systems and their phase equilibrium diagrams in greater detail.

3.9 Phase equilibrium diagrams (eutectic type)

Figure 3.8 shows a eutectic-type of phase equilibrium diagram and it can be seen that it is identical with the diagram produced for a sodium chloride and water solution (Fig. 3.7). That is, total solubility of the salt in water in the liquid state and total insolubility (crystals of ice and separate crystals of salt) in the solid state. In the general case of Fig. 3.8, the two components present are referred to as metal A and metal B. Although they are mutually soluble in the liquid state, both components retain their individual identities of crystals of A and crystals of B in the solid state.

Fig. 3.8 *Phase equilibrium diagram (eutectic type)*

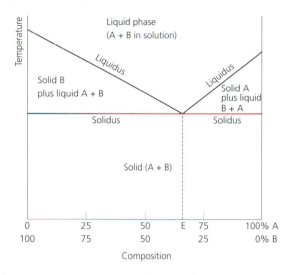

If you refer to Fig. 3.8, you can see that the line joining the points where solidification begins is referred to as the *liquidus* and that the line joining the points where solidification is complete is referred to as the *solidus*.

This type of equilibrium diagram gets its name from the fact that at one particular composition (E), the temperature at which solidification commences is a minimum for the alloying elements present. With this composition the liquidus and the solidus coincide at the same temperature, thus the liquid changes into a solid with both A crystals and B crystals forming instantaneously at the same temperature. This point on the diagram is called the *eutectic*, the temperature at which it occurs is the *eutectic temperature*, and the composition is the *eutectic composition*.

In practice, few metal alloys form simple eutectic type phase equilibrium diagrams. Exceptions to this are the cadmium–bismuth alloys and the thermal equilibrium diagram for such alloys is shown in Fig. 3.9. It can be seen that the eutectic composition occurs when the alloy consists of 40 per cent cadmium and 60 per cent bismuth. For this composition solidification occurs at just over 140 °C with both metals crystallising out of solution simultaneously. The eutectic structure is usually lamellar in form, as shown in Fig. 3.10. In this instance there are alternate layers or 'laminations' of cadmium and bismuth.

Let's now consider the cooling of an alloy consisting of 80 per cent cadmium and 20 per cent bismuth (a hyper-eutectic alloy).

- Above the liquidus there is there is a liquid solution of molten bismuth and molten cadmium.
- As the solution cools to the liquidus temperature, for the alloy under consideration, crystals of pure cadmium precipitate out (point A, Fig. 3.9). This increases the concentration of bismuth and reduces the concentration of cadmium present in the remaining solution. Thus the solidification temperature is reduced to that appropriate for this new ratio of cadmium and bismuth, and further crystals of pure cadmium precipitate out. This again reduces the percentage of cadmium present in the remaining

Fig. 3.9 *Cadmium–bismuth phase equilibrium diagram*

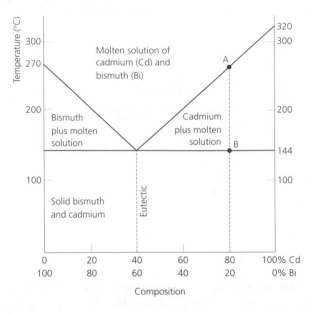

Fig. 3.10 *Lamellar structure of eutectic composition*

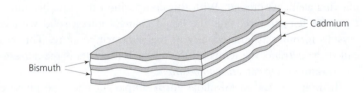

solution and the solidification temperature is further reduced with more pure cadmium crystals being precipitated out. This process repeats itself until the eutectic composition is reached (point B, Fig. 3.9).

- At the eutectic composition, crystals of cadmium and bismuth precipitate out simultaneously to form lamellar eutectic crystals of the two metals as shown in Fig. 3.10. Thus the final composition of the solid alloy will consist of crystals of pure cadmium in a matrix of crystals of eutectic composition.

Similarly for an alloy of 80 per cent bismuth and 20 per cent cadmium (hypo-eutectic), the amount of cadmium present in solution compared with the amount of bismuth present in solution will gradually increase as crystals of pure bismuth precipitate out until the eutectic composition is reached. Thus, in this instance the composition of the solid alloy will consist of crystals of pure bismuth in a matrix of crystals of eutectic composition.

Finally, for an alloy of 60 per cent bismuth and 40 per cent cadmium only crystals of eutectic composition will be present. These solid alloy compositions are shown in Fig. 3.11.

Fig. 3.11 *Solid composition of cadmium–bismuth alloys: (a) 20% Cd, 80% Bi; (b) 40% Cd, 60% Bi; (c) 80% Cd, 20% Bi*

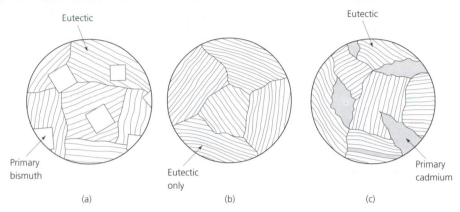

3.10 Phase equilibrium diagram (solid solution type)

It has already been stated that copper and nickel are not only mutually soluble in the liquid (molten) state, they are also mutually soluble in the solid state and they form a substitutional solid solution. The phase equilibrium diagram for copper–nickel alloys is shown in Fig. 3.12. Again, the line marked liquidus joins the points where solidification commences, whilst the line marked solidus joins the points where solidification is complete. This time there is no eutectic composition.

Fig. 3.12 *Copper–nickel phase equilibrium diagram*

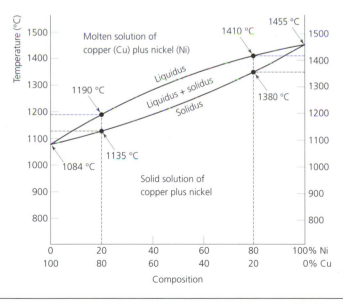

- For 100 per cent copper and 0 per cent nickel (pure copper) there is a single solidification temperature of 1084 °C. This is to be expected since, for a pure metal (in fact for any pure crystalline substance), the transition from liquid to solid takes place at a constant temperature.
- For an alloy of 80 per cent copper and 20 per cent nickel, Fig. 3.12 shows that solidification starts at 1190 °C and is complete at 1135 °C. Between the solidus and the liquidus is a solution of molten copper and nickel together with crystals of a solid solution of copper and nickel.
- For an alloy of 80 per cent nickel and 20 per cent copper, Fig. 3.12 shows that solidification starts at 1410 °C and is complete by 1380 °C.
- Finally, Fig. 3.12 shows that for 100 per cent nickel and 0 per cent copper (pure nickel) solidification occurs at the single temperature of 1445 °C.
- Below the solidus the alloy consists entirely of crystals of copper and nickel in solid solution.

3.11 Phase equilibrium diagram (combination type)

Many metals and non-metals are neither completely soluble in each other in the solid state, nor are they completely insoluble. Therefore they form a phase equilibrium diagram of the type shown in Fig. 3.13. In this system there are two solid solutions labelled α and β. The use of the Greek letters α, β, γ, etc., in phase equilibrium diagrams may be defined, in general, as follows:

- A solid solution of one component A in an excess of another component B, such that A is the solute and B is the solvent, is referred to as solid solution α.

Fig. 3.13 *Combination type phase equilibrium diagram*

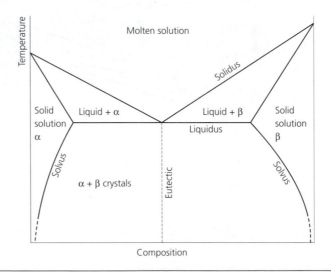

- A solid solution of the component B in an excess of the component A, so that **B** now becomes the solute and A becomes the solvent, is referred to as solid solution β.
- In a more complex alloy, any further solid solutions or intermetallic compounds which may be formed would be referred to by the subsequent letters of the Greek alphabet. That is, γ, δ, etc.

Tin–lead alloys (soft solders) are a typical example of the combination type of phase equilibrium diagram as shown in Fig. 3.14. Reference to the tin–lead phase equilibrium diagram shows that the α phase is a solid solution of 19.2 per cent tin in 80.8 per cent lead at the eutectic temperature, and that the β phase is a solid solution of 2.6 per cent lead in 97.4 per cent tin at the eutectic temperature. This diagram can be explained as follows:

- Above the liquidus ABC there is a homogeneous solid solution of molten tin and lead.
- For hypo-eutectic alloys, the solidus is the line ADB. Between the liquidus and the solidus the hypo-eutectic alloys will consist of the liquid solution of tin and lead plus crystals of the solid solution of α composition.
- Below the eutectic temperature the line separating the α phase from the α + β phase is called the solvus (see Fig. 3.13).
- For hyper-eutectic alloys, the solidus is the line BFC between the liquidus and the solidus, the hyper-eutectic alloys will consist of the liquid solution of molten lead and tin plus crystals of the β composition.
- Below the eutectic temperature, the line separating the α + β phase from the β phase is also called the solvus.

Fig. 3.14 *Tin–lead phase equilibrium diagram*

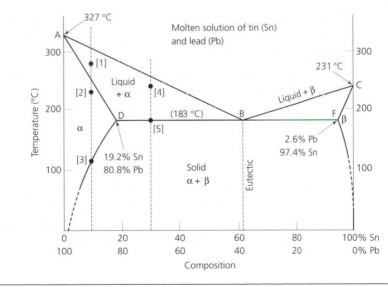

Describe the solidification of an alloy of composition 10 per cent tin and 90 per cent lead.

- Upon cooling from the molten state, where both metals are completely soluble in each other, to a temperature below the liquidus ([1] Fig. 3.14) the crystals of the α-phase solid solution start to grow.

- As in the previous diagrams, solidification is complete when the solidus is reached.

- The solid alloy will consist of crystals of the α phase in solid solution ([2] Fig. 3.14).

- The composition of this solid solution will be 19.2 per cent tin in 80.8 per cent lead, as previously stated.

- As the temperature falls, it will eventually meet the solvus ([3] Fig. 3.14). At this point the solid solution will be saturated with tin.

- Further cooling to room (ambient) temperature will result in the tin precipitating out to form the only other solid solution possible in this system, the β phase.

- Thus the final composition will consist of tin-rich crystals of the β phase dispersed through a matrix of crystals of low tin content α phase.

Describe the solidification of an alloy of composition 30 per cent tin and 70 per cent lead.

- Upon cooling from the molten state, where both metals are completely soluble in each other, to below the liquidus ([4] Fig. 3.14), the crystals of α phase will start to grow.

- This increases the concentration of tin and reduces the concentration of lead in the remaining molten solution.

- The solidification temperature is reduced to that appropriate for this new ratio, and the process repeats itself with more and more α phase solid solution being precipitated out until the eutectic composition is reached ([5] Fig. 3.14).

- At this point, crystals of both α- and β-phase solid solutions are precipitated out simultaneously to form lamellar eutectic crystals.

- Thus the final composition will consist of crystals of α-phase solid solution in a matrix of crystals of eutectic composition.

Examples 3.1 and 3.2 explain the behaviour of the various types of soft solder in common use. Tinman's solder has a composition of 60 per cent tin and 40 per cent lead. Since this is approximately the eutectic composition this solder has the lowest melting point and also solidifies instantly with no 'pasty' range. These factors, together with its relatively high tin content and low electrical resistance, accounts for its widespread use for soldered joints in the electronics industry.

On the other hand, a plumber requires a solder with a long pasty range which will solidify slowly and enable a *wiped joint* to be made. Plumber's solder has a composition of 80 per cent lead and 20 per cent tin so that there is a large temperature range between the liquidus and the solidus. At the same time the liquidus temperature is safely below that for pure lead so that there is no danger of melting the lead pipes or components being joined.

There are many other examples of binary alloys which could be quoted, but the three examples considered cover the three most common types of phase equilibrium diagrams. In Chapter 4 we will meet with the iron–carbon phase equilibrium diagram when we consider plain carbon steels.

3.12 Coring

So far, cooling above the liquidus has been assumed to be so slow that equilibrium is achieved as each change occurs. This rarely occurs in practice. By way of an example, the cooling of a copper–nickel alloy under equilibrium conditions will be considered in greater detail and the mechanism of solidification will then be compared with the same alloy cooled under production conditions.

Figure 3.15 shows part of the copper–nickel phase equilibrium diagram enlarged for clarity. It is convenient to consider an alloy of 70 per cent copper and 30 per cent nickel since solidification will conveniently centre on 1200 °C. When the molten alloy cools to the liquidus small dendrites of copper–nickel solid solution commence to form. If a line is drawn from T_1 on the liquidus parallel to the composition axis until it cuts the solidus, it is apparent that the composition of the solid solution will be 41 per cent copper and 53 per cent nickel. Since the overall composition of the alloy is still 70 per cent copper and 30 per cent nickel, the fact that the newly formed dendrites have 53 per cent nickel will result in the remaining molten solution having less than 30 per cent nickel.

As the alloy cools down to 1200 °C, the dendrites grow in size. A line drawn through T_2 parallel to the composition axis until it cuts the solidus indicates that the composition of the solid solution for this temperature will be 62 per cent copper and 38 per cent nickel. Thus between T_1 and T_2 the percentage of copper present in the dendrite has increased, whilst the percentage of nickel present in the dendrite has fallen. Since the line through T_2 cuts the liquidus at 78 per cent copper and 22 per cent nickel, this is the composition of the remaining molten alloy.

Solidification is complete at T_3, with the composition of the solid solution 70 per cent copper and 30 per cent nickel. The line from T_3 to the liquidus indicates that the last drop of molten alloy will have a composition of 87 per cent copper and 13 per cent nickel. Thus it is apparent that since the core of the crystal was formed under T_1 conditions, the crystal will have a nickel-rich core and a copper-rich case unless something can restore the balance.

Fig. 3.15 *Copper-rich Cu–Ni alloys*

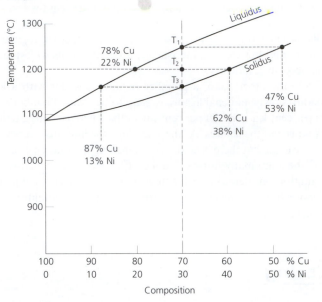

If the entire process is slow enough so that equilibrium within the crystal is maintained from the start, then *diffusion* will occur with copper atoms migrating into the core of the crystal and nickel atoms migrating into the case of the crystal. By the time cooling is complete, the composition should be uniform throughout with 70 per cent copper and 30 per cent nickel as shown in Fig. 3.16.

Fig. 3.16 *Crystal growth: (a) dendritic nucleus at liquidus temperature; (b) diffusion of copper and nickel as crystal commences to grow*

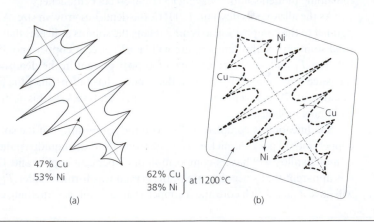

In phase equilibrium diagrams it is always assumed that cooling will be slow enough for equilibrium to be maintained. Under production conditions in the foundry, where cooling is more rapid than the ideal, there is insufficient time for diffusion to become complete and the nickel-rich core will become apparent when an etched specimen is examined under a microscope. The core of the crystal will have the outline appearance of the initial dendrite from which the crystal has grown. The result of this more rapid cooling is called *coring* and, since coring leads to lack of uniformity in the structure of the metal, this adversely affects its mechanical properties. Coring can largely be eliminated by heat treatment. The casting is heated to just below the solidus for the alloy concerned until diffusion is complete.

Once diffusion is complete the rate of cooling is irrelevant. However, over-fast cooling creates stresses in the metal. On the other hand, excessively long periods of heating and excessively slow cooling results in grain growth which may improve ductility but reduce mechanical strength and may cause machining problems as the metal will tend to tear and leave a poor surface finish rather than cut cleanly.

SELF-ASSESSMENT TASK 3.3

1. With reference to phase equilibrium diagrams, explain what is meant by the terms. You may use sketches to illustrate your answers:

   (a) eutectic
   (b) solid solution
   (c) intermetallic compound
   (d) solvus
   (e) solidus

EXERCISES

3.1 Discuss the main advantages and limitations of metal alloys compared with pure metals.

3.2 Describe what is meant by an 'intermetallic compound' and how its properties are likely to vary from those of a solid solution.

3.3 Sketch typical examples of the following phase equilibrium diagrams for binary alloys:

   (a) simple eutectic type
   (b) solid solution type
   (c) combination type

3.4 Draw the phase equilibrium diagram for cadmium–bismuth alloys. With reference to the diagram:

   (a) State the eutectic composition and describe the structure of the alloy in the solid state.
   (b) Explain in detail the cooling of an alloy of composition 90 per cent cadmium and 10 per cent bismuth from above the liquidus to below the solidus, and describe its structure in the solid state.

3.5 Draw the phase equilibrium diagram for copper–nickel alloys. With reference to the diagram:

(a) Explain in detail the cooling of an alloy of composition 60 per cent copper and 40 per cent nickel from above the liquidus to below the solidus, and describe its structure in the solid state.

(b) Explain why this series of alloys do not show a eutectic composition.

3.6 Draw the phase equilibrium diagram for tin–lead alloys.

(a) Indicate on the diagram:
   (i) the liquidus
   (ii) the solidus
   (iii) the solvus

(b) Explain in detail the cooling of an alloy of composition 60 per cent lead and 40 per cent tin from above the liquidus to below the solidus and describe its structure in the solid state.

(c) Explain briefly why plumber's solder has a high lead content, and why solder used for securing electronic components has a eutectic composition.

3.7 With reference to the copper–nickel phase equilibrium diagram, explain what is meant by:

(a) coring
(b) diffusion

3.8 With reference to the copper–aluminium phase equilibrium diagram, explain what occurs in the alloy during:

(a) solution treatment
(b) precipitation hardening

3.9 With the aid of sketches, explain what is meant by a 'family' of cooling curves and how these curves are used to construct phase equilibrium diagrams.

3.10 With the aid of a diagram, explain the difference between the terms 'hyper-eutectic solution' and 'hypo-eutectic solution'.

4 Plain carbon steels

The topic areas covered in this chapter are:

- The iron–carbon system.
- The iron–carbon phase equilibrium diagram.
- Cooling transformations for plain carbon steels with a eutectic composition.
- Cooling transformations for plain carbon steels with a hyper-eutectoid composition.
- Cooling transformations for plain carbon steels with a hypo-eutectoid composition.
- Critical change points.
- The effect of the carbon content on the properties of plain carbon steels.
- Plain carbon steels (BS 970).

4.1 Ferrous metals

Ferrous metals and alloys are based upon the metallic element iron. The name ferrous comes from the Latin name for iron which is *ferrum*. Iron is a soft, grey metal and it is rarely found in the pure state outside the laboratory. Engineers usually find it associated with the non-metal carbon, with which it forms solid solutions and the compound iron carbide. The carbon content is carried over from smelting process during which the iron is extracted from its ore.

Since all the ferrous materials used by engineers contain iron in association with carbon, it could be argued that all such materials are ferrous alloys. However, as pointed out in the previous chapter, the term ferrous alloy is reserved for those ferrous materials containing additional metallic alloying elements in sufficient quantities substantially to modify the properties of the material. Those 'alloys' containing only carbon as the main alloying element are referred to as wrought iron, plain carbon steels and plain cast irons depending upon the amount of carbon present and the way in which it is associated with the iron content. Since it is no longer widely used for engineering purposes, wrought iron will not be considered further in this text. Table 4.1 shows the relationship between the amount of carbon present and the resulting ferrous metal. It also gives some typical applications of those metals. Plain carbon steels will be considered in this chapter. Cast irons will be considered in Chapter 6. Alloy steels were introduced in Chapter 2 of this book and they will be considered in greater detail in *Engineering Materials*, Volume 2.

Table 4.1 Ferrous metals

Name	Group	Carbon content (%)	Some uses
Low-carbon steel	Plain carbon steel	0.1–0.15	Sheet for pressing out such shapes as motor car body panels. Thin wire, rod, and drawn tubes.
	Plain carbon steel	0.15–0.3	General purpose workshop bars, boiler plate, girders
Medium-carbon steel	Plain carbon steel	0.3–0.5 0.5–0.8	Crankshaft forgings, axles Leaf springs, cold chisels
High-carbon steel	Plain carbon steel	0.8–1.0 1.0–1.2 1.2–1.4	Coil springs, wood chisels Files, drills, taps and dies Fine-edged tools (knives, etc.)
Grey cast iron	Cast iron	3.2–3.5	Machine castings

4.2 The iron–carbon system

Figure 4.1 shows the iron–carbon phase equilibrium diagram. Strictly, it should be called the iron–iron carbide diagram but conventionally it is called the iron–carbon diagram and this latter name will be used in this text. If you compare Fig. 4.1 with the diagrams shown in Chapter 3, you can see that it is of the *combination type* where two substances are completely soluble in each other (miscible) in the liquid (molten) state but only partially soluble in each other in the solid state. Figure 4.1 is different from the example considered in Chapter 3 because of the structural changes resulting from iron being allotropic – that is, it can exist in more than one form. These structural changes take place at 910 °C and 1400 °C.

- Below 910 °C the iron forms body-centred-cubic crystals.
- From 910 °C to 1400 °C it forms face-centred-cubic crystals.
- Above 1400 °C it reverts to body-centred-cubic crystals.

These changes in lattice structure are accompanied by changes in volume as the atoms in the crystal lattice rearrange themselves. For example, when iron is heated, it expands uniformly with temperature until it reaches 910 °C whereupon it contracts slightly as the atoms rearrange themselves into a more compact lattice, after which the material continues to expand uniformly again, as shown in Fig. 4.2(a). A simple apparatus for demonstrating this phenomenon is shown in Fig. 4.2(b). These structural changes are accompanied by latent heat energy being taken in or given out. If an iron rod is cooled from above 910 °C in a darkened room, it will suddenly glow with increased brightness. The change from face-centred to body-centred crystals releases latent heat energy more rapidly than it can be dissipated and the temperature of the rod momentarily rises and, for a moment, it glows more brightly. This phenomenon is called *recalescence*.

Fig. 4.1 *Iron–carbon phase equilibrium diagram*

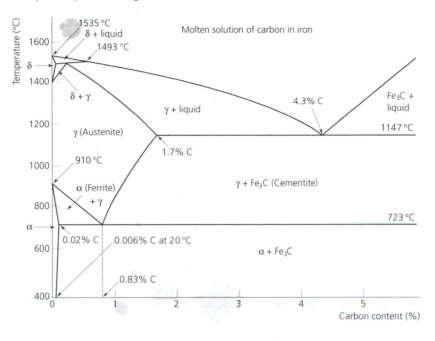

Fig. 4.2 *Effect of lattice change on volume: (a) change in volume as crystal lattices rearrange themselves; (b) method of demonstrating volume changes with temperature*

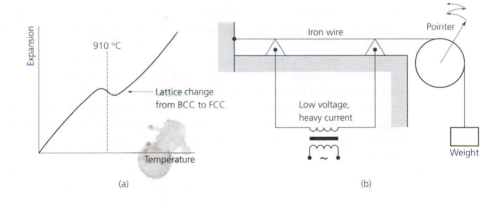

The iron–carbon phase equilibrium diagram appears to be very complex compared with those considered in Chapter 3. Fortunately this chapter is only concerned with the solid phases of the diagram, conventionally known as the 'steel section' of the full phase equilibrium diagram. The 'steel section' of the phase equilibrium diagram has been redrawn to larger scale in Fig. 4.3 to make it clearer.

Fig. 4.3 *Iron–carbon phase equilibrium diagram (steel section)*

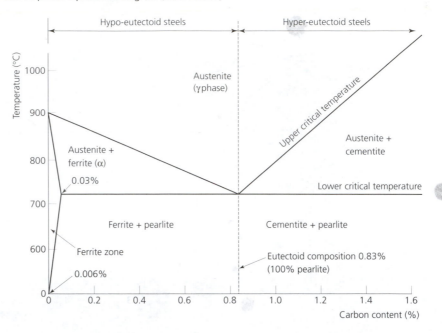

You can see from Fig. 4.3 that there are only three important phases:

- *Ferrite (α phase)*   This is a weak solution of carbon in body-centred-cubic crystals of iron. There is a maximum of 0.03 per cent carbon in solid solution at 723 °C, falling to 0.006 per cent carbon in solid solution at room temperature. (For all practical purposes it may be considered as 'pure' iron.) Ferrite is very soft, ductile and of relatively low strength.

- *Austenite (γ phase)*   This is a much more concentrated solid solution of carbon in iron than ferrite. Austenite is formed when carbon dissolves in face-centred-cubic crystals of iron in the solid state. The maximum amount of carbon which can be held in solution with iron in the solid state is 1.7 per cent at 1150 °C (see Fig. 4.4). Although this is the upper limit of carbon which can be present in plain carbon steels, for all practical purposes there is no advantage in increasing the carbon content beyond about 1.2 to 1.4 per cent.

- *Cementite (iron–carbide phase)*   An excess of carbon (C) combines with iron (Fe) to form iron carbide (Fe_3C). Each molecule of iron carbide contains three atoms of iron chemically combined with one atom of carbon. This is true up to the limit of 1.7 per cent carbon at room temperature, beyond which the excess carbon is precipitated out as 'free' or uncombined flakes of graphite.

Thus steels can be defined as:

Those alloys of iron and carbon in which the entire carbon content is combined with the iron in solid solution or as iron carbide (or both) and that no free carbon is present.

Since the maximum amount of carbon which can combine totally with the iron is 1.7 per cent, it follows that any carbon in excess of this figure will precipitate out as graphite flakes (graphite is an allotrope of carbon) and the resulting material will not be a steel but a cast iron (see Chapter 6).

The steel section of the iron–carbon phase equilibrium diagram is very similar to the combination type of equilibrium diagram shown in Chapter 3. In the combination type diagram there was one eutectic composition at which both alloying elements crystallised out simultaneously at the same temperature to form a lamellar structure. However, in the steel section of the iron–carbon phase equilibrium diagram such transformations occur in the solid state and the point at which ferrite and cemetite (iron carbide) precipitate out from the solid solution of austenite is called the *eutectoid* point.

Eutectoid points are similar to eutectic points but occur in the solid state. This point occurs at a temperature of 723 °C when 0.83 per cent carbon is present. The crystals which are precipitated out have a lamellar structure consisting of alternate layers of ferrite and cementite. This lamellar structure of ferrite and cementite is referred to as *pearlite* and it is the toughest structure which can exist in plain carbon steels. When the carbon content is 0.83 per cent the steel consists entirely of pearlite and it has maximum toughness. Steels with a carbon content below 0.83 per cent are called *hypo-eutectoid* steels, whereas steels with a carbon content above 0.83 per cent are called *hyper-eutectoid* steels.

4.2.1 *Cooling transformations for a steel with a eutectoid composition*

The transformations which occur during the cooling of a eutectoid composition (0.83 per cent carbon) steel are shown in Fig. 4.4(a). The steel commences to solidify directly into austenite at the liquidus (T_1) and solidification is complete when the solidus is reached at temperature T_2. The steel now consists entirely of γ-phase crystals of austenite. At 723 °C (T_3) the austenite suddenly changes into pearlite as shown. It remains as pearlite at all temperatures below 723 °C. Should the temperature rise above 723 °C, the structure will return to the solid solution of austenite again. These changes occur each time the steel is heated above 723 °C or cooled below it. Figure 4.4(b) shows a microphotograph of lamellar pearlite and the individual layers within the crystals are clearly shown.

4.2.2 *Cooling transformations for a steel with a hypo-eutectoid composition*

The transformations which occur during the cooling of a hypo-eutectoid steel are shown in Fig. 4.5(a). In this example the steel contains 0.5 per cent carbon. Again, the steel will commence to solidify at temperature T_1 and dendrites of body-centred-cubic (BCC) crystals of α-phase composition will begin to form (see also Fig. 4.1). This α phase continues to crystallise out of the residual molten steel until temperature T_2 is reached (1493 °C). At this temperature a peritectic reaction occurs between the α phase and the remaining molten steel to give austenite (γ phase) plus liquid. The temperature continues to fall and solidification is complete at temperature T_3. The steel now consists entirely of crystals of the solid solution γ-phase austenite. No further changes occur until the steel reaches temperature T_4.

Fig. 4.4 *A 0.83% carbon steel: (a) eutectoid transformations; (b) lamellar pearlite (×600)*

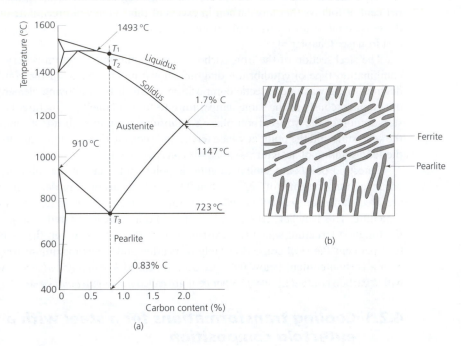

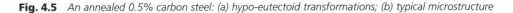

(a)

(b)

Fig. 4.5 *An annealed 0.5% carbon steel: (a) hypo-eutectoid transformations; (b) typical microstructure*

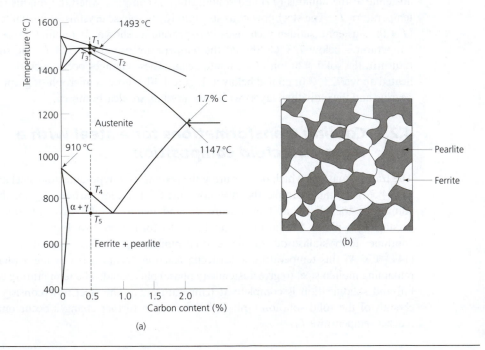

(a)

(b)

A detailed study of peritectic reactions is beyond the scope of this book. However, it can be said, simply, that during a peritectic reaction the two phases are already present in a heterogeneous mixture associate together to produce a third phase. At the same time one or both of the original phases will disappear entirely.

In this example a solid (α phase) reacts with the liquid phase to produce the γ phase (austenite). When a liquid phase reacts in this way it usually forms an envelope or coating around the new phase. This prevents further reaction except very slowly by diffusion. The term *peritectic* is derived from the Greek word 'peri' which means *around* and refers to the envelope or coating that forms around the newly created phase. Think of the word *perimeter* which is the line forming an *envelope* around the area of a circle, and literally means the distance 'measured around' the circle.

When the steel cools slowly below the temperature T_4, crystals of ferrite will start to grow in the austenite so that both α- and γ-phase crystals will be present. Since α-phase crystals of ferrite contain rather less than 0.03 per cent carbon in solid solution, the carbon content of the remaining phase, austenite, will increase progressively as more and more ferrite is formed until at 723 °C (the eutectoid temperature) the structure will contain ferrite (> 0.03 per cent carbon) and austenite (0.83 per cent carbon) which is the euctectoid composition. Thus at T_5 the austenite will change suddenly into the eutectoid composition of pearlite, and the final composition of the steel below T_5 will consist of crystals of ferrite and crystals of pearlite. Figure 4.5(b) shows a typical microstructure for an annealed 0.5 per cent carbon steel.

4.2.3 *Cooling transformations for a steel with a hyper-eutectoid composition*

The transformations which occur during the cooling of a hyper-eutectoid steel of 1.2 per cent carbon content are shown in Fig. 4.6(a). This time, solidification commences at temperature T_1 and is complete by the time the steel has cooled to temperature T_2. There is no peritectic reaction. The structure is entirely austenitic with crystals of the solid solution γ phase below temperature T_2. Upon cooling below temperature T_3 needles of primary cementite (iron-carbide) will start to precipitate out. Since cementite contains 1.7 per cent carbon, the remaining austenite will be less rich in carbon. Eventually, at 723 °C, the carbon content of the remaining austenite will have been reduced to 0.83 per cent and as the temperature falls below 723 °C (T_4) it will be transformed into crystals of pearlite with a eutectoid composition. Thus the final composition of the steel will consist of crystals of pearlite surrounded by bands of primary cementite at the crystal boundaries. Figure 4.6(b) shows a typical microstructure for an annealed 1.2 per cent carbon steel.

4.3 Critical change points

The construction of phase equilibrium diagrams from a family of cooling curves was explained in Chapter 3. The iron–carbon phase equilibrium diagram is also constructed from just such a family of cooling curves by connecting its *critical change points*. The change points are often referred to, simplistically, as the upper critical temperature (UCT) and the lower critical temperature (LCT). The critical change points, where changes in

Fig. 4.6 *An annealed 1.2% carbon steel: (a) hyper-eutectoid transformations; (b) typical microstructure*

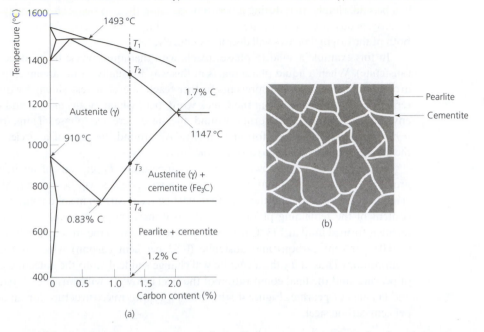

(a)

composition and structure occur, are also called *arrest points* since the time–temperature heating or cooling curve stops at these points as the latent heat energy associated with change is taken in during heating or given out during cooling, as shown in Fig. 4.7.

A_1 is the temperature at which the eutectoid transformations take place; that is the transformation of austenite into pearlite on cooling and vice versa on heating. For plain carbon steels A_1 is always constant and is 723 °C.

A_3 is the temperature above which hypo-eutectoid steels are wholly austenitic (γ phase).

A_{cm} is the temperature above which hyper-eutectoid steels are wholly austenitic (γ phase).

Due to what is known as *thermal inertia* the arrest points do not occur at exactly the same temperatures on heating curves as they do on cooling curves. Therefore the arrest points

Fig. 4.7 *Cooling curves for (a) hypo-eutectoid and (b) hyper-eutectoid carbon steels*

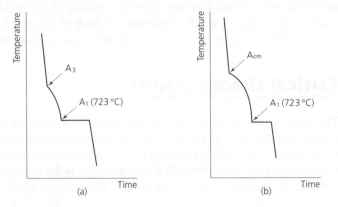

just described require further identification to indicate whether they were derived by heating or by cooling. This further notation makes use of the French word for heating which is *chauffage* and the French word for cooling which is *refroidissement*. Thus the critical change points on a time–temperature *heating curve* are called: Ac_1, Ac_3 and Ac_{cm}. Similarly the critical change points on a time–temperature *cooling curve* are called Ar_1, Ar_3 and Ar_{cm}. Their disposition on the phase equilibrium diagram for plain carbon steels is shown in Fig. 4.8.

Fig. 4.8 *Critical change points for carbon steels*

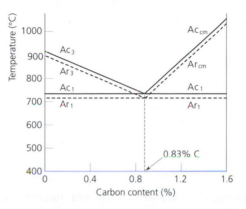

Since phase equilibrium diagrams are generally only used by engineers for determining heat treatment criteria (see Chapter 7), the cooling diagram based on Ar temperatures is the one usually quoted in engineering texts. This will be referred to again in Chapter 5. There is also an A_2 temperature lying between A_1 and A_3, and this is the temperature above which steels become non-magnetic when heated and below which they become magnetic again when cooled. Since this does not affect the mechanical properties of a steel, it is not normally included on the phase equilibrium diagram to avoid confusion. The A_2 temperature is often referred to as the *Curie point* after the French physicist who discovered it.

4.4 The effect of carbon on the properties of plain carbon steel

Figure 4.9 shows the effect of the carbon content upon the properties of plain carbon steels which have been cooled slowly enough to enable them to achieve phase equilibrium. It can be seen from Fig. 4.9 that low-carbon steels, consisting mainly of ferrite, are soft and ductile and relatively weak, reflecting the properties of the ferrite itself.

The increased amount of carbon in medium-carbon steels promotes the formation of cementite. This results in an increased presence of pearlite, making such steels stronger, tougher and harder, but not so ductile.

When the carbon content reaches approximately 0.83 per cent the steel consists entirely of pearlite. This is the eutectoid composition previously described and it produces plain carbon steel of maximum toughness and strength.

Fig. 4.9 *Properties of plain carbon steels*

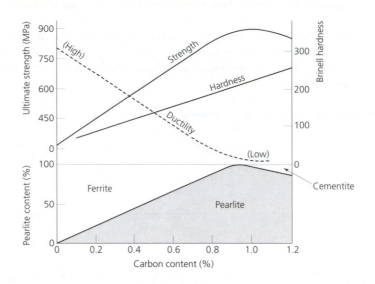

Increasing the carbon content still further increases the amount of cementite (iron–carbide) present in the steel. Since the maximum amount of combined cementite occurred at 0.83 per cent carbon content, where the composition of the steel is totally pearlitic, any increase in the carbon content results in the formation of excess cementite appearing around the crystal boundaries. This increases the hardness and wear resistance of the steel, but at the expense of reduced toughness, strength and ductility.

SELF-ASSESSMENT TASK 4.1

1. State briefly what is meant by the terms:

   (a) ferrite
   (b) pearlite
   (c) cementite
   (d) austenite

2. Describe the composition at room temperature of the following, and explain how their properties are influenced by their composition:

   (a) a low-carbon steel
   (b) a high-carbon steel

3. Sketch a cooling curve for a 0.5 per cent carbon steel. Identify and label the change points on the curve and explain what happens at each point.

4.5 Plain carbon steels

These are ferrous materials containing between 0.1 and 1.7 per cent carbon as the main alloying element. In addition, plain carbon steels contain the following elements by accident or by design.

Manganese	up to 1.0%
Phosphorus	up to 0.05%
Silicon	up to 0.3%
Sulphur	up to 0.05%

- *Manganese*   This is an essential constituent element since it ensures a sound ingot free from blow holes. Further, it combines with any sulphur present which would otherwise weaken the steel. In general, manganese raises the yield point (see Section 11.3) and increases the strength and toughness of the steel. However, it also increases the tendency of the steel to crack and distort when quench hardened (see Section 5.7) and, for this reason, the content should be kept below 0.5 per cent in medium- and high-carbon steels.
- *Phosphorus*   This is an impurity carried over from the iron ore. It forms compounds which make the steel brittle and, therefore, should be removed as far as possible during the refinement processes. It should not be present in excess of 0.05 per cent.
- *Silicon*   This is an impurity carried over from the iron ore. Its presence should be limited to between 0.1 and 0.3 per cent in the steels otherwise it can cause breakdown of the cementite which would result in weakness. Silicon has little direct effect upon the mechanical properties of plain carbon steels providing the amount present is limited to the percentage quoted above. It is often added to cast irons to prevent chill hardening (see Section 6.2). Silicon improves the magnetic properties of the 'soft' ferro-magnetic materials.
- *Sulphur*   This is an impurity carried over from the fuel used in the blast furnace to extract the iron from its ore. Sulphur tends to combine with the iron to form iron sulphide which greatly weakens the steel. For this reason the sulphur content must be kept below 0.05 per cent and there should always be at least 5 times as much manganese present as there is sulphur. Fortunately, sulphur has a greater affinity for manganese than it has for steel and will combine with the manganese in preference to the iron. Unlike iron sulphide which weakens the steel, manganese sulphide has no such adverse effect.

  Some free-cutting steels contain up to 0.2 per cent sulphur to improve their machinability at the expense of strength for lightly stressed turned parts. In this instance, the high sulphur content provides an extreme pressure lubrication between the chip and the tool and also causes the chips (swarf) to break up into easily disposable particles.

Table 4.2 gives the composition, properties and typical applications of some general purpose plain carbon steels, together with reference to their British Standard specification. In comparing the properties of these steels it should be noted that some have had their properties modified by heat treatment. The state of the steel is, therefore, also stated in the table. Although various specifications have been quoted in Table 4.2, the most widely used is BS 970. This specification is concerned with wrought steels and was originally published in six parts.

Table 4.2 Some plain carbon steels

| Type of steel | British Standards | Composition (%) | | Condition | Properties | | | | | Applications |
		C	Mn		YP (MPa)	UTS (MPa)	Elong. (%)	Impact (J)	Hardness (H_B)	
Low-carbon steel	BS 970.040A10	0.10	0.40	Process annealed after cold rolling	—	300	28	—	—	Car body panels produced by drawing and pressing.
	BS 15	0.20	0.20	As rolled	240	450	25	—	—	General purpose: mild steel. Welding quality, high tensile mild steel for building construction, etc.
	BS 968	0.20	1.50	As rolled	350	525	20	—	—	
Casting steel	BS 1504/161B	0.30	—	Annealed after casting to refine grain.	265	500	18	20	150	General purpose, medium strength castings for machining.

Medium-carbon steel	BS 970.080M40	0.40	0.80	Toughened by quenching from 850 °C, temper at 600 °C.	500	700	20	55	200	Axles, crankshafts, etc., under moderate stress.
	BS 970.070M55	0.55	0.70	Harden by quenching from 825 °C. Temper at 600 °C.	550	750	14	—	—	Gears and machine parts subject to wear.
	—	0.70	0.35	Quench harden from 790/810 °C in water. Temper at 150–300 °C as appropriate.	—	—	—	—	780	Hand chisels, cold sets, screwdriver blades, blacksmith's tools, etc.
	BS 4659: BW18	1.00	0.35	Quench harden from 760/780 °C in water. Temper at 150–300 °C as appropriate.	—	—	—	—	800	Taps, screwing dies, wood drills, press tools, hand (fitting) tools, files, measuring and marking out in instruments, etc.
	BS 4659: BW1C	1.20	0.35	Quench harden from 760/780 °C in water or oil. Temper at 150–300 °C as appropriate.	—	—	—	—	820	Fine edge tools, knives, files, surgical instruments.

Part 1 of the current edition of BS 970 has been prepared under the direction of the Iron and Steel Standards Committee and is identical with BS 970: Part 1: 1983 except for the deletion of requirements for quenched and tempered steels and steels for bright bar. This new standard, together with BS EN 10083-1, BS EN 10083-2 and BS 970: Part 3 supersedes BS 970: Part 1: 1983, which is now withdrawn. This new edition of BS 970 introduces technical changes but it does not reflect a full review or revision of the standard, which will be undertaken by the Iron and Steels Standards Committee in due course.

Appendix D of BS 970: Part 1: 1983 is not included in the new edition of the standard as it no longer reflects the current situation on steel grades.

European requirements for quenched and tempered steels are specified in BS EN 10083-1 and BS EN 10083-2 which are English language versions of EN 10083-1 and EN 10083-2 and are published simultaneously with BS 970: Part 1.

Work is continuing in Europe to prepare standards covering stainless steels, boron steels and surface quality. As the European standards are published, the corresponding tables and text will be withdrawn from BS 970: Part 1 until it is eventually withdrawn. The current edition of BS 970 still specifies Izod and Charpy impact test values for steels. It should be noted that only Charpy impact tests will be specified in all the new European standards.

The European requirements for stainless steels are specified in BS EN 10088-1, BS EN 10088-2 and BS EN 10088-3, and are published simultaneously with this amendment. These new standards should be used in preference to BS 970 whenever applicable.

BS 970: *Wrought steels* is currently published in four parts:

BS 970: Part 1: *General inspection and testing procedures and specific requirements for carbon, carbon manganese, alloy and stainless steels*

BS 970: Part 2: *Requirements for steels for hot formed springs*

BS 970: Part 3: *Bright bars for general engineering purposes*

BS 970: Part 4: *Valve steels*

This text is only concerned with plain carbon steels, some examples of which are listed in Table 4.2. Alloy steels will be considered in *Engineering Materials*, Volume 2.

Until 1970, steels in the United Kingdom were specified by a system of 'EN' numbers. This standardisation came about during the shortages of the Second World War and were variously described as standing for 'Economy Number' or 'Emergency Number'. On no account are they to be confused with the current EN European Standards.

The old EN numbers were a simple numerical listing of steels and gave no indication to the properties and composition of the steel. Between 1970 and 1972 the British Standards Institution revised BS 970: *Wrought steels*, into a more logical and informative coding system using six symbols for each grade of steel. This code, which is still used, is built up as follows.

- The first three symbols are a number code indicating the type of steel.

  000 to 199   Carbon and carbon–manganese steels. The numbers indicate the manganese content × 100.

  200 to 240   Free-cutting steels. The second and third numbers indicate the sulphur content × 100.

  250   Silicon–manganese valve steels.

  300 to 499   Stainless and heat-resisting steels.

  500 to 999   Alloy steels.

- The fourth symbol is a letter code:

  A   The steel is supplied to a chemical composition determined by chemical analysis of a batch sample.

  H   The steel is supplied to a hardenability specification.

  M   The steel is supplied to a mechanical property specification.

  S   The material is a stainless steel.

- The fifth and sixth symbols are a number code indicating the mean carbon content. The actual mean carbon content $\times 100$ is the code. Thus a steel of specification BS 970.040A10 would be interpreted as follows:

  BS 970   indicates the specification

  040   classifies the steel as 'Plain carbon' and indicates that the manganese content is 040/100 = 0.4 per cent

  A   indicates that the composition has been determined by batch analysis

  10   indicates that the carbon content is 10/100 = 0.1 per cent.

Some examples of the use of the six-symbol code applied to plain carbon steels are given in Table 4.3.

Table 4.3 *Applications of the six symbol code*

BS 970 specification	Description
070M26	A plain carbon steel with a composition of 0.26% carbon and 0.70% manganese. The letter 'M' indicates that the steel has to meet a prescribed mechanical property specification.
150M36	As above except that the composition for this steel is: 0.36% carbon, 1.5% manganese.
220M07	A low-carbon free-cutting steel with a composition of 0.07% carbon and 0.20% sulphur. Again the letter 'M' indicates that the steel has to meet a prescribed mechanical property specification.
070A20	A low-carbon steel with a composition of 0.20% carbon and 0.70% manganese. The letter 'A' indicates that the steel has to meet a prescribed chemical composition specification.

The final factor to be considered in the coding of wrought steels is the *limiting ruling section*. This will be considered in greater detail in Sections 5.10 and 5.11 but, briefly, it is the maximum diameter bar of given composition which, after appropriate heat treatment, will attain its specified mechanical properties.

For example, a plain carbon steel bar of composition 070M55 can attain condition 'R' (Table 4.4) after heat treatment providing it is not greater than 100 mm diameter. However, if it is to attain condition 'S' (Table 4.3), then its diameter must be limited to 63 mm. Thus, in the first instance, the limiting ruling section is 100 mm diameter and, in the second instance, the limiting ruling section is reduced to 63 mm diameter.

Table 4.4 Carbon and carbon–manganese steels (derived from BS 970)

Heat treatment condition symbol

Steel	P — Tensile strength: 550–700 MPa, Brinell hardness: 152–207					Q — Tensile strength: 620–770 MPa, Brinell hardness: 179–229					R — Tensile strength: 690–850 MPa, Brinell hardness: 201–255					S — Tensile strength: 770–930 MPa, Brinell hardness: 223–277					T — Tensile strength: 850–1000 MPa, Brinell hardness: 248–302				
	LRS	R_e	A	I	$R_{p0.2}$	LRS	R_e	A	I	$R_{p0.2}$	LRS	R_e	A	I	$R_{p0.2}$	LRS	R_e	A	I	$R_{p0.2}$	LRS	R_e	A	I	$R_{p0.2}$
070M20	19	355	20	41	340	13	415	16	34	400	—	—	—	—	—	—	—	—	—	—	—	—	—	—	—
070M26	29	355	20	41	325	19	415	16	34	400	—	—	—	—	—	—	—	—	—	—	—	—	—	—	—
080M30	63	340	18	34	310	29	400	16	34	370	—	—	—	—	—	—	—	—	—	—	—	—	—	—	—
080M36	—	—	—	—	—	—	—	—	—	—	13	465	16	34	450	—	—	—	—	—	—	—	—	—	—
080M40	—	—	—	—	—	63	385	16	34	355	19	465	16	34	450	—	—	—	—	—	—	—	—	—	—
080M46	—	—	—	—	—	—	—	—	—	—	29	450	16	—	415	13	525	14	—	510	—	—	—	—	—
080M50	—	—	—	—	—	100	370	16	—	340	63	430	14	—	400	29	495	14	—	465	13	570	12	—	555
070M55	—	—	—	—	—	—	—	—	—	—	100	415	14	—	385	63	480	14	—	450	19	570	12	—	55
120M19	100	355	18	28	325	29	450	16	47	415	19	510	16	34	495	—	—	—	—	—	—	—	—	—	—
150M19	150	340	18	27	310	63	430	16	54	400	29	510	16	41	480	—	—	—	—	—	—	—	—	—	—
120M28	—	—	—	—	—	100	415	16	41	385	29	510	16	34	480	13	510	16	34	555	—	—	—	—	—
150M28	—	—	—	—	—	150	400	16	47	370	63	480	16	41	450	19	570	14	34	555	—	—	—	—	—
120M36	—	—	—	—	—	100	415	18	41	385	29	510	16	34	480	—	—	—	—	—	—	—	—	—	—
150M36	—	—	—	—	—	150	400	18	47	370	63	480	16	41	450	29	555	14	41	525	13	635	12	34	620
216M28	63	355	20	34	325	19	430	18	41	415	13	495	16	54	480	—	—	—	—	—	—	—	—	—	—
212M36	100	340	20	34	310	63	400	18	47	370	29	480	16	34	450	—	—	—	—	—	—	—	—	—	—
225M36	—	—	—	—	—	63	400	18	34	370	29	480	16	34	450	—	—	—	—	—	—	—	—	—	—
216M36	100	340	20	34	310	63	400	18	34	370	29	465	16	34	430	—	—	—	—	—	—	—	—	—	—
212M44	—	—	—	—	—	100	400	18	34	370	63	465	16	34	430	13	540	14	27	525	13	600	12	27	585
225M44	—	—	—	—	—	—	—	—	—	—	100	450	16	34	415	29	525	14	27	495	—	—	—	—	—

Note: LRS = Limiting ruling section, A = Elongation (%), $R_{p0.2}$ = 0.2 % proof stress (MPa), R_e = Yield stress (MPa), I = Izod impact value (J).

EXERCISES

4.1 Explain how a plain carbon steel differs from wrought iron and cast iron in respect of their compositions and properties.

4.2 Explain why calculations for the linear expansion of a steel with increase in temperature are only valid up to the Ac_1 point.

4.3 A typical plain carbon steel is specified as BS 970.080M40. From this specification derive the composition of the steel.

4.4 If the heat treatment condition 'Q' is added to the specification in question 4.3, list the ruling section and the properties which could be expected of this steel.

4.5 Show by means of a diagram the effect of carbon content on the hardness, tensile strength and ductility of annealed plain carbon steels.

4.6 Sketch the 'steel section' of the iron–carbon phase equilibrium diagram for plain carbon steels and describe the cooling from the austenitic condition to room temperature for a:
   (a) hyper-eutectoid steel
   (b) hypo-eutectoid steel
   (c) steel of eutectoid composition

4.7 With reference to the iron–carbon phase equilibrium diagram:
   (a) Explain what is meant by the following change points:
       (i) Ar_1
       (ii) Ar_3
       (iii) Ar_{cm}
   (b) Explain the difference between Ar points and Ac points.
   (c) Explain what is meant by the 'Curie point'.

4.8   Since iron is allotropic:
  (a)   State the types of crystal it will form:
        (i)   above 1400 °C
        (ii) below 910 °C
        (iii) between 910 and 1400 °C
  (b)   Explain what is meant by 'recalescence', and describe how this phenomenon may be demonstrated.

4.9   Explain briefly what is meant by 'limiting ruling section' and how it is related to the hardenability of steels.

4.10 Explain why the 'Ar' points for steels are more likely to be of use than the 'Ac' points when heat treating steels.

5 Heat treatment of plain carbon steels

The topic areas covered in this chapter are:
- A review of the heat-treatment processes to be considered.
- Recrystallisation, cold working and hot working.
- Critical temperatures.
- Stress-relief annealing, spheroidising annealing and full annealing.
- Normalising.
- Quench hardening, quenching media, tempering and hardening faults.
- Mass effect.
- Case hardening and surface hardening.

5.1 Heat-treatment processes

Plain carbon steels and alloy steels are among the relatively few engineering materials which can be usefully heat treated in order to vary their mechanical properties. The other main group is the heat-treatable aluminium alloys (see Section 7.5). Steels can be heat treated because of the structural changes that can take place within solid iron–carbon alloys. The various heat-treatment processes appropriate to plain carbon steels are:

- Annealing.
- Normalising.
- Hardening.
- Tempering.

In all the above processes the steel is heated slowly to the appropriate temperature for its carbon content and then cooled. It is the rate of cooling which determines the ultimate structure and properties that the steel will have, providing that the initial heating has been slow enough for the steel to have reached phase equilibrium at its process temperature. Before describing the processes listed above it is necessary to establish some basic principles.

5.1.1 *Recrystallisation*

During cold-working processes, the grain of the metal becomes distorted and internal stresses are introduced into the metal. If the temperature of the cold-worked metal is now raised sufficiently, nucleation occurs and 'seed' crystals form at the grain boundaries at points of maximum internal stress. This is called *nucleation*. The more severe the cold working and the greater the internal stress, the lower will be the temperature at which nucleation occurs for a given metal. The principle of nucleation is shown in Fig. 5.1. The minimum temperature at which the reformation of the grain occurs is called the *temperature of recrystallisation*.

At temperatures above the recrystallisation temperature, the kinetic energy of the atoms on the edges of the distorted grains increases. This allows these edge atoms to move away and attach themselves to the newly formed nuclei which will then begin to grow into grains. This process continues until all the atoms of the original, distorted crystals have been transferred. Since, after severe cold working, more nuclei form than the number of original grains, the grain structure after recrystallisation is usually finer than the original grain structure before cold working. Thus a degree of grain refinement occurs.

Fig. 5.1 *Recrystallisation: (a) before working; (b) after cold working; (c) nucleation commences at recrystallisation temperature; (d) crystals commence to grow as atoms migrate from the original crystals and attach themselves to the nuclei; (e) after annealing is complete the grain structure is restored*

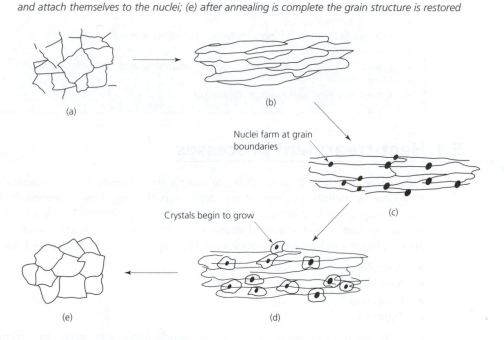

5.1.2 *Cold working*

This occurs when metal is bent, squeezed or stretched to shape *below* the temperature of recrystallisation. Examples of such processes are: pressing out car body panels, cold-drawing rods, wires and tubes, cold-heading rivets, and cold-rolling strip and sheet metal. Cold working results in distortion of the grain of the metal and, eventually, the metal becomes so stiff and brittle that it breaks. (This is what happens in a tensile test (see Section 11.2).) Metal that has become harder and stiffer as a result of cold working is said to be *work hardened*.

The metal must not be allowed to become excessively work hardened or it will be prone to fracture. Once work hardened, it needs to be annealed to restore its grain structure before further cold working is performed upon it. The heat-treatment process of annealing is described in Section 5.2.

5.1.3 *Hot working*

This occurs when metal is bent, squeezed or stretched to shape *above* the temperature of recrystallisation. Examples of such processes are: forging, hot rolling and extrusion. Since the process temperature is above the temperature of recrystallisation, the grains reform as fast as they are distorted by the processing. If the metal could be retained at this temperature, there would be no limit to the amount of hot working to which the metal could be subjected.

In practice there are strict limitations. The initial temperature has to be limited so that overheating and 'burning' of the metal does not occur or it will be excessively weakened or, in some cases, so that the melting point is not reached. Since the metal cools naturally once it has been removed from the furnace for processing, the finishing temperature of the process has to be carefully judged in order that:

- It is not too high so that subsequent *grain growth* occurs as this leads to a reduction in strength and to poor machining properties.
- It is not too low so that surface cracking occurs.

5.1.4 *Critical temperatures*

These temperatures, at which changes of state (phase changes) occur on phase equilibrium diagrams, have already been discussed in Chapters 3 and 4. For example, the iron–carbon phase equilibrium diagram shows the change from austenite to ferrite and pearlite commences, when cooling, at the Ar_3 line and is completed by the time the metal has cooled slowly to the Ar_1 line. Similarly, the change from austenite to pearlite and cementite (iron carbide) commences at the Ar_{cm} line and is completed by the time the metal has cooled to the Ar_1 line. The Ar_{cm}, Ar_3 and the Ar_1 lines connect the critical temperatures for the individual alloys of iron and carbon. Heat treatment pressures are related to these temperatures.

SELF-ASSESSMENT TASK 5.1

1. Explain why plain carbon steels can benefit from heat treatment.

2. Discuss the basic requirements for a heat-treatment process.

3. Explain the essential differences between cold working and hot working.

5.2 Annealing processes

All annealing processes are concerned with rendering steel soft and ductile so that it can be cold worked and/or machined. There are three basic annealing processes, and these are:

- *Stress-relief annealing* at subcritical temperatures (also known as 'process annealing' and 'interstage annealing').
- *Spheroidised annealing* at subcritical temperatures.
- *Full annealing* for forgings and castings.

The process chosen depends upon the carbon content of the steel, its pretreatment processing, and its subsequent processing and use. Figure 5.2(a) shows the temperature bands for the annealing processes superimposed on the iron–carbon phase equilibrium diagram. In all annealing processes the cooling rate is as slow as possible.

Fig. 5.2 *Annealing temperatures for plain carbon steels: (a) annealing temperatures; (b) spheroidised annealing*

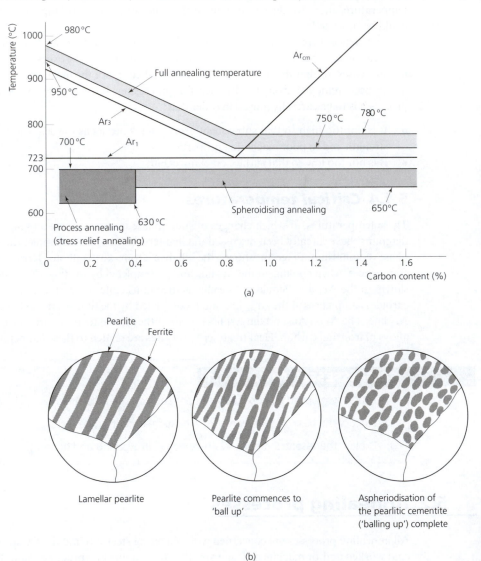

(a)

(b)

5.3 Stress-relief annealing

This process is reserved for steels below 0.4 per cent carbon content. Such steels will not satisfactorily quench harden (Section 5.7) but, as they are relatively ductile, they are frequently cold worked and become work hardened. Since the grain structure will have been severely distorted by the cold working, recrystallisation can commence at 500 °C but, in practice, annealing is usually carried out between 630 and 700 °C to speed up the process and limit grain growth.

The rate of cooling and the length of time for which the steel is heated depends upon the subsequent processing and use to which the material is going to be put. If further cold working is to take place then increased ductility and malleability will be required. This is achieved by prolonging the heating and slowing the cooling to encourage grain growth. However, if grain refinement and strength and toughness is of more importance, then heating and cooling should be more rapid.

5.4 Speroidising annealing

It has already been stated that crystals of pearlite have a laminated structure consisting of alternate layers of cementite and ferrite. When steels contain more than 0.4 per cent carbon are heated to just below the critical temperature (650–700 °C) the cementite in the crystals tends to 'ball up'. This is referred to as the aspheroidisation of pearlitic cementite and the process is shown diagramatically in Fig. 5.2(b). Since the temperatures involved are subcritical, no phase changes take place and spheroidisation of the cementite is purely a surface tension effect.

If the layers of cementite are relatively coarse prior to annealing, they take too long to break down and tend to form coarse globules (spheroids) of cementite. This, in turn, leads to impaired physical properties and machined surfaces with a poor finish. Thus, grain refinement by a quench treatment prior to aspheroidisation is recommended to produce fine globules of cementite. The process is most effective when it is used to soften plain carbon steels containing more than 0.4 per cent carbon and which have been either work hardened or quench hardened.

After speroidising annealing the steel can be cold worked and it will machine freely to a good surface finish. Furthermore, steel which has been subjected to spheroidising annealing will re-harden more uniformly and with less chance of cracking. As with any other annealing process, slow cooling is required after the heating cycle. It is usual to turn off the furnace and allow the furnace and the charge to cool down slowly together.

5.5 Full annealing

Plain carbon steels solidify at temperatures well above the temperatures with which heat-treatment processes are concerned. As a result, large castings, well insulated by the sand mould, take a very long time to cool down. Similarly, large forgings, although hot worked at temperatures well below their melting points, are nevertheless processed at temperatures substantially above their upper critical temperatures for relatively long periods of time. In both cases grain growth is excessive and the physical properties of the metal are impaired. The ferrite settles out along the crystal boundaries of the coarse grains of austenite and also within the grains to provide a mesh-like structure, as shown in Fig. 5.3. This is called a *Widmanstätten structure*.

If you refer back to Fig. 5.2(a) you can see that to render the steel useable it has to be reheated to approximately 50 °C above the Ar_3 line for hypo-eutectoid steels and approximately 50 °C above the Ar_1 line for hyper-eutectoid steels. This results in the formation of fine grains of austenite that transform into relatively fine grains of ferrite and pearlite or pearlite and cementite (depending upon the carbon content) as the steel is slowly cooled to room temperature, usually in the furnace.

Fig. 5.3   *Widmanstätten structure*

5.6 Normalising

The normalising temperatures for plain carbon steel are shown in Fig. 5.4. The process resembles full annealing except that, whilst in annealing the cooling rate is deliberately retarded, in normalising the cooling rate is accelerated by taking the work from the furnace and allowing it to cool in free air. Provision must be made for the free circulation of cool air, but draughts must be avoided.

In the normalising process, as applied to hypo-eutectoid steels, it can be seen that the process temperature is the same as for full annealing. After 'soaking' the steel at the process temperature to ensure conversion to fine grain austenite, the more rapid cooling associated with the normalising process results in the transformation of the fine grain austenite into fine grain ferrite and pearlite. The relatively rapid cooling avoids the grain growth associated with annealing.

Fig. 5.4 *Normalising temperatures for plain carbon steels*

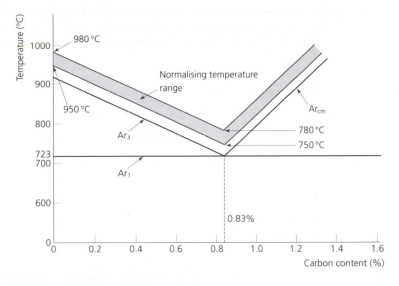

In the normalising process, as applied to hyper-eutectoid steels, it can be seen that the steel is heated to approximately 50 °C above the Ar_{cm} line. This ensures that the transformation to fine grain austenite corrects any grain growth or grain distortion that may have occurred previously. Again, the steel is cooled in free air and the austenite transforms into fine grain pearlite and cementite. The fine grain structure resulting from the more rapid cooling associated with normalising gives improved strength and toughness to the steel but reduces its ductility and malleability. The increased hardness and reduced ductility allows a better surface finish to be achieved when machining. (The excessive softness and ductility of full annealing leads to local tearing of the machined surface.) However, the level of ductility and malleability achieved by normalising is not sufficient for more than limited cold working.

Normalising is frequently used for stress relieving between rough machining and the finish machining of large castings and forgings to avoid subsequent 'movement' due to the slow release of internal stresses and loss of accuracy. At one time large castings and forgings were left outside to 'weather' for up to a year or more after rough machining to ensure that the workpieces became stabilised. Although highly successful, this procedure tied up an excessive amount of working capital and space and nowadays heat treatment is preferred as the work in progress is turned round more quickly.

5.7 Quench hardening

Figure 5.5 shows the temperature band from which plain carbon steels are cooled when they are quench hardened. It can be seen that this temperature band is the same as for full annealing. The band is not continued below 0.4 per cent carbon for, although some grain refinement and toughening occurs, no appreciable hardening takes place. If a plain carbon steel with a carbon content above 0.4 per cent is quenched (cooled very rapidly) from the appropriate temperature for its carbon content as shown in Fig. 5.5, there is not sufficient time for the equilibrium transformations previously described to take place and the steel becomes appreciably harder. The final hardness will depend solely upon the carbon content and the rate of cooling.

Fig. 5.5 *Hardening temperatures for plain carbon steels*

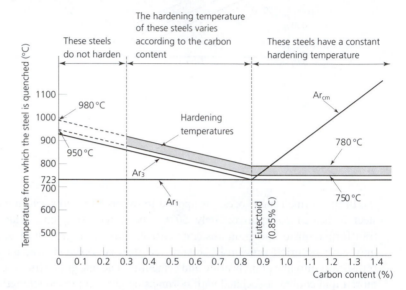

Let's refer to Fig. 5.6 in order to understand the reason for this increase in hardness. In the annealed condition metals can be formed by bending, stretching or squeezing them to shape. This is possible because the orderly arrangement of atoms in the crystals allows individual layers of atoms – called *slip planes* – to slide over each other as shown in Fig. 5.6(a). However, if distortion of the lattice occurs, as shown in Fig. 5.6(b), or particles of another material are introduced, as shown in Fig. 5.6(c), then slip cannot occur so easily and the metal becomes hard and brittle. It's like the difference between trying to slide two pieces of smooth paper over each other compared with trying to slide two sheets of glass paper over each other.

When a steel is heated to its hardening temperature it becomes austenitic. If it is cooled quickly, the equilibrium transformations into pearlite and ferrite or pearlite and cementite do not have time to take place. Instead, a structure called *martensite* is formed. This is the hardest structure that it is possible to produce in a plain carbon steel and, under a microscope, it appears as acicular (needle-shaped) crystals as shown in Fig. 5.7. Actually, these are sections through disc-shaped plates. What has happened is that the face-centred crystals of austenite have changed to body-centred crystals below the Ar_1 line as usual but, because of the rapid cooling, there has not been time for the cementite to form and the body-centred crystals are a *supersaturated solid solution* of carbon in iron (martensite). This so distorts the lattice structure that slip virtually becomes impossible and the steel becomes very hard and brittle.

Fig. 5.6 *Slip and hardness: (a) indentation is easy in a ductile material as slip occurs; this indicates that the material is soft; (b) distortion of the slip planes makes slip extremely difficult; this reduced the amount of indentation indicating that the material is hard; (c) particle hardening: the introduction of particles B distorts the slip planes and makes slip difficult*

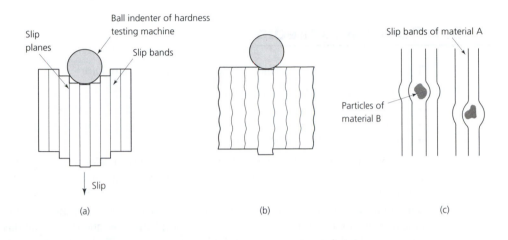

Fig. 5.7 *Martensitic structure: fine, needle-like crystals*

Large components do not cool as quickly as small components and may not achieve the *critical cooling rate* necessary for maximum hardness. The critical cooling rate is defined as the slowest cooling rate (quenching rate) which will produce a martensitic structure throughout the mass of the steel. If this cooling rate is not achieved some pearlite will be formed and the steel will be tougher but substantially less hard. However, there is no virtue in exceeding the critical cooling rate to any great extent. Once maximum hardness has been achieved any increase in the cooling rate will only result in cracking and distortion of the workpiece. Further, there is no particular advantage in heating hyper-eutectoid steels above their Ar_{cm} temperature when hardening them and, in practice, the hardening temperature for hyper-eutectoid steels is just above the Ar_1 temperature (see Fig. 5.5). Quenching hyper-eutectoid steels from this lower temperature helps to prevent grain growth, cracking and distortion.

The critical cooling rate can be substantially reduced by the addition of alloying elements to the steel. This enables thicker components to be hardened with less chance of cracking and distortion and is one of the most important reasons for using alloy steels. Alloy steels and their heat treatment will be dealt with in detail in *Engineering Materials*, Volume 2.

5.8 Quenching media

The most commonly used quenching media in order of severity are:

- Compressed air blast – least severe.
- Oil.
- Water.
- Brine (10 per cent solution) – most severe.

The choice of quenching bath depends upon the type of steel being treated and the resultant properties required. Brine, which is a solution of common salt and water, is only occasionally used to provide very rapid cooling for plain carbon tool steels and case-hardening steels where maximum hardness is required. Such severe quenching can lead to cracking in all but the simplest components, and plain water and quenching oils are most commonly used for plain carbon and alloy steels. To avoid cracking and distortion the quenching rate should be no greater than that needed to give the required properties in the workpiece. Water provides a quenching rate approximately three times as great as oil and is usually used for plain carbon steel.

Oil quenching is usually used with very high-carbon steels (1.2–1.4 per cent) and alloy steels. Air blast quenching is usually reserved for high-speed steel tools and components of small section. The alloy content of such steels is sufficiently high to reduce the critical cooling rate to the very low level required for air blast quenching.

As soon as the heated workpiece is plunged into the quenching bath it becomes surrounded by a blanket of vaporised quenching media. Since this vapour has a low thermal conductivity it slows the cooling process. Therefore the work must be constantly agitated in the quenching bath to disperse the vapour as it forms and keep the work in contact with the liquid. Agitation of the quenching bath also helps to keep its temperature uniform.

Care must be taken to ensure that distortion is kept to a minimum during quenching, and for this reason it is essential to dip long thin components vertically into the quenching bath. Figure 5.8 shows how cracking and distortion can occur both by incorrect design and incorrect quenching, and how these faults may be avoided.

5.9 Tempering

A quench-hardened plain carbon steel is hard, brittle and hardening stresses are present. In such a condition it is of little practical use and it has to be reheated, or *tempered*, to relieve the stresses and reduce the brittleness. Tempering causes the transformation of martensite into less brittle structures. Unfortunately, any increase in toughness is accompanied by some decrease in hardness. Tempering always tends to transform the unstable martensite back into the stable pearlite of the equilibrium transformations.

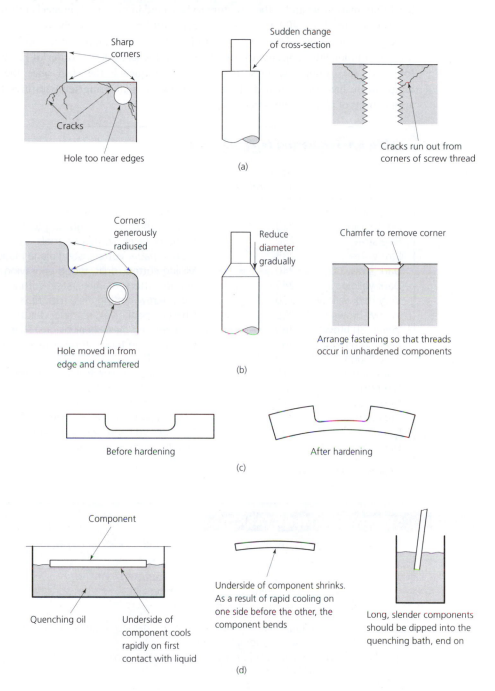

Fig. 5.8 *Causes of cracking and distortion: (a) incorrect engineering that promotes cracking; (b) correct engineering to reduce cracking; (c) distortion caused by an unbalanced shape being hardened; (d) how to quench long, slender components to avoid distortion*

Sharp corners

Cracks

Hole too near edges

Sudden change of cross-section

Cracks run out from corners of screw thread

(a)

Corners generously radiused

Hole moved in from edge and chamfered

Reduce diameter gradually

Chamfer to remove corner

Arrange fastening so that threads occur in unhardened components

(b)

Before hardening

After hardening

(c)

Component

Quenching oil

Underside of component cools rapidly on first contact with liquid

Underside of component shrinks. As a result of rapid cooling on one side before the other, the component bends

Long, slender components should be dipped into the quenching bath, end on

(d)

Tempering temperatures below 200 °C only relieve the hardening stresses, but above 220 °C the hard, brittle martensite starts to transform into a fine pearlitic structure called *secondary troostite* (or just 'troostite'). Troostite is much tougher although somewhat less hard than martensite and is the structure to be found in most carbon-steel cutting tools.

Tempering above 400 °C causes any cementite particles present to 'ball-up' giving a structure called *sorbite*. This is tougher and more ductile than troostite and is the structure used in components subjected to shock loads and where a lower order of hardness can be tolerated, for example springs. It is normal to quench the steel once the tempering temperature has been reached. Table 5.1 gives the tempering temperatures for various applications of plain carbon steel components.

Table 5.1 *Tempering temperatures*

Colour*	Equivalent temperature (°C)	Application
Very light straw	220	Scrapers, lathe tools for brass
Light straw	225	Turning tools, steel-engraving tools
Pale straw	230	Hammer faces, light lathe tools
Straw	235	Razors, paper cutters; steel plane blades
Dark straw	240	Milling cutters, drills, wood-engraving tools
Dark yellow	245	Boring cutters, reamers, steel-cutting chisels
Very dark yellow	250	Taps, screw-cutting dies, rock drills
Yellow-brown	255	Chasers, penknives, hardwood-cutting tools
Yellowish brown	260	Punches and dies, shear blades, snaps
Reddish brown	265	Wood-boring tools, stone-cutting tools
Brown-purple	270	Twist drills
Light purple	275	Axes, hot setts, surgical instruments
Full purple	280	Cold chisels and setts
Dark purple	285	Cold chisels for cast iron
Very dark purple	290	Cold chisens for iron, needles
Full blue	295	Circular and band saws for metals, screwdrivers
Dark blue	300	Spiral springs, wood saws

* Appearance of the oxide film that forms on a polished surface of the material as it is heated.

SELF-ASSESSMENT TASK 5.2

1. State which heat-treatment process you would choose (and give reasons for your choice) in order to:
   (a) prepare a cold-rolled steel sheet for severe cold working
   (b) stress relieve a forging prior to machining
   (c) toughen a quench-hardened steel

2. (a) Select a suitable quenching media for hardening a medium-carbon steel cold chisel. Give reasons for your choice.
   (b) Explain how you would quench the chisel to avoid warping and to ensure uniform and efficient cooling.

5.10 Mass effect

It has already been stated that the hardness of a plain carbon steel depends upon its carbon content and the rate of cooling from the hardening temperature. Obviously a thin component will cool more quickly than a thick component if both are quenched from the same temperature in the same quenching bath.

In a thick component the heat will be trapped at the centre so that the core of the component cools more slowly than the outer layers. This leads to a variation in hardness across a section of the component, as shown in Fig. 5.9(a). This variation in hardness is referred to as *mass effect*.

Plain carbon steels have a highly critical cooling rate and, therefore, large sections cannot be fully hardened throughout, as shown in Fig. 5.9(b). Thus plain carbon steels are said to have a *poor hardenability*. On the other hand, a 3 per cent nickel steel containing only 0.3 per cent carbon will harden uniformly throughout its section because it has a relatively low critical cooling rate. Such a steel is said to have *good hardenability*. Therefore, hardenability can be defined as the ease with which uniform hardness is obtained in a material.

Fig. 5.9 *Mass effect on hardenability*

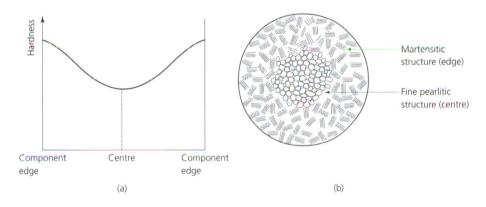

Martensitic structure (edge)

Fine pearlitic structure (centre)

Component edge — Centre — Component edge

(a)

(b)

Hardness and hardenability should not be confused. It has already been stated that a 1.0 per cent plain carbon steel has poor hardenability compared with a 3 per cent nickel steel containing only 0.3 per cent carbon. However, because of its higher carbon content the 1.0 per cent plain carbon steel will have a very much higher surface hardness.

Lack of uniformity of structure and hardness in steels with poor hardenability characteristics can seriously affect their mechanical properties. For this reason it is necessary to specify the maximum diameter or *limiting ruling section* for which the stated mechanical properties can be achieved under normal heat-treatment conditions. One of the main reasons for adding alloying elements such as nickel and chromium to steels is to reduce the mass effect and increase the ruling section for which prescribed properties can be achieved.

5.11 The 'Jominy' (end-quench) test

This test is used to determine the hardenability of steels. It involves heating a specimen to the hardening temperature appropriate for its carbon content so that it is fully austenitic, and then quenching it by spraying a jet of water against its lower end. Details of the test and the specimen are shown in Fig. 5.10. The specimen cools very rapidly at the quenched end and progressively less rapidly towards the opposite (shouldered) end. A flat is ground along the side of the cold specimen and its hardness is tested every 3 mm from the quenched end. The hardness is plotted against distance from the quenched end to give a hardenability curve, as shown in Fig. 5.11 which compares a plain carbon steel with an alloy steel.

Fig. 5.10 *Jominy end-quench test*

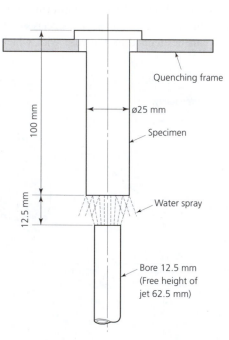

5.12 Case hardening

Often components require a hard case to resist wear and a tough core to resist shock loads. These two properties do not exist in a single steel since, for toughness, the core should not exceed 0.3–0.4 per cent carbon whilst, to give adequate hardness, the surface of the component should have a carbon content of approximately 1.0 per cent. The usual solution to this problem is case hardening.

Fig. 5.11 *Hardenability curve*

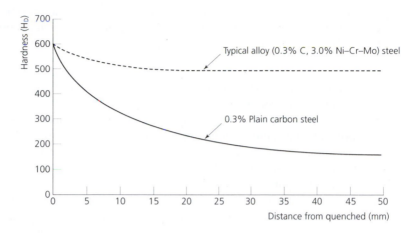

This is a process by which carbon is added to the surface layers of a low-carbon plain or low-alloy steel component to a carefully regulated depth, after which the component goes through successive heat-treatment processes to harden the case and refine the core. Thus the process has two distinct steps, as shown in Fig. 5.12.

● Carburising (the addition of carbon).
● Heat treatment (hardening and core refinement).

Fig. 5.12 *Case hardening: (a) carburising; (b) after carburising; (c) after quenching component from a temperature above 780°C*

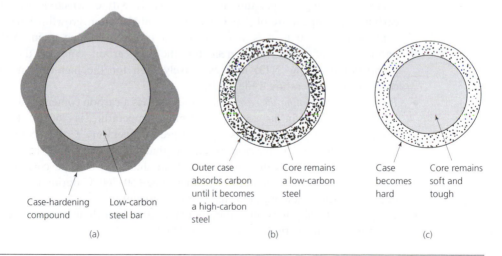

5.12.1 *Carburising*

Carburising makes use of the fact that low-carbon steels (approximately 0.1 per cent carbon) absorb carbon when heated to the austenitic condition. Various carbonaceous materials are used in the carburising process.

- *Solid media* Solid media such as bone charcoal or charred leather, together with an energiser such as sodium and/or barium carbonate. The energiser makes up to 40 per cent of the total composition.
- *Fused salts* Fused salts such as molten sodium cyanide, together with sodium carbonate and varying amounts of sodium and barium chloride are used in special salt-bath furnaces. Since cyanide is a deadly poison and represents from 20–50 per cent of the total content of the molten salts, stringent safety precautions must be taken in its use. The components to be carburised are immersed in the molten salts.
- *Gaseous media* Gaseous media are increasingly used now that 'natural' gas (methane) is widely available. Methane is a hydrocarbon gas containing organic compounds of carbon which are readily absorbed into steel. The methane gas is often enriched by the vapours given off when oil is 'cracked' by heating it in contact with platinum which acts as a catalyst.

It is a fallacy to suppose that carburising hardens steel. Carburising merely adds carbon to the outer layers and leaves the steel in a fully annealed condition with a coarse grain structure. Therefore, additional heat-treatment processes are required to harden and refine the case and to refine and toughen the core.

5.12.2 *Heat treatment*

Let's consider Fig. 5.13. This will clarify the following descriptions of the heat-treatment processes that follow carburising.

- *Refining the core* Since the core has a content of less than 0.3 per cent carbon, the correct annealing temperature is approximately 870 °C which is well below the carburising temperature of 950 °C which caused the grain growth. After raising the component to 870 °C it is water quenched to ensure a fine grain. Although the temperature of 870 °C is correct for the low-carbon core of the component (temperature A Fig. 5.13) it is excessively high for the high-carbon case of the component (temperature B Fig. 5.13).
- *Refining and hardening the case* Since the case has a carbon content of approximately 1.0 per cent carbon its correct hardening temperature is 760 °C. Therefore the component is reheated to this temperature (temperature C Fig. 5.13) and again quenched. This hardens the case and ensures that it has a fine grain. The temperature of 760 °C is too low to cause grain growth in the hyper-eutectic core providing the component is heated rapidly through the range 650–760 °C during the reheating and quenched without soaking at the hardening temperature.
- *Tempering* Ideally it is advisable to relieve any quenching stresses present in the component by tempering it at about 200–220 °C.

Fig. 5.13 *Case-hardening temperatures*

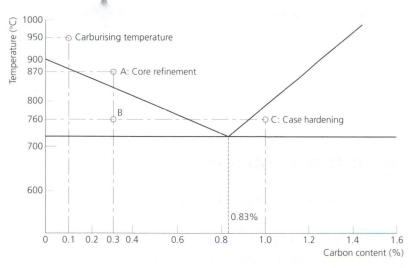

The above heat-treatment sequence is used to give ideal results in stressed components. However, in the interests of speed and economy the process is often simplified where components are lightly stressed or where alloy steels are used having less critical grain growth and quenching characteristics. In such circumstances the tempering process is left out. Sometimes heat treatment is limited to simply quenching immediately after carburising whilst the components are still at the carburising temperature.

5.13 Localised case hardening

It is often not desirable to harden a component all over. For example, it is undesirable to case harden screw threads. Not only would they be extremely brittle, but any distortion occurring during carburising and hardening could only be corrected by expensive thread-grinding operations. Various means are available for avoiding the local infusion of carbon during the carburising process.

- Heavily copper plating those areas to be left soft. The layer of copper prevents intimate contact between the component and the carbon, thus preventing carburisation. Note that copper plating cannot be used for salt-bath treatment, as cyanide dissolves the copper from the component.
- Encasing the areas to be left soft in fire clay. Mostly used when pack-carburising.
- Leaving surplus metal on the component. This is machined off, together with the infused carbon, between carburising and hardening. Although more expensive because of the extra handling involved, this is the surest way of leaving local soft areas. This is shown in Fig. 5.14.

Fig. 5.14 *Localised case hardening*

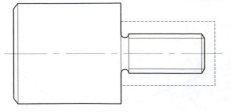

Surplus metal is left on the blank during carburising. Additional carbon is then removed during screw-cutting so that thread remains soft after heat treatment

5.14 Surface hardening

5.14.1 *Flame hardening*

Localised surface hardening can also be achieved in medium- and high-carbon steels and some cast irons by rapid local heating and quenching. Figure 5.15(a) shows the principle of flame hardening. A carriage moves over the workpiece so that the surface is rapidly heated by an oxy-acetylene or an oxy-propane flame. The same carriage carries the water-

Fig. 5.15 *Surface hardening: (a) flame hardening (Shorter process); (b) induction hardening*

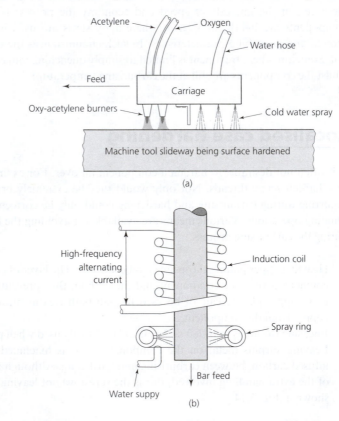

quenching spray. Thus the surface of the workpiece is heated and quenched before its core can rise to the hardening temperature. This process is often used for hardening the slideways of machine tools, for example, lathe bed-ways.

5.14.2 *Induction hardening*

Figure 5.15(b) shows how the same surface-hardening effect can be produced by high-frequency electromagnetic induction. The induction coil surrounding the component is connected to a high-frequency alternating current generator. This induces high-frequency eddy currents in the component causing it to become hot. When the hardening temperature has been reached, the current is switched off and a water spray quenches the component. The induction coil can be made from copper tube which also carries the quenching water or oil. This technique is often used for hardening gear teeth. The induction coil can be tailored to suit the profile of the component. The depth of the case can be controlled by the frequency of the alternating current. The higher the frequency, the nearer to the surface of the component will be the eddy currents resulting in a shallower depth of heating and, therefore, a shallower depth of hardening.

5.14.3 *Nitriding*

This process is used to put a hard, wear-resistant coating on components made from special alloy steels, for example, drill bushes. The alloy steels used for this process contain either 1.0 per cent aluminium, or traces of molybdenum, chromium and vanadium. Nitrogen gas is absorbed into the surface of the metal to form very hard nitrides. The process consists of heating the components in ammonia gas at between 500 and 600 °C for upwards of 40 hours. At this temperature the ammonia gas breaks down and the atomic nitrogen is readily absorbed into the surface of the steel. The case is applied to the finished component. No subsequent grinding is possible since the case is only a few micrometres thick. However, this is no disadvantage since the process does not affect the surface finish of the component and the process temperature is too low to cause distortion. Some of the advantages of nitriding are:

- Cracking and distortion are eliminated since the processing temperature is relatively low and there is no subsequent quenching.
- Surface hardnesses as high as 1150 H_D are obtainable with 'Nitralloy' steels.
- Corrosion resistance of the steel is improved.
- The treated components retain their hardness up to 500 °C compared with the 220 °C for case-hardened plain carbon and low-alloy steels.

SELF-ASSESSMENT TASK 5.3

1. Explain when 'case hardening' would be a more appropriate choice of process than 'through hardening'. Give an example where this is so.

2. With reference to the quench hardening of plain carbon steel, state what is meant by:
   (a) the critical cooling rate
   (b) mass effect

3. Explain the essential differences between case hardening and flame hardening by the Shorter process.

5.1 Briefly explain the following terms as applied to the heat treatment of plain carbon steels:
(a) quench hardening
(b) tempering
(c) full annealing

5.2 (a) Describe how a chisel made from 0.8 per cent carbon steel may be hardened and tempered.
(b) How can the heat-treatment temperatures be judged visually in this example?
(c) Why should a long, slender component be quenched with its axis vertical?

5.3 With reference to the quench hardening of plain carbon steels, explain what is meant by the term 'critical cooling rate'.

5.4 With the aid of diagrams explain in detail what is meant by:
(a) recrystallisation
(b) cold working
(c) hot working

5.5 With reference to plain carbon steels, explain in detail the essential differences between the following, and give one example in each instance where the process would be used:
(a) stress-relief (process) annealing
(b) spheroidising annealing
(c) full annealing

5.6 With reference to the steel section of the iron–carbon phase equilibrium diagram, explain the following:
(a) How can a hypo-eutectoid steel be heat treated to give maximum toughness? Describe the micro-constituents present in the steel after heat treatment.
(b) How can a hyper-eutectoid steel be heat treated so that it is suitable for cutting metal? Describe the micro-constituents present in the steel after heat treatment.

5.7 With reference to the iron–carbon phase equilibrium diagram, explain how a low-carbon steel can be case hardened by the pack-carburising process so that the components have a hardened and tempered case and a toughened, fine-grain core.

5.8 With the aid of diagrams:
(a) Explain how selected areas (e.g. screw threads) of a case-hardened component can be left soft for subsequent machining.
(b) Explain the principles of surface hardening by the following methods, and give an example where each of these processes would be appropriate:
 (i) flame hardening
 (ii) induction hardening

5.9 Explain fully what is meant by the following, as applied to plain carbon steels and how the 'Jominy end-quench test' can be used to assess hardenability:
  (a) limited ruling section
  (b) mass effect
  (c) hardenability

5.10 Discuss the advantages and limitations of the 'nitriding process' for applying a wear-resistant surface to engineering components.

6 Cast irons and their heat treatment

The topic areas covered in this chapter are:

- The iron–carbon system for cast irons.
- Alloying elements and impurities.
- Heat treatment of grey cast iron.
- Malleable cast iron.
- Spheroidal graphite cast iron.
- Alloy cast irons.
- Specifications, properties and uses of cast irons.

6.1 The iron–carbon system for cast irons

Cast iron is the name given to those ferrous metals containing more than 1.7 per cent carbon. It is similar in composition to crude pig iron as produced by the blast furnace. Unlike steel, it is not subjected to an extensive refinement process. After the pig iron has been remelted in a cupola furnace ready for casting, selected scrap iron and scrap steel are added to the melt to give the required composition. Although more expensive, casting ingots can be purchased already formulated to a guaranteed composition and purity where high-grade and alloy cast irons are required.

Since pig iron and scrap are cheap, and because there is no expensive refinement process, cast iron is a useful low-cost material compared with steel. It is widely used where castings of high rigidity, resistance to wear and compressive strength are required. Further, cast iron is easy to machine, it has a high fluidity which makes it easy to cast into intricate shapes, and it has a melting point between 1147 and 1250 °C, which is substantially lower than the melting point for mild steel.

Let's refer back to the iron–carbon phase equilibrium diagram (Fig. 4.1). This shows that the cast irons lie to the right of the 'steel section' of the diagram. Figure 6.1 shows the 'cast iron section' of the phase equilibrium diagram in greater detail. Since the maximum amount of carbon which can be held in solid solution as austenite is only 1.7 per cent, it is obvious that in all cast irons there will be surplus carbon. This surplus carbon can combine with the iron to form cementite (iron carbide) or it can be precipitated out as flake graphite (graphite is an allotrope of carbon). In the foundry, cementite is referred to as 'combined carbon' and flake graphite is known as 'free carbon'.

Figure 6.1 shows that there is a eutectic when there is 4.3 per cent carbon present. At this composition the molten metal solidifies at 1147 °C into austenite (γ phase) and cementite. Unless cooling is very rapid, graphite will be precipitated out due to the instability of the cementite as a result of some of the impurities present (particularly silicon). As cooling proceeds, further graphite is precipitated out from the austenite. At 723 °C, providing cooling is slow enough, the remaining austenite (γ phase) changes into ferrite (α phase). Thus at room temperature the composition consists of ferrite plus large flakes of graphite together with fine flakes of graphite formed by the decomposition of the cementite after solidification.

Fig. 6.1 *Cast iron section of the iron–carbon phase equilibrium diagram*

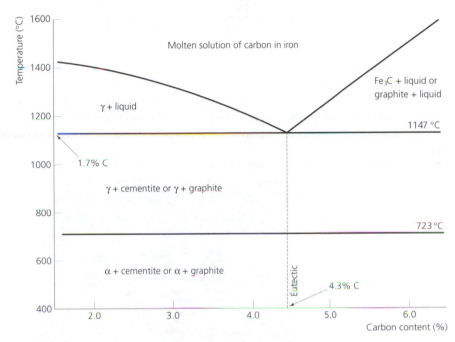

If cooling is sufficiently fast to prevent phase equilibrium being achieved, the austenite will change to ferrite and pearlite at the eutectoid temperature of 723 °C as the cementite essential for the formation of pearlite will be retained. With even faster cooling, the structure will consist of flake graphite in a matrix which is entirely pearlitic, as shown in Fig. 6.2. Ferritic and pearlitic cast irons containing free graphite flakes are called *grey cast irons* because of the grey appearance of their surfaces when freshly fractured. As in steel, increasing the amount of pearlite present enhances the toughness and hardness of the cast iron.

With even faster cooling (chilling), a different type of structure is likely to be formed. When the solidus temperature of 1147 °C is reached, the structure will consist entirely of austenite and cementite. On further cooling the austenite changes into pearlite, thus at room temperature the composition of the iron is pearlite and cementite. This type of cast

Fig. 6.2 *Pearlitic grey cast iron (source: The Castings Development Centre)*

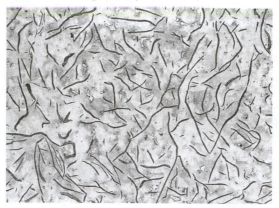

iron is known as *white cast iron* due to its white appearance when freshly fractured. It derives its appearance from the white crystals of cementite. A microphotograph of a typical white cast iron structure is shown in Fig. 6.3. The hardness of the cementite in white cast iron makes it difficult to machine and its use is limited mainly to wear-resistant components and as a basic material for conversion to white-heart malleable cast iron (see Section 6.4).

Fig. 6.3 *White cast iron (source: The Castings Development Centre)*

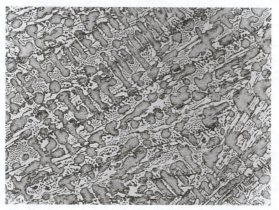

6.2 Alloying elements and impurities

Cast irons are not solely alloys of iron and carbon as the phase equilibrium diagram would suggest, but complex alloys in which impurities such as sulphur, phosphorus and alloying elements such as silicon and manganese have a significant influence on the properties of the casting. Although complex alloys, such cast irons are still referred to as common cast irons, the term *alloy cast irons* being reserved for those cast irons containing substantial amounts of such metallic elements as chromium, nickel, etc.

6.2.1 *Silicon*

This element is used to soften cast irons by promoting the formation of flake graphite at the expense of cementite. The silicon content is increased in irons used for light or thin components which might chill harden by cooling too quickly in the mould and becoming hard and brittle. The addition of significant amounts of silicon can reduce the eutectic composition down to 3.5 per cent carbon. Thus, at a constant rate of cooling, the addition of silicon to a cast iron having 3 per cent carbon will have the following effects.

- Ferritic grey cast iron is produced with 3 per cent silicon.
- Ferritic/pearlitic cast iron is produced with 2 per cent silicon.
- Pearlitic cast iron is produced with 1.5 per cent silicon.
- White cast iron is produced with no silicon.

Providing sufficient silicon has been added to break down the cementite into flake graphite (3 per cent silicon in this example) there is no benefit in adding extra silicon. In fact, the presence of excess silicon will lead to increased hardness and brittleness.

6.2.2 *Sulphur*

Sulphur is an impurity carried over from the fuel used in the blast furnace during the extraction of the iron from its ore. The presence of sulphur in cast irons, even in small quantities, has the effect of stabilising the cementite and preventing the formation of flake graphite. Thus, sulphur hardens the cast iron. It also causes embrittlement due to the formation of iron sulphide (FeS) at the grain boundaries.

6.2.3 *Manganese*

The addition of this element in small quantities is essential in all ferrous metals as it combines with any residual sulphur present to form manganese sulphide (MnS). Unlike ferrous sulphide, manganese sulphide is insoluble in the molten iron and floats to the top of the melt to join the slag. Thus by removing the sulphur, the manganese indirectly softens the cast iron and also removes a source of embrittlement. Increasing the manganese content beyond that required to neutralise the sulphur has the effect of stabilising the cementite and causing hardness in the iron, just as the sulphur did, but without any embrittlement. Thus it is important to balance the amount of manganese, sulphur and silicon present with great care. Manganese also promotes grain refinement and increases the strength of the cast iron.

6.2.4 *Phosphorus*

Like sulphur, this is a residual impurity from the extraction process. It is present in cast iron as iron phosphide (Fe_3P). This phosphide forms a eutectic with ferrite in grey cast irons, and with ferrite and cementite in white cast irons. Since these eutectics melt at only 950 °C, high phosphorus content cast irons have great fluidity. Cast irons containing 1 per cent phosphorus are thus very suitable for the production of thin section castings and highly intricate ornamental castings. Unfortunately phosphorus, like sulphur, causes hardness and embrittlement in the cast iron and the amount present must be kept to a minimum in castings where shock resistance and strength are important. The composition and applications of some typical grey cast irons are summarised in Table 6.1.

Table 6.1 *Typical grey cast irons*					
Composition (%)					
C	Si	Mn	S	P	Applications
3.30	1.90	0.65	0.08	0.15	Motor vehicle brake drums
3.25	2.25	0.65	0.10	0.15	Motor vehicle cylinder blocks
3.25	1.75	0.50	0.10	0.35	Medium machine castings
3.25	1.25	0.50	0.10	0.35	Heavy machine castings
3.60	1.75	0.50	0.10	0.80	Light and medium spun cast water pipes
3.50	2.75	0.50	0.10	0.9	Ornamental castings requiring maximum fluidity but only low strength

6.3 Heat treatment of grey cast iron

It is virtually impossible to arrange for all the parts of a casting to cool and solidify at the same rate. It is equally impossible to achieve the optimum cooling rate for a given composition of cast iron in any given casting. Therefore grey iron castings are frequently subjected to heat treatment to relieve stresses, refine the grain structure and improve their machinability.

6.3.1 *Annealing*

Grey iron castings can be heated to just above the Ac_1 temperature (approximately 760 °C), soaked at that temperature until equilibrium structures have been achieved and then cooled very slowly. The soaking time will depend upon the mass and thickness of the casting. This promotes the precipitation of flake graphite and breaks down any excess cementite which may be forming 'hard spots' or chills. These hard spots result from over-rapid cooling of thin sections due to the chilling effect of the mould. Annealing also removes internal stresses in the composition resulting from uneven cooling in the casting process.

6.3.2 *Quench hardening*

Ferritic grey cast irons can be heated to just above the Ac_1 temperature (approximately 760 °C) and quenched in water or oil. This prevents phase equilibrium being attained and results in the formation of pearlite rather than ferrite and flake graphite, thus giving increased toughness and hardness. Great care has to be taken in the quenching process to avoid cracking the castings. For this reason it is not usual to attempt to attain a white iron structure. To relieve the stresses created by quenching it is usual to temper the castings at 450–475 °C. Quench hardening grey iron castings is not a common practice.

6.3.3 *Stress relieving*

Most iron castings have internal stresses when they are released from the mould. These stresses not only reduce the strength of the casting, they may also cause it to warp and distort during machining. Traditionally, machine beds and frames which were required to be dimensionally and geometrically stable were 'weathered' (see also Section 5.6); that is, they were rough machined and left out of doors for a period of time varying from months to years. The continual changes in temperature caused diffusion of the internal structure coupled with continual expansion and contraction. This resulted in the release of the internal stresses and allowed the castings to warp to their final shape so that, after finish machining, no further distortion took place.

Nowadays, castings are more likely to be stress relieved by soaking them in a furnace at (approximately) 550 °C for a period ranging from several hours to several days, depending upon the size of the casting. Very slow cooling follows the heating. Thus the temperature for the stress relief of iron castings is very much lower than that required for the full annealing of steel castings and forgings (see Section 5.5).

SELF-ASSESSMENT TASK 6.1

1. Explain why grey cast iron contains 'free' carbon and how this affects its properties.

2. Describe the effects of the impurities, sulphur and phosphorus, on the properties of grey cast iron.

3. Explain why grey iron castings are sometimes annealed.

4. Explain why large and complex castings are stress relieved after rough machining and state the processes that are available for stress relieving.

6.4 Malleable cast iron

Malleable cast irons are produced from white cast irons by a variety of heat-treatment processes, depending upon the final composition and structure required. As their name implies, malleable cast irons have increased malleability and ductility, increased tensile strength, and increased toughness.

6.4.1 *Black-heart process*

In this process the white iron castings are heated in airtight boxes out of contact with air at 850–950 °C for 50–170 hours, depending upon the mass and thickness of the castings. The effect of this prolonged heating is to break down the iron carbide (cementite) of the white

---

cast iron into small rosettes of graphite. The final structure is of ferrite and fine carbon particles, as shown in Fig. 6.4. The name 'black-heart' comes from the darkened appearance of the iron, when fractured, resulting from the formation of free graphite. The relationship of the process to the iron–carbon phase equilibrium diagram is shown in Fig. 6.5.

Fig. 6.4 *Black-heart malleable cast iron (source: The Castings Development Centre)*

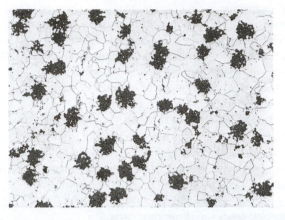

Fig. 6.5 *Black-heart transformations*

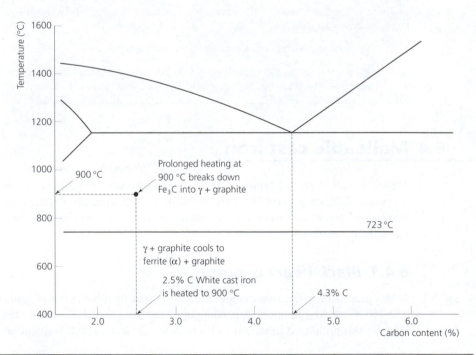

6.4.2 White-heart process

In this process the castings are packed into airtight boxes with iron oxide in the form of high-grade ore. They are then heated to about 1000 °C for between 70 and 100 hours, depending upon the mass and thickness of the castings. The ore oxidises the carbon in the castings and draws it out, leaving a ferritic structure near the surface and a pearlitic structure near the centre of the casting. There will also be some fine rosettes of graphite. White-heart castings behave much as expected of a mild steel casting, but with the advantage of a very much lower melting point and higher fluidity at the time of casting. Figure 6.6 shows a typical white-heart structure.

Fig. 6.6 *White-heart malleable cast iron (source: The Castings Development Centre)*

6.4.3 Pearlitic process

This process is similar to the black-heart process inasmuch as the castings are heated to 850–950 °C for 50–170 hours in a non-oxidising environment. As in the black-heart process the iron carbide (cementite) breaks down into austenite and free graphite. However, in the pearlitic cast iron process, rapid cooling prevents the austenite changing into ferrite and graphite, and a pearlitic structure is produced instead. Since this 'pearlitic cast iron' also has a fine grain resulting from the rapid cooling, it is harder, tougher and has a higher tensile strength than black-heart cast iron. However, there is a marked reduction in malleability and ductility. Pearlitic malleable irons can be produced by increasing the manganese content of the melt from 1.0–1.5 per cent. This inhibits the production of free graphite and encourages the formation of cementite and pearlite. Figure 6.7 shows a typical pearlitic cast iron.

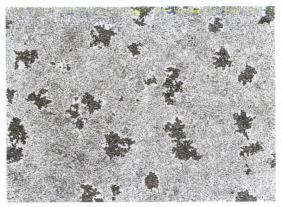

6.5 Spheroidal graphite cast iron

Spheroidal graphite (SG) cast iron goes under a variety of names, including nodular iron, ductile iron, high duty iron, etc. When traces of the metals magnesium or cerium are added to ordinary grey cast iron, the graphite flakes become redistributed throughout the mass of the metal as fine spheroids of graphite, as shown in Fig. 6.8. The flakes of graphite in common grey cast iron leave voids in the metal with sharp corners similar to cracks. Stress concentrations occur at these sharp corners from which cracks are propagated. The rounded nodules of carbon in spheroidal graphite cast iron do not create stress concentrations to the same extent, and this greatly enhances the strength of the castings. It also reduces the likelihood of fatigue failure.

Fig. 6.8 *Speroidal graphite cast iron (source: The Castings Development Centre)*

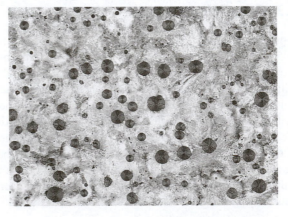

6.6 Alloy cast irons

The alloying elements in cast irons are similar to those in alloy steels.

- *Nickel*   This is used for grain refinement, to add strength, and to promote the formation of free graphite. Thus it toughens the casting.
- *Chromium*   This stabilises the combined carbon (cementite) present and thus increases the hardness and wear resistance of the casting. It also improves the corrosion resistance of the casting, particularly at elevated temperatures. As in alloy steels, nickel and chromium tend to be used together because they have certain disadvantages when used separately which tend to offset their advantages. However, when used together the disadvantages are overcome, whilst the advantages are retained.
- *Copper*   This is used very sparingly as it is only slightly soluble in iron; however, it is useful in reducing the effects of atmospheric corrosion.
- *Vanadium*   This is used in heat-resisting castings as it stabilises the carbides and reduces their tendency to decompose at high temperatures.
- *Molybdenum*   This dissolves in the ferrite and, when used in small amounts (0.5 per cent), it improves the impact strength of the casting. It also prevents 'decay' at high temperatures in castings containing nickel and chromium. When molybdenum is added in larger amounts it forms double carbides, increases the hardness of castings with thick sections, and also promotes uniformity of the microstructure.
- *Martensite*   Martensitic cast irons contain between 4 and 6 per cent nickel and approximately 1 per cent chromium, for example, 'Ni-hard' cast iron. It is naturally martensitic in the cast state but, unlike alloys with rather less nickel and chromium, it does not need to be quench hardened thus reducing the possibility of cracking and distortion. It is used for components which need to resist abrasion. It can only be machined by grinding.
- *Austenite*   Austenitic cast irons contain between 11 and 20 per cent nickel and up to 5 per cent chromium. These alloys are corrosion resistant, heat resistant, tough and non-magnetic. Since the melting temperatures of alloy cast irons can be substantially higher than those for common grey cast irons, care must be taken in the selection of moulding sands and the preparation of the surfaces of the moulds. Increased venting of the moulds is also required as the higher temperatures cause more rapid generation of steam and gases. The furnace and crucible linings must also be suitable for the higher temperatures and the inevitable increase in maintenance costs is also a significant factor when working with high-alloy cast irons.

The *growth* of cast irons is caused by the breakdown of pearlitic cementite into ferrite and graphite at approximately 700 °C. This causes an increase in volume. This increase in volume is further aggravated by hot gases penetrating the graphite cavities and oxidising the ferrite grains. This volumetric growth causes warping and the setting up of internal stresses leading to cracking, particularly at the surface.

Therefore, where castings are called upon to operate at elevated temperatures, alloy cast irons should be used. A low-cost alloy is 'Silal', which contains 5 per cent silicon and a relatively low carbon content. The low carbon content results in a structure which is composed entirely of ferrite and graphite with no cementite present. Unfortunately 'Silal' is rather brittle because of the high silicon content. A more expensive alloy is 'Nicrosilal'. This is an austenitic nickel–chromium alloy which is much superior in all respects for use at elevated temperatures.

SELF-ASSESSMENT TASK 6.2

1. Briefly describe the difference between the white-heart and black-heart processes for making cast irons malleable.
2. How do malleable cast irons and spheroidal graphite cast irons differ?
3. Discuss the advantages and limitations of malleable and SG cast irons with common grey cast iron.

6.7 Properties and uses of white cast irons

White cast irons are little used except as a basis for malleable cast irons (Section 6.4). They are wear resistant but lack ductility and they break easily since they are very brittle. They are virtually unmachinable except by grinding. The composition and properties of a typical white cast iron may be listed as follows:

Composition		*Properties*	
Carbon	2.5%	Elongation	none
Silicon	0.8%	Tensile strength	250–450 MPa
Manganese	0.4%	Hardness	$400\,H_B$
Sulphur	0.08%	Phosphorus	0.1%

6.8 Properties and uses of grey cast irons

These are the most widely used cast irons and they vary in composition according to specific applications. Table 6.1 has already listed some typical examples together with their composition and uses. The tensile strength of such cast irons varies between 150 and 350 MPa.

Cast irons requiring high fluidity are used for very intricate mouldings for decorative iron work and architectural tracery. They lack mechanical strength since the high fluidity is obtained by ensuring that the melt has a high silicon content (2.5–3.5%) and a high phosphorus content (approximately 1.5%). It is the high phosphorus content which reduces the strength of the metal.

General-purpose engineering irons have to have reasonable mechanical strength. Ideally they should have fine graphite flakes in a matrix of pearlite. Such a cast iron would combine good mechanical properties with good machinability. The silicon content would be dependent upon the thickness and mass of the casting, varying between 2.5 per cent for thin sections and about 1.5 per cent for castings with thick sections. The phosphorus content would be kept low to improve toughness and shock resistance, although up to 0.8 per cent may be present to improve fluidity. Sulphur must be kept below 0.1 per cent to avoid segregation, hard spots and embrittlement.

Local hardening of grey iron castings (e.g. slideways) can be achieved by chilling. This is done by introducing 'chills' or metal plates into the mould just behind the surface layer of sand to promote rapid cooling and the formation of cementite. The hardness occurs only at the surface of the casting and the core remains grey and tough. See also flame hardening (Section 5.14).

Large heavy castings do not require a high silicon content as they naturally cool slowly and there is adequate time for any cementite to break down into ferrite and flake graphite. A typical composition would be 1.2–1.5 per cent silicon, 0.5 per cent phosphorus and 0.1 per cent sulphur. Table 6.2 (derived from BS 1452) lists the essential properties of some typical grey cast irons.

Table 6.2 *Properties of grey cast iron*

Grade	UTS (MPa)	0.1% proof stress (MPa)	Compressive strength (MPa)	0.1% compressive proof stress (MPa)
150	150	98	600	195
180	180	117	672	234
220	220	143	768	286
260	260	169	864	338
300	300	195	960	390
350	350	228	1080	455
400	400	260	1200	520

Grade	Shear strength (MPa)	Modulus of elasticity (GPa) Tension	Modulus of elasticity (GPa) Compression	Modulus of rigidity (GPa)
150	173	100	100	40
180	207	109	109	44
220	253	120	120	48
260	299	128	128	51
300	345	135	135	54
350	403	140	140	56
400	460	145	145	58

Note: For grey cast iron, hardness is not related to tensile strength but varies with casting section thickness and materials composition.

6.9 Specifications for grey iron castings

BS 1452 specifies the requirements for seven grades of grey iron castings. These are: 150, 180, 220, 260, 300, 350 and 400. As well as designating the specification, these numbers refer to the tensile strength of a standard test piece in MPa (remember that $1\,\text{MPa} = 1\,\text{MN/m}^2 = 1\,\text{N/mm}^2$) taken from the same melt as the castings being manufactured. The standard does not specify the chemical composition of the iron or its processing in the foundry. It only specifies the properties, test conditions and quality. How these are attained are left to the discretion of the foundry in consultation with the customer. Basically, the customer need only specify the standard and the grade, e.g. BS 1452 grade 220, in which case the foundry will then use its discretion in achieving this specification. However, the customer may also specify or require:

- A mutually agreed chemical composition.
- Casting tolerances, machining locations.
- Test bars and/or test certificates.
- Whether testing and inspection is to be carried out in the presence of the customer's representative.
- Any other requirement, such as hardness tests and their locations, non-destructive tests and quality assurance.

Throughout BS 1452 the accent is on the quality of the finished casting and the methods of testing to ensure that the specified quality and properties have been achieved. For example, the frequency of testing is specified as follows:

- Grades 150 and 180: one test bar for up to 10 tonnes of castings.
- Grades 220 and 260: one test bar for up to 5 tonnes of castings.
- Grades 300, 350 and 400: one test bar for up to 1 tonne of castings.

Where a single casting exceeds the above figures, there shall be one test bar per casting. The test bar should be sufficient to manufacture at least three test pieces to allow for retesting.

BS 1452 also specifies that tensile testing (Section 11.2) shall be in accordance with BS 18: Part 2 and that the test piece shown in Fig. 6.9 shall be machined from a cast cylindrical test bar 30 mm diameter and 230 mm minimum length. At least three test pieces should be cast so that in the event of the first test piece failing, two further tests can be made. Should either of these retests fail, then the whole melt and any castings made from it is deemed to have failed the requirements of BS 1452.

Fig. 6.9 *Machined tensile test pieces for grey cast iron with (a) plain and (b) screwed ends (source: BS 1452)*

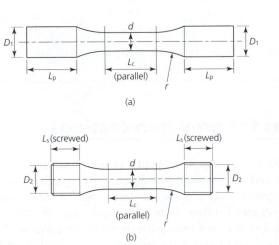

Gauge diameter d
  Nominal value .... 20 mm
  Machining tolerance* .... ± 0.5 mm
Minimum parallel length L_c .... 55 mm
Minimum radius r
  Nominal value .... 25 mm
  Machining tolerance .... −0/+5 mm
Plain ends
  Minimum diameter D_1 .... 25 mm
  Minimum length L_p .... 65 mm
Screwed ends
  Minimum diameter at
  root of thread D_2 .... 25 mm
  Minimum length L_s .... 30 mm

* If it is desired to calculate the tensile strength on the basis of the nominal diameter, the machining tolerance shall be ±0.10 mm. *Note*: With screwed ends, any form of thread may be used provided that the diameter at the root of the thread is not less than that specified.

Although the BS grades of grey cast iron are based upon minimum tensile strengths obtained from test pieces machined from 30 mm diameter cast test bars, the strength of the castings produced will vary according to their mass, section thickness and consequential rate of cooling. Thin sections of low mass will cool quickly and show increased tensile strength and hardness, whilst thicker sections will cool more slowly and show a reduced tensile strength and hardness. This effect is shown in Fig. 6.10.

Fig. 6.10 *Variations of tensile strength with section thickness – for example: an iron which will give a tensile strength of 220 N/mm² in a 30 mm diameter test bar should give a tensile strength of approximately 220 N/mm² in the centre of a casting whose ruling section is 15 mm, and approximately 147 N/mm² where the cross-section is 100 mm (source: BS 1452)*

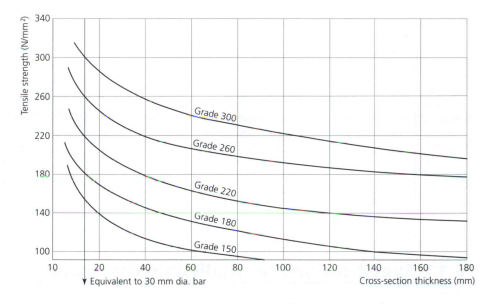

Many other properties of the grey cast irons specified in BS 1452 have to be considered and the full specifications are comprehensively listed in Appendix B of that specification.

6.10 Properties and uses of malleable cast irons

Malleable irons exploit the excellent casting properties of cast iron during the casting process, after which they are converted by heat-treatment processes into a composition and structure whose properties resemble that of a low-carbon steel. This results in castings which are stronger and much less brittle than ordinary cast irons and are widely used in the automobile, agricultural machinery and machine tool industries for the manufacture of small and medium sized stressed components.

Malleable iron castings are also used in the electrical industry for conduit fittings, switch gear cases and components. The properties and applications of some typical malleable cast irons (derived from BS 6681) are listed in Table 6.3.

Table 6.3 _Properties and uses of malleable cast iron_

| Type of cast iron | Condition | Properties | | | Applications |
		UTS (MPa)	Elong. (%)	Hardness (H_B)	
Black-heart malleable	Annealed	300–350	6–12	150 max.	Wheel hubs, brake drums, conduit fittings, control levers and pedals.
White-heart malleable	Annealed ·	340–480	3–15	230 max.	Wheel hubs, bicycle and motor cycle frame fittings. Gas, water, and steam pipe fittings.
Pearlitic malleable	Normalised	450–700	3–6	150–290	Gears, couplings, camshafts, axle housings, differential housings and components.

6.11 Specifications for malleable cast irons

BS 6681 supersedes BS 309, 310 and 3333 for specifying the requirements of malleable cast irons. These are specified by a capital letter designating the type of iron:

W White-heart malleable cast iron
B Black-heart malleable cast iron
P Pearlitic malleable cast iron

The letter is followed by two figures designating the minimum tensile strength in MPa of a 12 mm diameter test piece. For this purpose the tensile strength is divided by 10, e.g. if the tensile test returned a result of 320 MPa, the two-figure designation would be 32. Finally, there are two figures representing the minimum elongation percentage on the specified gauge length (see Fig. 6.11 and Table 6.4). Thus a complete designation for a malleable cast iron could be:

W35-04   White-heart malleable cast iron with a minimum tensile strength of 350 MPa, and a minimum elongation of 4 per cent.

Table 6.4 _Dimensions of tensile test bar_

(See Fig. 6.11)

Diameter (d) (mm)	Tolerance on diameter (mm)	Nominal cross-sectional area (S₀) (mm²)	Gauge length (L₀ = 3d) (mm)	Minimum parallel length (L_c) (mm)	Radius at shoulder (r) (mm)	Preferred shank dimensions (mm) Diameter (D)	Length (L)
9	±0.7	83.6	27	30	6	13	40
12	±0.7	113.1	36	40	8	16	50
15	±0.7	176.7	45	50	8	19	60

Note: Where the test bar is tested in the unmachined state, the tensile strength is calculated from the measured diameters of each test bar. For this purpose, the diameter is obtained by taking the average of measurements taken in the same plane at right angles to each other.
Source: BS 6681.

Fig. 6.11 _Tensile test piece for malleable cast iron (source: BS 6681)_

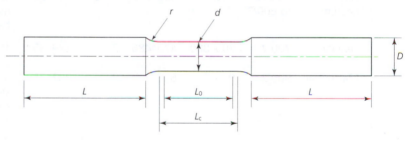

As for grey cast irons the specification is not concerned with the chemical composition except for stating that the phosphorus content shall not exceed 0.12 per cent. The composition and manufacturing processes are left to the discretion of the foundry in consultation with the customer. The melt and the castings made from it will satisfy the requirements of BS 6681 providing the test results and the general quality of the castings meet the specifications laid down.

At least one test bar should be produced from each melt and at least one test bar for each 2000 kg of castings produced. Two additional test bars from each batch should be kept in reserve for retests (see grey cast iron). Since malleablising is a heat-treatment process, the test bars should be heat treated alongside the castings to which they refer.

BS 6681 sets out the full range of mechanical properties for each designated grade of malleable cast iron and a comprehensive testing programme to ensure that the specifications are met. These details are beyond the scope of this chapter and the specification itself should be consulted.

6.12 Properties and uses of spheroidal and nodular graphite cast irons

Speroidal graphite cast irons are now used for quite highly stressed components in the automobile industry. By designing components with this material in mind, spheroidal graphite castings can, in some instances, replace steel forgings at much lower cost; for example: camshafts, crankshafts and differential gear carriers for motor vehicles. The properties and some further applications for typical spheroidal graphite (SG) cast irons are listed in Table 6.5.

Table 6.5 *Properties of spheroidal graphite cast iron*						
		Properties				
Type of cast iron	Grade	UTS (MPa)	0.2% PSI (MPa)	Elong. (%)	Hardness (H_B)	Applications
Ferritic	350/22	350–420	220–270	12–22	160–212	Water main pipes, hydraulic cylinder
Ferritic/ pearlitic	450/10 to 600/3	450–600	320–370	3–10	160–269	and valve bodies, machine vice handles.
Pearlitic	700/2	700–800	420–480	2	229–352	These grades will surface harden
Martensitic	900/2	900	600	2	302–359	and can replace steel forgings for such stressed applications as automobile engine camshafts and crankshafts.

6.13 Specifications for spheroidal and nodular graphite cast irons

BS 2789 specifies the requirements for spheroidal or nodular graphite cast irons. Again, the standard does not specify the chemical composition of the iron, its method of manufacture or any subsequent heat treatment. The standard is solely concerned with the properties, testing and quality control of the finished castings. How this is attained is left to the discretion of the foundry in consultation with the customer. It is a very comprehensive standard and it is only possible to review briefly some of its more important points within the scope of this chapter. The standard itself should be consulted for more detailed study.

This revised standard includes requirements for tensile strength, elongation, 0.2 per cent proof stress and, for two grades of iron, resistance to impact. The standard covers the majority of commercial applications and includes the requirements for a total of nine grades of iron including two new grades: 900/2 which increases the range of mechanical properties available to designers and grade 450/10 which has a higher proof stress to tensile stress ratio than the previous grades. The grades specified are:

900/2   This grade has a tempered martensitic structure.

800/2   These grades have a mainly pearlitic matrix, characterised by high
700/2   tensile strength but at the expense of lower ductility and resistance to
600/3   impact.

500/7   These intermediate grades have ferritic/pearlitic matrices combining
450/10   strength with reasonable ductility and impact strength.

420/12   This grade has a mainly ferritic matrix of moderately high tensile
          strength, but with substantial ductility and impact resistance.

400/18   These grades have wholly ferritic matrices with even greater ductility
350/20   and even higher resistance to impact.

To interpret these grades, the first three figures indicate the minimum tensile strength in MPa and the final figure (after the /) indicates the minimum elongation percentage. The addition of the letter L followed by a number in the case of 400/18L20 and 350/22L40 indicates that the impact strength must be attained at low temperatures, that is at $-20\,°C$ and $-40\,°C$ respectively.

The standard specifies the shape and dimensions of test pieces for tensile testing and it also specifies how the test bars are derived; that is, whether they are cast separately or whether they are cast onto the main casting or onto a runner bar. If they are 'cast on', then they must not be separated from the main casting until they have cooled below $500\,°C$.

6.14 Composition, properties and applications of some alloy cast irons

Although much more expensive because of the high cost of the alloying elements which they contain and the increased foundry costs in casting them, they are widely used for special applications where their properties can be exploited. By suitable alloying, their properties can be tailored to give high strength, high corrosion resistance, high wear resistance and high temperature resistance, whilst retaining the improved casting properties of cast iron compared with cast steel. They can also be cast to intricate shapes which cannot be achieved by steel forgings whilst exhibiting almost comparable mechanical properties. The composition, properties and some typical applications of alloy cast irons are listed in Table 6.6.

Table 6.6 Typical alloy cast irons

Type of cast iron	Composition (%)								Properties		Applications
	C	Si	Mn	S	P	Ni	Cr	Other	UTS (MPa)	Hardness (H_B)	
Chromidium	3.2	2.1	0.8	0.05	0.17	—	0.32	—	275	230	Cylinder blocks, brake drums and discs, clutch casings, differential carriers, etc.
Wear and shock resistant	2.9	2.1	0.7	0.05	0.10	1.75	0.10	0.8 Mo 0.15 Cu	450	300	Crankshafts for automobile diesel and petrol engines. High strength. Good shock and vibration damping properties.
Ni-hard	2.8	1.3	—	—	—	21.0	2.0	—	—	60	Martensitic iron of great hardness and wear resistance, ore crushing jaws, abrasive material handling components.
Ni-resist	2.9	2.1	1.0	0.05	0.1	15.0	2.0	6.0 Cu	215	130	A corrosion resistant alloy suitable for valve and pump bodies handling sulphur and chloride solutions.
Silal	2.5	5.0	—	—	—	—	—	—	215	—	High temperature resistance, suitable for exhaust manifolds, furnace components, etc.

1. Explain why cast irons are lower in cost, weight for weight, compared with steels.

2. The British Standards for cast irons only specify the properties to be achieved.

   (a) How are these properties achieved?
   (b) Who is responsible for the composition of the melt?

3. State the rules for producing test pieces for the various types of cast iron and sizes of castings.

EXERCISES

6.1 In addition to iron and carbon, list the other elements that may be found in a grey cast iron and describe their effect on the properties of a grey cast iron.

6.2 Describe what is meant by the term 'chilling' in connection with the casting process, and explain how 'chilling' affects the properties of a grey iron casting.

6.3 Explain what is meant by 'growth' in a cast iron and state how this can be controlled.

6.4 (a) With reference to BS 1452 list the following properties of a grade 220 grey cast iron:
     (i) tensile strength
     (ii) 0.1 per cent proof stress
     (iii) compressive strength
     (iv) shear strength
   (b) Why does this standard **not** give the composition for each grade of cast iron listed in it?

6.5 Briefly explain the effect of the following alloying elements on cast irons:
   (a) chromium
   (b) copper
   (c) vanadium

6.6 Discuss in detail the following malleablising processes as applied to cast irons, together with typical properties (BS 6681: 1986) and applications:
   (a) white-heart process
   (b) black-heart process
   (c) Pearlitic processes

6.7 Discuss the effects on the structure, and properties of the following heat-treatment processes as applied to cast irons:
   (a) Annealing
   (b) stress relieving

6.8 (a) Sketch the cast iron section of the iron–carbon phase equilibrium diagram and label the more important compositions and temperatures.

(b) With reference to the above diagram, distinguish between the formation of ferritic grey cast iron and pearlitic grey cast iron.

6.9 Describe, with the aid of diagrams, the essential difference between grey cast iron and spheroidal graphite (SG) cast iron and compare their advantages and limitations.

6.10 State the composition of a suitable cast iron for each of the following applications, giving reasons for your choice:

(a) a lightly stressed ornamental casting

(b) an internal combustion engine crankshaft

(c) a machine tool bed

(d) a mains water pipe

(e) brake discs

(f) a furnace flue damper

7 Non-ferrous metals, their alloys and their heat treatment

The topic areas covered in this chapter are:

- Non-ferrous metals in general.
- Aluminium and its alloys.
- Copper and its alloys.
- Aluminium bronze alloys.
- Cupro-nickel alloys.
- Magnesium alloys.
- Zinc alloys.
- Tin–lead alloys – soft solders and white bearing metals.

7.1 Non-ferrous metals

The term 'non-ferrous metals' refers to the thirty-eight metals other than iron that are known to man. The non-ferrous metals which are most commonly used by engineers are listed in Table 7.1. Two of the most important non-ferrous metals are aluminium and copper. They not only form the bases of many important alloys, but they are widely used in their own right as pure metals.

A list of non-ferrous metals would not be complete without mention of the 'new metals' listed below. Although known for many years, these metals have only been available in bulk for engineering applications since the Second World War. Further, it is only comparatively recently, with the development of supersonic aircraft and the nuclear power industry, that there has been a large-scale commercial demand for these materials.

- *Niobium, tantalum* and *zirconium* are used in atomic reactor components.
- *Tellurium* is used instead of lead in free-cutting alloys (higher strength).
- *Titanium* is used in supersonic aircraft and rockets as it has a higher strength/weight ratio than aluminium alloys and retains its strength at the elevated temperatures encountered in supersonic flight.
- *Beryllium* is used as an alloying element with copper to make instrument springs which are resistant to fatigue and corrosion. Beryllium–copper alloys can be quench hardened like steel and are used to provide 'non-sparking' tools for use in oil fields, chemical installations, gas rigs, etc.

Table 7.1 Common non-ferrous metals

Metal	Density (kg/m³)	Melting point (°C)	Properties	Typical uses
Aluminium	2700	660	Lightest of the commonly used metals. High electrical and thermal conductivity. Soft, ductile and low tensile strength 93 MPa.	The base of many engineering alloys. Lightweight electrical conductors.
Copper	8900	1083	Soft, ductile and low tensile strength 232 MPa. Second only to silver in conductivity, it is much easier to joint by soldering and brazing than aluminium. Corrosion resistant.	The base of brass and bronze alloys. It is used extensively for electrical conductors and heat exchangers, such as motor car radiators.
Lead	11 300	328	Soft, ductile and very low tensile strength. High corrosion resistance.	Electric cable sheaths. The base of 'solder' alloys. The grids for 'accumulator' plates. Lining chemical plant. Added to other metals to make them 'free-cutting'.
Silver	10 500	960	Soft, ductile and very low tensile strength. Highest electrical conductivity of any metal.	Widely used in electrical and electronic engineering for switch and relay contacts.
Tin	7300	232	Resists corrosion.	Coats sheet mild steel to give 'tin plate'. Used in soft solders. One of the bases of 'white metal' bearings. An alloying element in bronzes.
Zinc	7100	420	Soft ductile and low tensile strength. Corrosion resistant.	Used extensively to coat sheet steel to give 'galvanized iron'. The base of die-casting alloys. An alloying element in brass.

▶

Metal	Density (kg/m^3)	Melting point (°C)	Properties	Typical uses
Chromium	7500	1890	Resists corrosion. Raises strength but lowers ductility of steels. Improves heat-treatment properties.	Used as an alloying element in high-strength and corrosion-resistant steels. Used for electroplating.
Cobalt	8900	1495	Improves wear-resistance and 'hot hardness' of high-speed steels.	Used as an alloying element in 'super' high-speed steels and in permanent-magnet alloys.
Manganese	7200	1260	High affinity for oxygen and sulphur. Soft and ductile.	Used to de-oxidise steels and to offset the ill-effects of the impurity sulphur. Larger amounts improve wear resistance.
Molybdenum	9550	2620	A heavy, heat-resistant metal that alloys readily with other metals.	Used as an alloying element in high-strength nickel–chrome steels to improve mechanical and heat-treatment properties. It reduces mass effect and temper-brittleness.
Nickel	8900	1458	A strong, tough, corrosion-resistant metal widely used as an alloying element.	Used as an alloying element to improve the strength and mechanical properties of steel. Tends to unstabilise the carbon during heat treatment, and chromium has to be added to counteract this effect in medium- and high-carbon steels. Used for electroplating.

These 'new metals' are very expensive compared with the more conventional engineering materials and they are only used where their special properties can be fully exploited.

The pure non-ferrous metals are used mainly where their properties of corrosion resistance and high electrical and thermal conductivity can be exploited. They are not widely used as structural materials in mechanical engineering because of their relatively low strengths. However, as will be shown later in this chapter, their mechanical properties can be greatly enhanced by alloying these metals together.

7.2 Aluminium

Pure aluminium is a weak, ductile material with a low density (2.3 g/mm^3 compared with 7.9 g/mm^3 for iron). Its electrical conductivity is second only to copper, as is its thermal conductivity. It is also highly resistant to atmospheric corrosion. Because of its low strength it is of little use as a structural material, and for such purposes aluminium alloys are preferred.

7.2.1 *High-purity aluminium*

Aluminium is used where corrosion resistance or high electrical or high thermal conductivity is required. The impurities present are less than 0.5 per cent. Aluminium has a high affinity for atmospheric oxygen and a film of aluminium oxide quickly forms on a freshly cut surface. This film is virtually homogeneous and prevents further corrosion taking place. The oxide film is also virtually transparent, so the aluminium retains its surface appearance for a long time. Unfortunately aluminium reacts violently with alkalis to give off hydrogen, so that care has to be taken not to subject it to caustic degreasing compounds. For this reason high-purity aluminium is *unsuitable* for marine environments where the salt spray is highly alkaline. Special alloys have been developed for marine use. Pure aluminium is used for lining food-processing vessels, and also for decorative architectural purposes both internally and externally.

Aluminium is a good conductor of electricity. It is second only to copper with a conductivity of approximately two-thirds of that metal. However, because of its lower density, it is a better conductor on a weight for weight basis. For this reason it is used for the conductors in the overhead grid system, where aluminium conducting strands are laid up over a high-tensile steel core to form a composite cable. Aluminium also has a high thermal conductivity, but its use in lightweight heat exchangers is limited because of the difficulties encountered in soldering, brazing or welding it on a production basis.

7.2.2 *Commercial purity aluminium*

Commercially pure aluminium contains 1.0–0.5 per cent impurities and these have the effect of strengthening the metal at the expense of reducing its corrosion resistance and electrical conductivity. Table 7.2 compares the properties of high-purity and commercial-purity aluminium. It can be seen from the table that the mechanical properties are heavily influenced by the amount of cold working that the metal has received. By controlling the amount of cold working, varying degrees of strength and hardness can be produced. These are said to be the different *tempers* in which the metal is available. Since only a few of the non-ferrous alloys (and none of the non-ferrous metals) can be hardened by heat-treatment processes, the modification of the mechanical properties by cold working becomes an important consideration. The properties, uses and forms of supply of aluminium are summarised in Fig. 7.1.

Table 7.2 *Properties of aluminium*

Type	Condition	UTS (MPa)	Elong. (%)	Hardness (H_B)
High purity	Annealed	45	60	15
(99.99 % Al)	Half-hard	82	24	22
	Full-hard	105	12	30
Commercial	Annealed	87	43	22
purity	Half-hard	120	12	35
(99.0 % Al)	Full-hard	150	10	42

Note: High-purity aluminium has superior electrical conductivity and corrosion-resistance properties.

Fig. 7.1 *Aluminium: properties, uses and forms of supply*

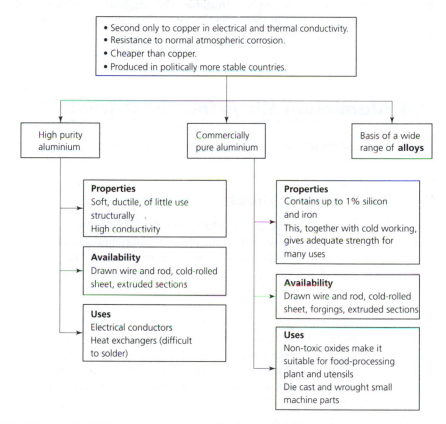

7.3 Aluminium alloys

Aluminium alloys can be divided into four categories:

1. Non-heat-treatable wrought alloys
2. Non-heat-treatable casting alloys
3. Heat-treatable wrought alloys
4. Heat-treatable casting alloys

- *Non-heat-treatable alloys*   These, as their name implies, do not respond significantly to heat-treatment processes beyond annealing and stress relief after casting or cold working.
- *Heat-treatable alloys*   These, as their name implies, do respond to heat treatment and in particular to the processes known as *solution treatment* and *precipitation hardening* (see Section 7.5).
- *Casting alloys*   These are alloys which can be used successfully for casting by a variety of processes, including sand casting and die casting. They may be heat treatable or non-heat treatable.
- *Wrought alloys*   These alloys whose mechanical properties allow them to be formed by a variety of processes, including forging, rolling, extrusion and drawing. They may be heat treatable or non-heat treatable. The non-heat-treatable wrought alloys can have their mechanical properties enhanced considerably by a combination of hot working and cold working.

7.4 Aluminium alloys (non-heat-treatable)

Let's now consider these alloy groups in rather more detail.

7.4.1 *Casting alloys*

The non-heat-treatable casting alloys are essentially binary alloys containing aluminium and silicon. It can be seen from the phase equilibrium diagram shown in Fig. 7.2(a) that silicon is only partially soluble in aluminium below the eutectic composition of approximately 12 per cent silicon. Thus the diagram is of the combination type. The addition of silicon to aluminium increases its fluidity and improves its general casting properties.

The microstructure for the eutectic composition shows a coarse eutectic structure of the α phase solid solution plus silicon. In this alloy the solid solution is of silicon in aluminium. The hyper-eutectic alloys show coarse silicon crystals in a matrix of eutectic structure. This combination of coarse eutectic structure and silicon crystals results in poor mechanical properties and embrittlement which can be overcome by a process known as *modification*.

Fig. 7.2 *Aluminium–silicon phase equilibrium diagram: (a) unmodified; (b) modified*

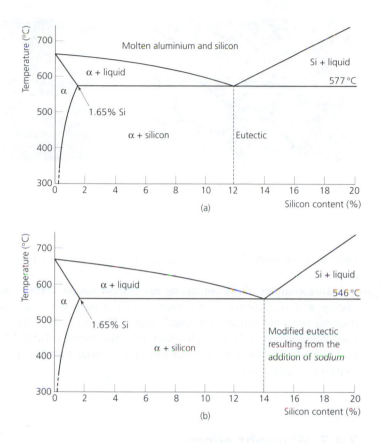

Modification consists of adding between 0.005 and 0.15 per cent metallic sodium to the melt immediately before casting. The effect is to delay precipitation of the silicon when the normal eutectic temperature is reached so that, at the commencement of nucleation, undercooling results in rapid crystallisation and a fine-grain structure with improved mechanical properties. It also raises the eutectoid composition to about 14 per cent silicon and lowers the eutectic temperature to 546 °C, as shown in Fig. 7.2(b). The effect of modification on a 13 per cent silicon alloy is shown in Fig. 7.3. The unmodified alloy (Fig. 7.3(a)) is, at 13 per cent silicon, a hyper-eutectic alloy and shows the typical coarse-grain structure which results in the poor mechanical properties previously mentioned. The modified alloy (Fig. 7.3(b)) is now at 13 per cent silicon. It is a hypo-eutectic alloy and shows the finer grain structure of α-phase solid solution in a fine eutectic matrix. This results in the cast alloy having increased strength, high ductility and good corrosion resistance. The low melting point and narrow freezing range of alloys only just below the eutectic composition makes them

Fig. 7.3 *Aluminium–silicon cast alloy structures (13% silicon): (a) unmodified; (b) modified*

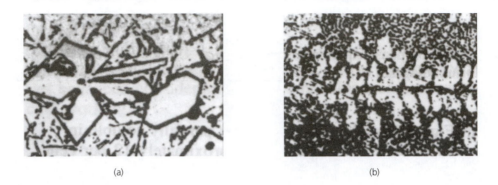

(a)   (b)

suitable for pressure die casting where rapid solidification improves the productivity of the process.

More complex alloys containing aluminium–silicon–copper are used for both sand casting and die casting, whilst aluminium–magnesium–manganese alloys are of high strength but only suitable for sand casting. However, these latter alloys are extremely corrosion resistant and are widely used for marine castings. Table 7.3 lists a number of non-heat-treatable aluminium alloys suitable for casting, together with their properties and some typical applications.

7.4.2 *Wrought alloys*

Typical wrought alloys contain either traces of manganese or traces of magnesium. A typical non-heat-treatable wrought alloy contains approximately 1.0–1.5 per cent manganese. This increases the tensile strength without materially affecting the excellent ductility of pure aluminium. This alloy is corrosion resistant and widely used for kitchen utensils, corrugated sheeting for roof-decking panels for the building trade, and for drawn aluminium tubing.

Another important non-heat-treatable wrought-aluminium alloy contains magnesium. This alloy is highly corrosion resistant, particularly in marine environments, and is widely used in shipbuilding. Unfortunately, the low melting point of aluminium–magnesium alloys, compared with steel, produces problems where the alloy is used for fire-resistant bulkheads. Figure 7.4 shows the relationship between tensile strength and magnesium content. It can be seen that the strength increases significantly as the magnesium content is increased to a practical limit of 7 per cent. There is no significant reduction in the ductility of this alloy as its strength increases.

Table 7.4 lists a number of non-heat-treatable wrought aluminium alloys, together with their properties and some typical applications.

Table 7.3 Non-heat-treatable, cast aluminium alloys

Type	Composition (%)			Condition	Properties			Applications
	Cu	Mn	Si		0.1% PS (MPa)	UTS (MPa)	Elong. (%)	
BS 1490/LM2	1.6	—	10.0	Chill cast	84	224	2.5	Gravity die casting. General-purpose alloy for lightly stressed parts not subjected to mechanical shock.
BS 1490/LM4	3.0	—	5.0	Sand cast	70	150	2	Sand castings; gravity and pressure die castings. General-purpose alloy where mechanical properties are of secondary importance.
				Chill cast	80	170	3	
BS 149/LM6	—	—	11.5	Sand cast	55	170	7	Sand castings; gravity and pressure die castings. Excellent foundry properties. One of the most widely used aluminium alloys when modified. Sumps. Gear boxes, radiators, large castings.
				Pressure die cast	85	215	4	
Birmalite	—	0.35	10.0	Sand cast	98	105	Nil	Sand castings and gravity die castings. Maintains its strength. Up to 300 °C, thus used for low-duty pistons and cylinder heads.
				Chill cast	154	168	0.5	

Fig. 7.4 *Effect of magnesium content on the tensile strength of annealed aluminium–magnesium alloys*

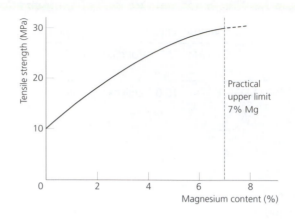

Table 7.4 *Non-heat-treatable, wrought aluminium alloys*

| Type | Composition (%) | | Condition | Properties | | | Applications |
	Mg	Other		0.1% PS (MPa)	UTS (MPa)	Elong. (%)	
BS 1470/7:N3	—	Cu 0.15 Si 0.6 Fe 0.75 Mn 1.2	Annealed Hard	45 170	110 200	34 4	Metal boxes, milk bottle caps, food containers, cooking utensils, roofing sheets, panelling of road transport vehicles and railway coaches.
BS 1470/7:N4	2.5	Cu 0.15 Mn 0.5 Si 0.6 Fe 0.75	Annealed $\frac{3}{4}$ hard	75 215	185 265	24 4	Marine superstructures, lifeboats, panelling subjected to marine environments, chemical plant, panelling for road transport vehicles and railway rolling stock.

| Type | Composition (%) | | Condition | Properties | | | Applications |
	Mg	Other		0.1% PS (MPa)	UTS (MPa)	Elong. (%)	
BS 1470/4:N6	5.0	Cu 0.15 Si 0.6 Fe 0.75 Mn 1.0	Annealed $\frac{1}{4}$ hard	125 215	265 295	18 8	Shipbuilding and applications requiring high strength and corrosion resistance.
BS 1470/4:N7	7.0	Cu 0.15 Si 0.6 Fe 0.75 Mn 1.0	Extruded	175	308	35	Roofing supports in mines and similar applications requiring high strength and corrosion resistance.

7.5 Aluminium alloys (heat treatable)

These are aluminium alloys containing copper which, together with other alloying elements, enables the alloy to respond to heat treatment. Reference back to Section 3.2 shows that solubility varies with temperature and that when the temperature of a saturated solution is lowered, the excess solute will precipitate out of solution. A similar effect can occur in solid solutions and advantage is taken of this in the heat treatment of some non-ferrous alloys, notably those containing aluminium and copper, and those containing aluminium and magnesium.

7.5.1 *Solution treatment*

Figure 7.5 shows part of the aluminium–copper phase equilibrium diagram. If an aluminium alloy containing 4 per cent copper is cooled from the molten condition to 500 °C (T_1), the alloy consists of crystals of α-phase solid solution of copper in aluminium. The solid solution does not become saturated until the solvus is reached at temperature T_2. If cooling continues slowly (equilibrium conditions) the crystals of solid solution will remain saturated but some precipitation will occur. At room temperature (T_3) the structure of the alloy will consist of the α-phase solid solution containing a coarse precipitate of the copper–aluminium intermetallic compound $CuAl_2$. In this condition the alloy will be soft and relatively weak.

Fig. 7.5 *The solution treatment of aluminium–copper alloys*

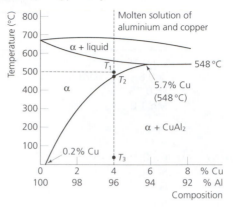

If, on the other hand, the alloy is quenched from the temperature T_1 so that it cools quickly, there is no time for equilibrium conditions to be achieved. This is because it requires the grouping together of atoms by diffusion for precipitation to occur, and this is a relatively slow process. The result of quenching from temperature T_1 is to prevent precipitation so that the grains consist of a *supersaturated solid solution* of α-phase alloy at room temperature. This is referred to as *solution treatment* and is the way in which such aluminium alloys may be 'annealed' or softened ready for cold working. In this condition the alloy will have a fine-grain structure and will be somewhat harder, stronger and tougher than when cooled under equilibrium conditions, but still very ductile.

7.5.2 *Precipitation hardening*

The solution-treated alloy can retain its supersaturated solid solution of α-phase grains at room temperature. However, this is not a stable condition and precipitation of the copper–aluminium intermetallic compound $CuAl_2$ will occur with elapse of time. The precipitate will be in the form of fine particles evenly distributed through the mass of the metal to give greater strength and hardness than that resulting from the coarse precipitation attained by cooling under equilibrium conditions. When precipitation hardening occurs naturally over a period of about 4 days it is referred to as *natural ageing*. When the precipitation process is speeded up by reheating the alloy to about 165 °C for about 10 hours, it is referred to as *precipitation hardening* or *artificial ageing*. Both terms mean the same thing.

The precipitation process chosen has an appreciable effect on the strength and hardness of the alloy, as can be seen from Fig. 7.6. Since solid solutions tend to be soft and ductile (the alloy is softened by 'solution treatment') and, since intermetallic compounds tend to be hard and brittle, it is the presence of the fine particles of the $CuAl_2$ precipitate that increases the strength and hardness of the precipitation age-hardened alloy. The greater the amount of the intermetallic precipitate present, the harder and stronger will be the alloy (see also Fig. 5.6).

To retard the precipitation process after solution treatment, components such as rivets made from 'duralumin' alloy are kept under refrigerated conditions in order to avoid the possibility of cracking whilst being cold worked (e.g. cold headed).

Fig. 7.6 *Effects of time and temperature on the precipitation hardening of aluminium alloys*

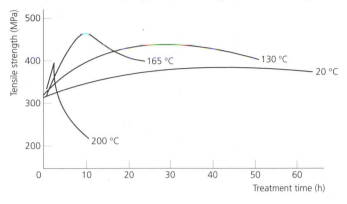

7.5.3 *Casting alloys*

These are usually complex alloys containing copper or nickel or both in significant amounts plus other alloying elements in lesser amounts. A number of these alloys contain up to 4 per cent copper, whilst others contain up to 2 per cent nickel. They are softened and grain refined after casting by solution treatment and hardened and strengthened by precipitation hardening when intermetallic compounds such as $CuAl_2$ and $NiAl_3$ will be present. These intermetallic compounds make the castings less ductile. Examples of these heat-treatable casting alloys are given in Table 7.5, together with their composition and some typical applications.

7.5.4 *Wrought alloys*

These are complex alloys of aluminium together with such alloying elements as copper, magnesium, silicon and zinc. One of the most popular of the heat-treatable wrought alloys already mentioned is 'duralumin'. It is strong and tough, yet it can be cold worked and easily machined after solution treatment. Table 7.6 lists examples of heat-treatable wrought alloys, together with their composition and some typical applications.

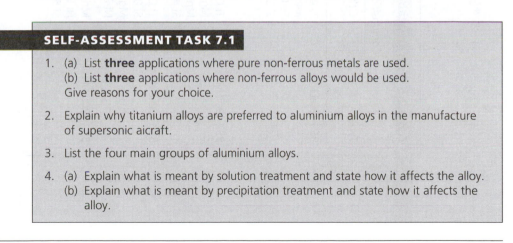

SELF-ASSESSMENT TASK 7.1

1. (a) List **three** applications where pure non-ferrous metals are used.
    (b) List **three** applications where non-ferrous alloys would be used.
    Give reasons for your choice.

2. Explain why titanium alloys are preferred to aluminium alloys in the manufacture of supersonic aicraft.

3. List the four main groups of aluminium alloys.

4. (a) Explain what is meant by solution treatment and state how it affects the alloy.
    (b) Explain what is meant by precipitation treatment and state how it affects the alloy.

Table 7.5 Heat-treatable, cast aluminium alloys

Type	Composition (%)		Condition	Properties			Applications
	Cu	Other		0.1% PS (MPa)	UTS (MPa)	Elong. (%)	
BS 1490:LM4	3.0	Ti 0.2 Mn 0.5 Fe 0.8 Si 5.0	Solution treated at 520 °C for 6 hours. Precipitation hardened at 170 °C for 12 hours.	252	294	1	General purpose alloy for sand casting; gravity and pressure die casting. Withstands moderate stress, shock and hydraulic pressure.
BS 1490:LM8	—	Ti 0.15 Mg 0.5 Mn 0.5 Si 4.5	Solution treated at 465 °C for 8 hours. Precipitation hardened at 165 °C for 10 hours.	—	280	2	Good casting properties and corrosion resistance. Mechanical properties can be varied by heat treatment.
BS 1490:LM14	4.0	Ti 0.2 Si 0.3 Mg 1.5 Ni 2.0	Solution treated at 510 °C. Precipitation hardened in boiling water for 2 hours.	215	280	—	Pistons and cylinder heads for liquid and air cooled engines. A good general purpose alloy.
BS 1490:LM16	1.2	Mn 0.5 Ni 0.25 Si 5.0	Solution treated at 520 °C for 12 hours; water quenched. Precipitation hardened at 150 °C for 10 hours.	182	231	1	Suitable for castings of intricate shape. High pressure tightness: suitable for valve bodies and cylinder heads. Also used for water jackets and cylinder blocks.

Table 7.6 Heat-treatable, wrought aluminium alloys

Type	Composition (%)		Condition	Properties			Applications
	Cu	Other		0.1% PS (MPa)	UTS (MPa)	Elong. (%)	
DTD 372	—	Ti 0.2 Si 0.5 Mg 0.6	Solution treated at 520 °C; quenched. Precipitation hardened at 170 °C for 10 hours.	168	224	18	Good corrosion resistance. Extruded sections such as glazing bars and window sections. Windscreen and sliding roof sections for automobile body-building industry.
BS 1470/7:H114	4.0	Si 0.5 Mn 0.7 Mg 0.8	Solution treated at 480 °C; quenched. Age hardened at room temperature for 4 days.	280	400	10	General purpose alloy suitable for stressed parts in aircraft and other structures. The original 'Duralumin'.
BS 1470/7:H730	—	Mn 0.7 Mg 1.0 Si 1.0	Solution treated at 510 °C; quenched. Precipitation hardened at 175 °C for 10 hours.	150	250	20	Structural members for road and rail vehicles and shipbuilding. Architectural work. Ladders and scaffold tubes. High electrical conductivity: overhead powerlines.
DTD 5074	1.6	Ti 0.3 Mg 2.5 Zn 6.2	Solution treated at 465 °C; quenched. Precipitation hardened at 120 °C for 24 hours.	590	650	11	Strongest commercial alloy. Highly stressed aircraft components. Military equipment.

7.6 Copper

This is one of the few non-ferrous metals which has sufficient strength to be used unalloyed. Very pure copper has a density of 8.93 g/mm³. It is very ductile and can be readily drawn into rods, wires and tubes. It has excellent electrical and thermal conductivity, and has good corrosion resistance. Like aluminium it reacts with atmospheric oxygen to form a thin, homogeneous oxide film on a freshly cut surface and this prevents further oxidation. High-conductivity copper has a purity better than 99.9 per cent. This is called cathode copper because the refined copper forms the cathode of an electrolytic cell, as shown in Fig. 7.7

Fig. 7.7 *Electrolytic cell for the refinement of copper*

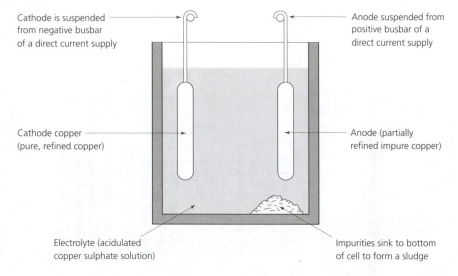

Cathode is suspended from negative busbar of a direct current supply

Anode suspended from positive busbar of a direct current supply

Cathode copper (pure, refined copper)

Anode (partially refined impure copper)

Electrolyte (acidulated copper sulphate solution)

Impurities sink to bottom of cell to form a sludge

The impure copper forms the anode and the electrolyte is an acidulated solution of copper sulphate. On the passage of a direct electric current through the cell, metallic copper from the electrolyte is deposited on the cathode. This upsets the chemical balance of the electrolyte, which then dissolves copper from the anode to restore the balance. The process continues as long as the current flows and the copper deposited upon the cathode has a very high degree of purity. Any impurities are precipitated to the bottom of the electrolytic cell as a dross (sludge) which has to be removed periodically.

Copper produced by this method of refinement is used for electrical conductors and heat exchangers.

For structural purposes, high-conductivity copper is insufficiently strong because of its high degree of purity. To increase the strength, impurities in the form of copper oxides are allowed to form in the metal. This is a form of particle reinforcement as discussed in Section 9.4. Such coppers are referred to as tough pitch and are used for general-purpose sheets rods and tubes. Tough pitch copper (which has been 'fire-refined') does not have such good electrical conductivity or corrosion resistance as electrolytically refined, high-purity, cathode copper.

Pure copper is a difficult material to machine to a good surface finish. However, the addition of traces of the metal tellurium or the non-metal sulphur produces a free-cutting copper with only slightly impaired ductility and conductivity. For example, the addition of 0.5 per cent tellurium forms the chip-breaking compound copper-telluride, whilst maintaining an electrical conductivity of 90 per cent that of high-conductivity copper. Similarly the addition of 0.4 per cent sulphur produces the chip-making compound copper sulphide, whilst maintaining an electrical conductivity of 95 per cent that of high-conductivity copper. However, the use of sulphur, although cheaper, substantially reduces the strength of the copper. To prevent gassing and porosity during welding, traces of phosphorus are added to the copper. The phosphorus combines with any dissolved oxygen present, and copper treated in this manner is said to be phosphorus deoxidised. It should be noted that the presence of only 0.04 per cent phosphorus results in a 25 per cent reduction in electrical conductivity. Therefore phosphorus deoxidised copper is unsuitable for electrical conductors. Figure 7.8 shows the relationship and some typical uses of the more commonly available forms of copper, whilst Table 7.7 lists some typical properties.

Fig. 7.8 *Copper: properties, uses and forms of supply*

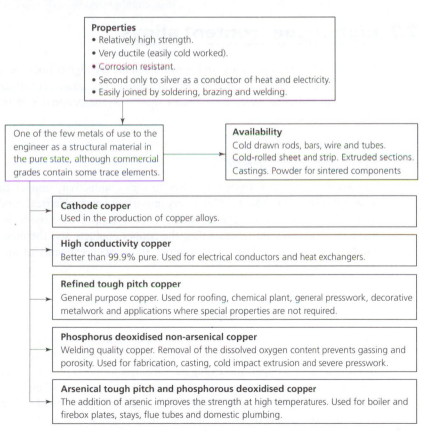

Table 7.7 Properties of copper

Description	Purity	Oxygen content	Condition	UTS (MPa)	Elong. (%)	Hardness (H_B)
Electrolytic tough pitch high-conductivity copper	99.90% (min.)	0.05%	Annealed	220	50	45
			Hard	400	4	115
Fire refined tough pitch high-conductivity copper	99.85% (min.)	0.05%	Annealed	220	50	45
			Hard	400	4	115
Oxygen-free high-conductivity copper	99.05% (min.)	—	Annealed	215	60	45
			Hard	340	6	115
Phosphorus de-oxidised copper	99.85% (min.)	O_2 nil P 0.013–0.05%	Annealed	210	60	145
			Hard	320	4	115
Arsenical copper	99.20% (min.)	O_2 0.05% 0.3–0.5 As	Annealed	220	50	45
			Hard	400	4	115

7.7 High copper content alloys

The first group of copper alloys to be considered are the high copper content alloys. That is, the additional alloying elements represent only a very small percentage of the total; yet these small additions make a significant change to the properties of the alloy compared with pure copper.

7.7.1 Silver copper

The addition of only 0.1 per cent silver to high-conductivity copper raises its annealing temperature by up to 150 °C. This is very important where electrical conductors have to be soldered to hard-drawn copper contacts. If pure copper were used, the heat required to make the soldered joint would soften the copper contacts and the increased rate of wear would render them useless – for example, the copper segments of an electric motor or generator commutator.

7.7.2 Cadmium copper

Like silver, cadmium has little effect upon the conductivity of the copper. Again, like silver, cadmium also raises the annealing temperature. In addition, however, it also strengthens and toughens the copper, increasing its resistance to metal fatigue. As cadmium copper is substantially free from oxygen, it is not susceptible to 'gassing' when it is braze welded.

Cadmium copper is used for low- and medium-voltage overhead transmission cables where its high conductivity and high strength enables it to be used over relatively long spans. It is also used for traction purposes, e.g. the overhead conductors for electrified railway systems. Because of its resistance to metal fatigue, annealed cadmium copper is also recommended for aircraft wiring. Here its flexibility, combined with its resistance to metal fatigue, reduces the risk of failure due to vibration-induced metal fatigue.

7.7.3 *Chromium copper*

This is one of the few non-ferrous alloys which can be heat treated to improve its mechanical properties. A typical alloy containing 0.5 per cent chromium can be quenched from 100 °C. This leaves the alloy in a soft and ductile condition with a rather low electrical conductivity. However, if the metal is reheated to 500 °C for approximately two hours, its mechanical and electrical properties are restored.

Since the properties of chromium copper depend upon heat treatment rather than upon cold working, it can be used in cast as well as in wrought forms. Similarly, components can be formed from annealed sheets or extruded rod and then hardened after manipulation and machining.

7.7.4 *Tellurium copper*

This has already been introduced as a free-cutting copper alloy in Section 7.6. Tellurium forms stable compounds with copper and the addition of 0.5 per cent makes the copper as machinable as free-cutting brass yet leaves the electrical conductivity relatively unaffected. Tellurium copper has a high corrosion resistance and is used extensively for high-duty electrical contacts in machines and switchgear for use in hostile environments such as mines, ships and chemical plant. The addition of traces of nickel and silicon allows it to be heat treated in a similar manner to chromium copper, but at the expense of some loss of conductivity.

7.7.5 *Beryllium copper*

Beryllium copper is used where mechanical rather than electrical properties are required. Beryllium copper can be softened by heating it to 800 °C and quenching in water. In this condition the material can be extensively cold worked and machined. Subsequent heat treatment consists of heating the alloy to between 300 and 320 °C for upwards of two hours. The resulting mechanical properties will depend to some extent upon the degree of cold working that took place between the first and second treatments.

Beryllium copper is used widely for instrument springs, flexible bellows, corrugated diaphragms (aneroid barometers and altimeters) and the bourdon tubes of pressure gauges.

Hand tools made from beryllium copper are almost as strong and hard wearing as those made from steel but, since they will not strike sparks from other metals or from flint stones, such tools are widely used in hazardous locations where there is a high risk of fire or explosion – for example, mines, oil refineries, oil rigs, chemical plant, etc. The high cost of this alloy precludes its use for more conventional applications.

7.8 Brass alloys

These are alloys of copper and zinc. They tend to give low-strength, porous castings which depend upon a combination of hot working and cold working to consolidate the metal and improve their mechanical properties. The more common brasses are listed in Table 7.8.

Table 7.8 *The brass alloys*

| Name | Composition (%) | | | Applications |
	Cu	Z	Other	
Cartridge brass	70	30	—	Most ductile of the copper–zinc alloys. Widely used in sheet metal pressing for severe deep drawing operations. Originally developed for making cartridge cases, hence its name.
Standard brass	65	35	—	Cheaper than cartridge brass and rather less ductile. Suitable for most engineering processes.
Basis brass	63	37	—	The cheapest of the cold working brasses. It lacks ductility and is only capable of withstanding simple forming operations.
Muntz metal	60	40	—	Not suitable for cold working but hot works well. Relatively cheap due to its high zinc content, it is widely used for extrusion and hot-stamping processes.
Free-cutting brass	58	39	Pb 3.0	Not suitable for cold working but excellent for hot working and high-speed machining of low-strength components.
Admiralty brass	70	29	Sn 1.0	This is virtually cartridge brass plus a little tin to prevent corrosion in the presence of salt water.
Naval brass	62	37	Sn 1.0	This is virtually Muntz metal plus a little tin to prevent corrosion in the presence of salt water.

The change in properties with change in composition is more readily understood by reference to the copper–zinc phase equilibrium diagram shown in Fig. 7.9. Although more complex than any phase equilibrium diagram considered so far, it is fairly easy to understand if taken section by section. As in previous diagrams the α phase consists of a solid solution, which in this case is a solid solution of zinc in copper. Like all solid solutions it is ductile and suitable for cold working.

When the amount of zinc present exceeds that required to saturate the α-phase solid solution, a β phase is introduced. This new phase is tougher, stronger and harder than the α phase, but considerably less ductile. The β phase can only exist down to 454 °C below which it is modified to β' phase. This modified phase lowers the ductility and malleability still further. If even more zinc is added an even more brittle and less ductile phase is introduced called γ brass. Commercial brasses usually contain only α or $\alpha + \beta'$ phases. The effect of these phases on the properties of a range of brasses is shown in Fig. 7.10, and these properties should be compared with the applications previously listed in Table 7.8.

Fig. 7.9 *Copper–zinc phase equilibrium diagram*

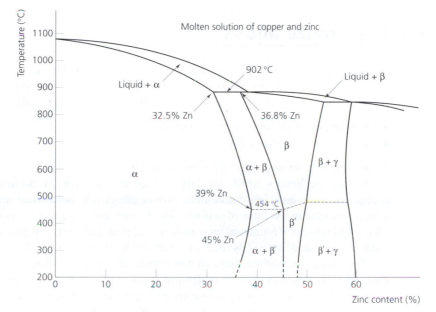

Fig. 7.10 *Effect of composition on the properties of brass*

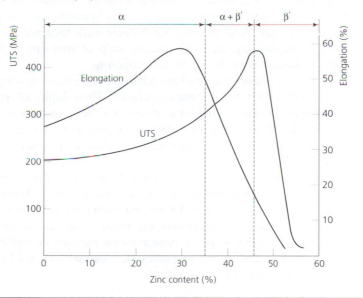

The $\alpha + \beta'$ brasses are largely used for hot working (hot stamping and extrusion) above 454 °C. This is because the β phase that exists above 454 °C is much more malleable and ductile than the β' phase that exists at room temperature. The addition of lead or tellurium to α or $\alpha + \beta'$ (duplex) brasses gives them free-cutting properties so that they machine easily. These additional alloying elements have little effect upon the strength, hardness and ductility of the brass.

7.9 Tin–bronze alloys

These are alloys of copper and tin together with a *deoxidiser*. The deoxidiser is essential to prevent the tin content oxidising during casting or hot working. Oxidation of the tin would result in a weak, 'scratchy' bronze. Two deoxidisers are commonly used:

- Phosphorus in the 'phosphor–bronze' alloys.
- Zinc in the 'gunmetal' alloys.

Some typical tin–bronze alloys are listed Table 7.9.

Unlike the brasses which are largely used in the wrought condition (rod, sheet, stampings, etc.), only low tin content bronze alloys can be worked and most bronze components are in the form of castings. Tin–bronze alloys are more expensive than brass alloys, but they are stronger, more corrosion and wear resistant, and give sound, pressure-tight castings which are widely used for steam and hydraulic valve bodies and mechanisms.

The phase equilibrium diagram for copper–tin alloys is shown in Fig. 7.11. This diagram is too complex to warrant detailed study at this stage, but it should be noted that up to about 10 per cent tin (depending upon temperature) only the α-phase solid solution of tin in copper is present. Thus, as for all solid solutions, these low tin content bronze alloys are ductile and can be cold worked. As previously stated, care must be taken to prevent oxidation as this will weaken the alloy and make it 'scratchy' and brittle. In these low tin–bronze alloys the deoxidising agent is always phosphorus and the resulting alloy is a phosphor-bronze. After the alloy has become work hardened due to cold working (cold rolling, wire drawing, etc.), the resulting strip or wire can be formed into instrument springs, electrical contacts and similar components.

If the tin content is increased much beyond 10 per cent, the alloy will show considerable amounts of the δ phase at room temperature. These duplex alloys of $\alpha + \delta$ phase are too brittle to be cold worked and are used solely for the production of castings. They consist of fine particles of the hard δ phase dispersed throughout a matrix of tough, ductile α phase. This results in a low-friction, hard-wearing surface, and a sound pressure-tight casting free from porosity.

These casting alloys may be deoxidised with traces of phosphorus so as to leave not more than a residue of 0.04 per cent in the finished casting. Alternatively, the phosphorus content may be increased up to 1.0 per cent when increased fluidity, strength and corrosion resistance are required. Increasing the phosphorus content to this level results in a reduction in toughness and a corresponding increase in brittleness, together with an improvement in anti-friction properties. Bearing materials and bearings are dealt with in detail in *Engineering Materials*, Volume 2.

Table 7.9 Tin–bronze alloys

Name	Composition (%)					Application
	Cu	Z	P	Sn	Pb	
Low-tin bronze	96	—	0.1–0.25	3.9–3.75	—	This alloy can be severely cold worked to harden it so that it can be used for springs where good elastic properties must be combined with corrosion resistance, fatigue resistance and electrical conductivity, e.g. contact blades.
Drawn-phosphor-bronze	94	—	0.1–0.5	5.9–5.5	—	This alloy is used in the work-hardened condition for turned components requiring strength and corrosion resistance, such as valve spindles.
Cast phosphor-bronze	rem.	—	0.03–0.25	10	—	Usually cast into rods and tubes for making bearing brushes and worm wheels. It has excellent anti-friction properties.
Admiralty gunmetal	88	2	—	10	—	This alloy is suitable for sand casting where fine-grained, pressure-tight components such as pump and valve bodies are required.
Leaded-gunmetal (free-cutting)	85	5	—	5	5	Also known as 'red brass', this alloy is used for the same purposes as standard, admiralty gunmetal. It is rather less strong but has improved pressure tightness and machining properties.
Leaded (plastic) bronze	74	—	—	2	24	This alloy is used for lightly loaded bearings where alignment is difficult. Due to its softness, bearings made from this alloy 'bed in' easily.

Fig. 7.11 *Copper–tin phase equilibrium diagram (copper-rich alloys)*

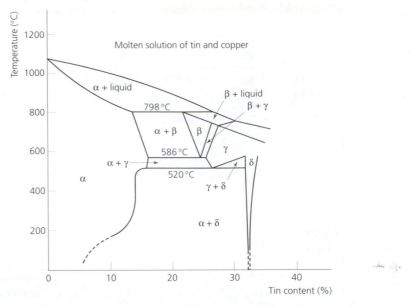

Zinc may be used in casting alloys as a deoxidiser and, as it replaces a substantial amount of the tin content, it reduces the cost of the alloy which is an important factor with large castings. As has already been stated, alloys containing zinc as a deoxidiser are referred to as 'gunmetal'. Such alloys give excellent pressure-tight, corrosion-resistant castings but they do not have the anti-friction properties of the more expensive phosphor–bronzes. Gunmetals are used for such applications as hydraulic and steam valve body castings and castings for marine use. The alloy gets its name from the fact that it was originally used for casting canon barrels for the Admiralty.

The more important phases found in the copper–tin alloy system may be summarised as follows:

α phase   Tin (Sn) in solution in excess copper (Cu)
β phase   Copper (Cu) in solution in excess tin (Sn)
δ phase   Hard, brittle intermetallic compound $Cu_{31}Sn_8$

7.10 Aluminium–bronze alloys

These are alloys of copper and aluminium. In aluminium–bronze alloys the predominant metal present is copper and the aluminium is only the alloying element, unlike the heat-treatable aluminium alloys described earlier in this chapter. Aluminium–bronze alloys are more expensive than 'tin–bronze' alloys but are more corrosion resistant at high temperatures. They are also more ductile and can be cold worked into tubes for boilers and condensers in steam and chemical plant.

Figure 7.12 shows the copper–aluminium (aluminium–bronze) phase equilibrium diagram. Once again it is the α-phase solid solution of aluminium in copper which produces

Fig. 7.12 *Copper–aluminium phase equilibrium diagram (copper-rich alloys)*

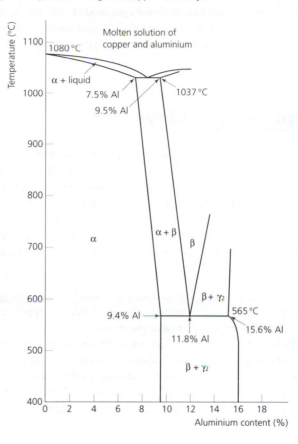

the ductile, cold-working alloys which can be rolled into sheets and drawn into tubes. Over about 10 per cent aluminium, duplex alloys consisting of $\alpha + \beta$ phases are produced above 565 °C and duplex alloys of $\alpha + \gamma_2$ phases are produced below 565 °C. The γ_2 phase is a hard brittle intermetallic compound Cu_9Al_4. Thus alloys containing this phase are unsuitable for cold working, but are excellent for casting.

These duplex aluminium–bronze alloys can also be hot worked over 565 °C when they consist of the more malleable $\alpha + \beta$ phases. Aluminium–bronze alloys containing 10 per cent aluminium are one of the few non-ferrous alloys which can be heat treated to be like steel.

For example, quenching this alloy from 900 °C produces a structure which consists entirely of the β' phase. This phase is analogous with martensite in steel since it is very hard and brittle and has acicular (needle-like) crystals. To increase the toughness of the alloy with some small loss of hardness, like steel, it can be tempered by reheating it to about 400 °C and quenching it. Slow cooling under equilibrium conditions from above 565 °C anneals the alloy and restores the normal $\alpha + \gamma_2$ phases at room temperature.

Non-ferrous metals, their alloys and their heat treatment   **179**

Aluminium–bronzes are highly corrosion resistant particularly at high temperatures since a homogeneous film of aluminium oxide forms on the surface of the metal. This film is also 'self-healing' if damaged. Table 7.10 lists some examples of aluminium–bronze alloys together with their composition and some typical applications.

7.11 Cupro-nickel alloys

Copper and nickel form alloys in which the two metals are wholly soluble (miscible) in each other, both in the liquid and in the solid state. Because of this, their phase equilibrium diagram was considered in some depth in Section 3.10. Since only the α phase is present over the entire range of alloys, they all have high strength and ductility and are equally suitable for hot working and cold working. These are relatively expensive copper-based alloys, but their strength, ductility and corrosion resistance make them highly suitable for such applications as high-duty boiler and condenser tubes, bullet envelopes and resistance wires.

- *Monel metal* This also contains traces of iron and manganese, has exceptionally high corrosion resistance particularly at elevated temperatures and is widely used for chemical plant and marine applications.
- *Nickel silvers* These contain up to 20 per cent zinc which reduces the cost and imparts a silvery appearance when highly polished. The nickel–silver alloys can be readily cold worked and are widely used for domestic table cutlery as an alternative to stainless steel which is more difficult to work to shape. Table 7.11 lists some cupro-nickel alloys, together with their composition and some typical applications.

SELF-ASSESSMENT TASK 7.2

1. Describe the main differences between high-conductivity copper and tough pitch copper, and suggest a typical application for each metal.

2. In what way are the high-copper content alloys different from the brass and bronze alloys.

3. List the main alloying elements in:

   (a) gunmetal
   (b) brass
   (c) phosphor–bronze

4. Give a typical example where the following alloys would be used in preference to a tin–bronze:

   (a) Aluminium–bronze
   (b) Cupro-nickel

Table 7.10 Copper–aluminium (aluminium-bronze alloys)

Composition (%)				Condition	Properties				Applications
Al	Fe	Cu	Other		0.1% PS (MPa)	UTS (MPa)	Elong. (%)	Hardness (H_D)	
5	—	rem.	Mn or Ni up to 4%	Annealed / Hard	112 / 532	350 / 700	70 / 4	80 / 200	Cold worked for decorative purposes such as imitation jewellery. Tubes for engineering applications, resistant to corrosion and oxidation.
8	—	rem.	Fe, Mn and Ni up to 2%	Hot worked	140	392	45	100	Chemical engineering plant suitable for use at moderately elevated temperatures.
10	5	80	Ni 5%	Hot forged	420	658	20	215	General engineering forgings combining strength and corrosion resistance – can be heat treated.
9.5	2 5	rem.	Ni and Mn up to 1.0%	As cast	168	476	30	115	General purpose alloy for both die-casting and sand-casting.
12	—	rem.	Fe, Mn and Ni from 5–8%	As cast	434	504	3	250	A hard, rigid alloy containing the γ_2 phase. Used where heavy compressive loads are involved. Good wear resistance.

Table 7.11 Cupro-nickel alloys

| Composition (%) | | | Properties | | | | | | Applications |
Cu	Ni	Other	Condition	0.1% PS (MPa)	UTS (MPa)	Elong. (%)	Hardness (H_D)		
80	20	—	Annealed	98	308	45	75		Very high ductility and corrosion resistance, will withstand severe cold working.
			Hard	420	490	5	165		
70	30	—	Annealed	98	322	45	80		Used for condenser and heat-exchanger tubes where high corrosion resistance is required.
			Hard	490	588	5	175		
60	40	—	Annealed	—	350	45	90		'Constantan' electrical resistance wire – high specific resistance and low temperature coefficient. Also used in thermocouples.
			Hard	—	588	5	190		
29	68	Fe 1.25 Mn 1.25	Annealed	196	490	45	120		'Monel Metal'. Good mechanical properties combined with excellent corrosion resistance properties. Used for chemical plant.
			Hard	518	658	70	220		

7.12 Magnesium alloys

Magnesium is the lightest of the engineering metals with a density of only $1.7\,g/mm^3$. Its electrical and thermal conductivity is about 60 per cent that of high-conductivity copper. It has a high affinity for oxygen and burns in air with a fierce white flame. It was widely used at one time for flash photography. Because of this fire risk, and because of its low tensile strength, magnesium is only used as an alloying element or as the basis of a range of ultra-lightweight alloys.

Although not as strong as aluminium alloys, magnesium alloys have a much lower density, so that in many instances their strength-to-weight ratio can actually be superior. The magnesium alloys fall into two categories:

- Casting alloys.
- Wrought alloys.

Magnesium alloys contain aluminium, zinc, zirconium and manganese, together with 'rare-earth' metallic elements in some instances. Like the aluminium alloys, magnesium alloys can be annealed by solution treatment. The alloy is heated at 380 °C for eight hours, when the temperature is raised to 410 °C for a further sixteen hours. Magnesium alloys can be precipitation hardened by heating at 190 °C for some ten to twelve hours.

The melting and casting of magnesium alloys is difficult and dangerous in view of the flammability of magnesium if accidentally overheated. These alloys are generally melted under a flux containing calcium and sodium fluorides to exclude atmospheric oxygen. As the molten metal is pored into the mould it is dusted with flour of sulphur. The sulphur burns on contact with the hot metal and blankets it with sulphur dioxide gas which excludes atmospheric oxygen without which the molten alloy cannot ignite. The presence of sulphur dioxide gas makes efficient fume extraction and good ventilation most important. Table 7.12 lists some cast and wrought magnesium alloys together with their composition, properties and some typical applications.

7.13 Zinc alloys

Pure zinc is an interesting metal. Its boiling point is so low that it is the only commercial metal which can be refined by distillation. It has a density of $7.1\,g/mm^3$ and a melting point of only 420 °C. The pure metal is relatively weak, but is widely used as a coating on steel to prevent corrosion by atmospheric attack and zinc-coated low-carbon steel sheet is known as 'galvanised iron' (see Sections 13.9 and 13.13).

Zinc-based alloys are used almost entirely for pressure die casting. They are widely used for such components as car door handles, carburettor and fuel pump bodies, and other lightly stressed components. Zinc pressure die castings can be machined, but they cannot be soldered or welded. Zinc-based alloys are popular for die casting as they have the following properties:

- High fluidity, which enables complex castings with thin sections to be made.
- Low melting point, which reduces die wear.

Table 7.12 Magnesium alloys

Type	Composition (%) (remainder Mg)						Condition	Properties			Applications
	Al	Mn	Zn	Zr	Th	Rare earths		0.1% PS (MPa)	UTS (MPa)	Elong. (%)	
Casting alloys	10.0	0.3	0.7	—	—	—	Chill-cast	115	200	2	Lightweight castings for the aircraft and high-performance car industry. For example: landing wheels, road wheels, crank cases, and miscellaneous coatings.
	—	—	4.0	0.7	—	1.2	As cast	95	170	5	
	—	—	—	0.7	3.0	—	Heat treated	130	215	4	
	—	—	—	—	—	—	Heat treated	100	210	8	
Wrought alloys	—	1.5	—	—	—	—	Rolled	95	200	5	Petrol tanks, oil tanks and other lightly stressed sheet metal components. Lightweight sections.
	—	1.0	—	—	3.0	—	Rolled	215	280	10	
	6.0	0.3	1.0	—	—	—	Forged	155	280	8	Lightweight forgings for the aircraft industry such as airscrew blades and undercarriage components.
	—	—	—	—	—	—	Extruded	140	215	8	
	—	—	3.0	0.7	—	—	Forged	170	265	8	
	—	—	—	—	—	—	Extruded	215	310	8	

- Unlike molten aluminium, they do not react with and errode the die steel.
- Narrow freezing range, which results in rapid solidification and allows high rates of production.
- Easy polishing and electroplating properties.
- Adequate strength for small components.

A typical zinc-based alloy marketed under the trade name of 'Mazak' contains:

Aluminium	4%
Copper	2.7%
Zinc	remainder

The zinc must be better than 99.9 per cent pure, otherwise even minute traces of impurities such as cadmium, tin and lead will lead to intercrystalline brittleness, swelling, corrosion and pitting of the electroplated surface. It was the lack of high-purity zinc which gave zinc alloy die castings their poor reputation for quality in the early days of the process. Zinc pressure die castings should have a strength of about 320 MPa and high rigidity.

The corrosion resistance of zinc-based die castings can be increased by the 'chromate passivation' process. The castings are cleaned and immersed in a solution of dilute sulphuric acid and sodium bichromate for not more than one minute. They are then washed and dried. The resultant passive film prevents corrosion in damp atmospheric conditions when the die casting has not been electroplated.

7.14 Tin–lead alloys

The soft solders (tin–lead alloys) were discussed in Section 3.10. For the full range of soft-solder alloys see BSEN 29453. Table 7.13 lists some standard, tin–lead alloy soft solders and Table 7.14 list some of their typical applications. The 'white' bearing metal alloys are also based on tin and some examples and their applications are listed in Table 7.15. Bearing metals and bearings will be dealt with in greater detail in *Engineering Materials*, Volume 2.

The non-ferrous metals and alloys discussed in this chapter are only intended as a representation of the range available. Any one group could be a study in its own right.

SELF-ASSESSMENT TASK 7.3

1. Compare the advantages and limitations of magnesium-based alloys with the aluminium-based alloys.

2. Suggest **two** applications where magnesium alloys would be suitable.

3. Explain why zinc-based die-cast toys made before the Second World War are now found to be weak and brittle.

4. Explain why a soft solder with approximately 60 per cent tin and 40 per cent lead is ideal for use in assembling electronic circuit board components.

Table 7.13 Tin–lead alloys (soft solders)

Group	Alloy number	Alloy designation	Melting or solidus/liquidus temperature (°C)	Chemical composition (%)										Sum of all impurities except Sb, Bi and Cu
				Sn	Pb	Sb	Cd	Zn	Al	Bi	As	Fe	Cu	
Tin–lead alloys	1	S-Sn63Pb37	183	62.5–63.5	rem	0.12	0.002	0.001	0.001	0.10	0.03	0.02	0.05	0.08
	1a	S-Sn63Pb37E	183	62.5–63.5	rem	0.05	0.002	0.001	0.001	0.05	0.03	0.02	0.05	0.08
	2	S-Sn60Pb40	183–190	59.5–60.5	rem	0.12	0.002	0.001	0.001	0.10	0.03	0.02	0.05	0.08
	2a	S-Sn60Pb40E	183–190	59.5–60.5	rem	0.05	0.002	0.001	0.001	0.05	0.03	0.02	0.05	0.08
	3	S-Pb50Sn50	183–215	49.5–50.5	rem	0.12	0.002	0.001	0.001	0.10	0.03	0.02	0.05	0.08
	3a	S-Pb50Sn50E	183–215	49.5–50.5	rem	0.05	0.002	0.001	0.001	0.05	0.03	0.02	0.05	0.08
	4	S-Pb55Sn45	183–226	44.5–45.5	rem	0.50	0.005	0.001	0.001	0.25	0.03	0.02	0.08	0.08
	5	S-Pb60Sn40	183–235	39.5–40.5	rem	0.50	0.005	0.001	0.001	0.25	0.03	0.02	0.08	0.08
	6	S-Pb65Sn35	183–245	34.5–35.5	rem	0.50	0.005	0.001	0.001	0.25	0.03	0.02	0.08	0.08
	7	S-Pb70Sn30	183–255	29.5–30.5	rem	0.50	0.005	0.001	0.001	0.25	0.03	0.02	0.08	0.08
	8	S-Pb90Sn10	268–302	9.5–10.5	rem	0.50	0.005	0.001	0.001	0.25	0.03	0.02	0.08	0.08
	9	S-Pb92Sn8	280–305	7.5–8.5	rem	0.50	0.005	0.001	0.001	0.25	0.03	0.02	0.08	0.08
	10	S-Pb98Sn2	320–325	1.5–2.5	rem	0.12	0.002	0.001	0.001	0.10	0.03	0.02	0.05	0.08

Note: The temperatures given under the heading 'Melting or solidus/liquidus temperature' are for information purposes and are not specified requirements for the alloys; all single figure limits are maxima.

Table 7.14 *Typical uses of soft-solders (tin–lead alloys)*

Alloy number	Typical uses
1 1a 2	Soldering of electrical connections to copper; soldering of electronic assemblies; hot dip coating of ferrous and non-ferrous metals; high quality sheet metal work; capillary joints including light gauge tubes in copper and stainless steel; manufacture of electronic components; machine soldering of printed circuits.
2a	Hand and machine soldering of electronic components; can soldering.
3 3a 4 5	General engineering work on copper, brass and zinc; can soldering.
6 7	Jointing of electric cable sheaths. 'wiped' joints.
8 9 10	Lamp solder; dipping solder; for service at very low temperatures (e.g. less than −60 °C).

Table 7.15 *Tin–lead alloys (white metals for bearings)*

Composition (%)					Properties and applications
Sn	Sb	Cu	Pb	P	
93	3.5	3.5	—	—	Big-end bearings for light- and medium-duty, high-speed internal combustion engines.
86	10.5	3.5	—	—	Main bearings for light- and medium-duty, high-speed internal combustion engines.
80	11.0	3.0	6.0	—	General-purpose, heavy-duty bearings. Lead improves plasticity where alignment is a problem.
60	10.0	28.5	1.5	—	Heavy-duty marine reciprocating engines, electrical machines.
40	10.0	1.5	48.5	—	Low-cost, general-purpose, medium-duty bearing alloy.

7.1   State the main alloying elements in:
(a)   brass
(b)   duralumin
(c)   gunmetal
(d)   phosphor-bronze

7.2   State an example of each of the following types of aluminium alloy and, in each case, state the composition and a typical application:
(a)   non-heat-treatable wrought alloy
(b)   non-heat-treatable casting alloy
(c)   heat-treatable wrought alloy
(d)   heat-treatable casting alloy

7.3   State the effect on pure copper of adding the following alloying elements:
(a)   tellurium
(b)   beryllium
(c)   cadmium
(d)   arsenic

7.4   (a)   Sketch the aluminium–silicon phase equilibrium diagram and explain in detail the changes which take place as a 6 per cent silicon alloy is cooled from the molten state to room temperature.
(b)   Discuss the need for 'modification' of aluminium–silicon alloys by the addition of metallic sodium, and show the effect of such modification on the aluminium–silicon phase equilibrium diagram.

7.5   Discuss in detail the differences in composition, properties and applications of:
(a)   high-conductivity copper
(b)   fire-refined tough pitch copper
(c)   oxygen-free high-conductivity copper
(d)   phosphorus deoxidised copper

7.6   (a)   State the composition, properties and typical applications of the aluminium alloy known as 'duralumin'.
(b)   Describe in detail, with reference to the aluminium–copper phase equilibrium diagram, the softening of duralumin by solution treatment, and the subsequent hardening of duralumin by natural ageing (precipitation).

7.7   Sketch the copper–zinc phase equilibrium diagram and refer to it when describing the differences in composition, properties and typical applications of:
(a)   the α brasses
(b)   the duplex brasses $(\alpha + \beta')$

7.8   Sketch the copper–tin phase equilibrium diagram and refer to it when describing the differences in compositions, properties and typical applications of:
(a)   the α-phase bronze alloys
(b)   the duplex bronze alloys $(\alpha + \delta)$

7.9 Describe the composition, properties and a typical application of:
- (a) a magnesium alloy
- (b) a zinc alloy suitable for die casting

7.10 Select a suitable non-ferrous metal or alloy for each of the following applications, giving reasons for your choice in each instance:
- (a) an electrical conductor to be jointed by soft soldering
- (b) a die-cast frame for a domestic electricity meter
- (c) an internal combustion engine piston
- (d) a white-metal bearing lining for the main bearings of an automobile engine crankshaft

8 Polymeric (plastic) materials

The topic areas covered in this chapter are:

- The history of plastics.
- Additives.
- General properties of polymeric materials.
- Properties and applications of elastomers.
- Properties and applications of typical thermoplastics.
- Properties and applications of typical thermosetting plastics.

8.1 The history of plastics

Before looking at polymeric (plastic) materials in more detail, let's quickly look at the history of these diverse and important and interesting materials. They do not exist naturally and their existence had to await the development of the chemical engineering industry and, in particular, the oil refining industry which provides the raw materials for many of these plastics.

The history of the plastics industry can be traced back to circa 1854 when Alexander Parkes of Birmingham, England, discovered that nitrocellulose could be mixed with camphor to form a material he called 'Parkesene'. He developed this first man-made plastic material as a substitute for carved ivory and boneware (e.g. napkin rings and knife handles). This new material could be quickly and cheaply moulded to shape whereas the traditional materials had to be hand carved and turned to shape. There was a lot of opposition by the traditional trades to this new material and it could have sunk into oblivion.

However, renamed 'Celluloid', it was adopted by John Hyatt of Colorado, USA, in 1868 for the manufacture of cheap billiards balls for his saloons in place of the more expensive and traditional ivory billiards balls. Unfortunately the balls were not unknown to explode on impact and had a reputation for starting the occasional gunfight!

Celluloid was a relatively unstable and highly flammable material and was the cause of many serious accidents. It was widely used for moulding brush backs, combs, toys and similar small and cheap articles. It was also used as cinema film and still camera film. Because of its instability and flammability, much valuable archive film has been lost.

Nowadays it has been supplanted by the much safer and more durable *cellulose acetate* 'safety film'.

The second major plastic to emerge was phenol–formaldehyde resin in 1907. It was called 'Bakelite' after the German chemist Baekeland who patented the process for its manufacture. Its development fortunately coincided with the expansion of the electrical engineering industry which immediately adopted it for the manufacture of moulded insulators and instrument cases. The plastics industry, as we know it today, had arrived.

Unfortunately Bakelite products could only be produced in dark colours such as black and brown. It was not until it was found that the phenol could be replaced by a substance called 'urea' that the light coloured urea–formaldehyde resin was developed. This could be produced in a whole range of colours by the addition of pigments.

In the 1930s it was discovered that another substance, 'melamine', could be used in place of urea and produced a stronger material which did not absorb as much water and did not stain as readily. It was able to withstand higher temperatures and was, and still is, widely used for tableware and cooking utensils.

As mass production continued to expand, the real potential of the manufacture of different polymeric (plastic) materials in which long-chain molecules (polymers) were produced from single molecules (monomers) began to be seen. With the steady development of this concept of producing new materials by chemical synthesis, a future large-scale industry was born. Although many synthetic materials were initially only produced in relatively small quantities, the demands of the Second World War resulted in the large-scale development of the materials we know today – namely, such materials as polystyrene, polyvinylchloride (PVC) and polymethylmethacrylate (Perspex).These and a steadily expanding range of materials reached the domestic market by 1945.

8.2 Additives

The principles underlying the structure and behaviour of polymeric materials were considered in Chapter 2. This chapter will now consider the properties and uses of these materials in greater depth.

It is not usual to use pure resin to mould a plastic product. The appearance and performance of most plastic and elastic polymers can be improved by the use of various additives which will now be considered.

8.2.1 *Plasticisers*

These are added to polymeric materials to reduce their rigidity and brittleness and improve their flow properties whilst being formed or moulded. There are two main groups of plasticisers:

Primary plasticisers
These are used to reduce the Van der Waal's forces between adjacent molecular chains by introducing monomers whose polar groups partially neutralise those of the polymer groups and allow greater mobility between adjacent polymer chains.

Secondary plasticisers

These are monomers of a compatible but inert material without polar groups which may be added to provide mechanical separation of the polymer chains in the same way that a lubricant separates a shaft from its bearing. Separation of the polymer chains in this manner reduces the Van der Waal's forces of attraction between them, as shown in Fig. 8.1. Secondary plasticisers may, themselves, be divided into two groups according to the method of application.

Fig. 8.1 *Plasticisers: (a) strong Van der Waal's forces link the polymer chains; (b) a secondary plasticiser separates the chains, weakening the Van der Waal's forces*

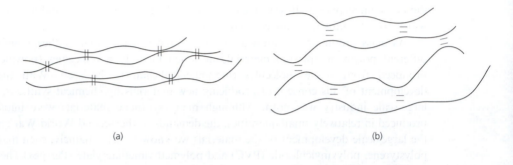

(a)   (b)

- *Internal plasticisation* Small amounts of plasticiser are added during polymerisation. The additive produces bulky side groups which achieve separation of the polymer chains as they form during the polymerisation process (secondary plasticisation). For example, polyvinyl chloride (PVC), which is a rigid and rather brittle plastic material, can be made flexible for raincoats, garden hosepipes and the insulation of electric cables by the addition of 15 per cent vinyl acetate as a secondary plasticiser during polymerisation.
- *External plasticisation* This is the more common method of plasticisation. The plasticiser, in the form of a low-volatility liquid solvent, is added after polymerisation. It disperses throughout the plastic, filling the voids between the polymer chains acting as a lubricant. Again, this is secondary plasticisation as the plasticiser separates the polymer chains and weakens the Van der Waal's forces as previously explained.

Only the amorphous zones are lubricated in this manner as highly crystalline polymers will not absorb sufficient solvent plasticiser. Although it is difficult, effectively, to plasticise high-crystallinity polymers, the presence of a plasticiser helps to reduce the amount of crystallinity present by interfering with the formation of orderly arrays in the polymer chains. By reducing the amount of crystallinity present, the plasticiser also reduces the glass transition temperature (T_g). Figure 8.2 shows the effect of adding a plasticiser to polyvinyl chloride.

Fig. 8.2 *Effect of a plasticiser on the mechanical properties of PVC*

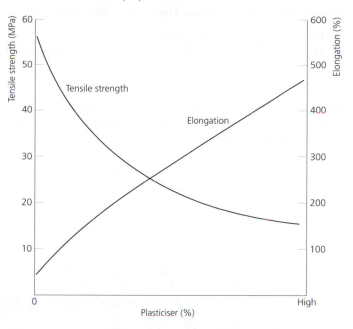

8.2.2 *Fillers*

These have a considerable influence on the properties of mouldings produced for any given polymeric material. They improve the impact strength and reduce shrinkage during moulding. Typical fillers together with their properties are listed in Table 8.1. Fillers are essential in thermosetting moulding powders and may be present in quantities up to 80 per cent by weight. They are less frequently found in thermoplastics, where their presence is normally limited to 25 per cent by weight. The exceptions are thermoplastic floor tiles which may contain up to 40 per cent calcium carbonate by weight.

Table 8.1 *Fillers*

Filler material	Properties
Glass fibre	Good electrical insulation properties
Wood flour; calcium carbonate	Low cost, high bulk, low strength
Asbestos	Heat resistant (no longer recommended; health hazard)
Aluminium powder	High mechanical strength
Shredded paper Shredded cloth Mica granules	Good strength, combined with reasonable electrical insulation properties

The selection of a filler is usually determined by the properties it can impart to the plastic product and also its cost. However, all fillers must:

- Have a low moisture absorption rate.
- Not adversely affect the colour or surface finish of the product.
- Not cause abrasive wear in the processing equipment.
- Be capable of being 'wetted' by the resin.

Figure 8.3 shows the effect of various filler materials on the stress–strain curve for a typical phenolic resin, whilst Fig. 8.4 shows the effect of various filler materials on the elastic modulus (rigidity) of a typical polyester resin.

Fig. 8.3 *Effect of fillers on the mechanical properties of a phenolic thermoset*

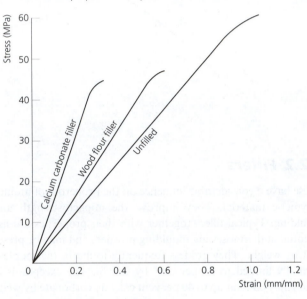

Fig. 8.4 *Effect of fillers on the elastic modulus of an isophthalic polyester*

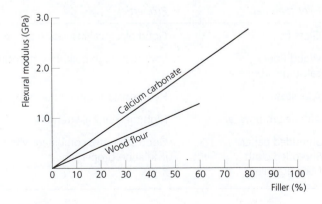

8.2.3 *Stabilisers*

Degradation of polymeric materials occurs when they are exposed to heat, sunlight and weathering. Such degradation is usually accompanied by colour change, deterioration in mechanical properties, cracking and surface crazing. Since these are environmental changes, they will be dealt with more fully in Section 13.15. Sometimes such degradation is encouraged where biodegradable plastics are acceptable, since this assists in their safe disposal. However, stabilisers are used where plastics have to withstand environmental attack over a long period of time – for example, window frames and roof guttering.

8.2.4 *Colorants*

These can be subdivided into dyestuffs, organic pigments and inorganic pigments. Dyestuffs are usually aromatic organic chemicals which are soluble in a variety of solvents. They absorb light selectively to produce their characteristic colours. They are suitable for colouring transparent and translucent polymers such as the acrylics, cellulosics and polystyrene. Unfortunately organic dyestuffs have only a limited colour stability when exposed to sunlight, and they may also degrade at the moulding temperatures of some high-temperature polymers.

Organic pigments are opaque and cannot be used to colour transparent plastic materials. They are usually found in opaque plastic products. Their light and heat stability is markedly superior to that for dyestuffs. Inorganic pigments based on metal oxides and salts have the greatest opacity and superior light and heat stability. Examples are:

- Titanium oxide are used in non-toxic white plastics and paints.
- Iron oxides are used to provide yellows, tans and reds.
- Cadmium salts give brighter yellows and reds but are toxic.
- Carbon black is used as an ultraviolet radiation absorber and for the production of black products.

8.2.5 *Antistatic agents*

These are included to increase surface conductivity so that static charges can leak away. This prevents the attraction of dust particles and reduces the risk of explosion in hazardous environments caused by the spark associated with an electrical discharge. It also prevents electric shocks which can occur when synthetic fabric materials are handled in very dry climates.

8.3 General properties of polymeric materials

The properties of polymeric materials can vary widely, but they all have certain properties in common:

8.3.1 *Electrical insulation*

All polymeric materials exhibit good electrical insulation properties. However, their usefulness in this field is limited by their low heat resistance and their softness. Thus they are useless as formers on which to wind electric radiator elements, and as insulators for use out of doors where their relatively soft surface would soon be roughened by the weather. Dirt collecting on this roughened surface would then provide a conductive path, causing a short circuit.

8.3.2 *Strength/weight ratio*

Polymeric materials vary in strength considerably. Some of the stronger, such as nylon, compare favourably with the weaker metals. All polymeric materials are much lighter than any of the metals used for engineering purposes. Therefore, properly chosen and proportioned, their strength/weight ratio compares favourably with many light alloys and they are steadily taking over engineering duties which, until recently, were considered the prerogative of metal.

8.3.3 *Corrosion resistance*

All polymeric materials are inert to most inorganic chemicals and can be used in environments which are hostile even to the most corrosion-resistant metals. The synthetic rubbers, which are a product of polymer chemistry, are superior to natural rubber (polyisoprene) since they are not attacked by oils and greases. The degradation of polymers will be considered more fully in Section 13.15.

SELF-ASSESSMENT TASK 8.1

1. Explain why polymeric materials are popularly called 'plastics' when many are elastic and others are brittle.

2. Explain why 'additives' are added to plastic materials.

3. List **three** general properties of plastic materials and explain how these result in the fundamental differences between plastics and metals.

8.4 Properties and applications of elastomers

Elastomers are substances which permit extreme reversible extensions to take place at normal temperatures. Natural rubber is an obvious and very important elastomer. Because elastomers are cross-linked like thermosets, their polymer chains cannot slide over each other under applied loads to take on a permanent set. However, unlike true thermosets which have very many cross-links, elastomers have relatively few cross-links. Further, the helical molecular chain of an elastomer uncoils when stressed, as shown in Fig. 8.5, and recovers when the load is removed in a similar manner to a coil spring.

Fig. 8.5 *Behaviour of an elastomer under stress: (a) unstressed elastomer chain; (b) stressed elastomer chain*

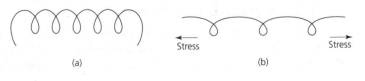

(a)   (b)

Many components do not require great strength but they do require softness, flexibility and reversible elongation. Thus the rubbers (natural and synthetic elastomers) are ideal for such applications as:

- Resilient floor coverings.
- Weatherstripping.
- Footwear.
- Vehicle tyres.
- Joint sealants.
- Expansion joints.
- Anti-vibration mountings.

Like other polymers, elastomers become brittle below their glass transition temperatures. The T_g for most synthetic rubbers is about $-20\,°C$ but for silicones and for natural rubber it is about $-60\,°C$.

The elastomers are usually addition polymerised as thermoplastics and then cross-linked (vulcanised) with sulphur at approximately every five-hundredth carbon atom. Increased vulcanisation increases the cross-linking and this, in turn, increases the stiffness and reduces the elongation percentage of the material. For example, fully vulcanised, natural rubber becomes a rigid, brittle thermoset called ebonite.

The uses to which elastomers (rubbers) may be put by engineers may be classified as follows:

Vibration insulation and isolation
- Shock absorbers.
- Anti-vibration machine and engine mountings.
- Sound insulation.

Distortional systems
- Correctives for misalignment: such as flexible couplings.
- Changing shapes: such as belts, flexible hose, covered rollers and tyres.
- Seals and gaskets.
- Rubber hydraulics (forming tools).

Protective systems
- Protection against abrasion.
- Protection against corrosion.
- Electrical insulation.
- Protective clothing: gloves, aprons, boots.

8.4.1 *Commercial elastomers*

Some typical groups of elastomers used in engineering are as follows:

Acrylic rubbers
These are derived from the same family of polymeric materials as 'Perspex' but, in order to give them the properties associated with elastomers, they are not so heavily cross-linked. This group of rubbers have excellent resistance to oils, oxygen, ozone and ultraviolet radiation and they are used as the basis for the latex paints developed for motor vehicles.

Butyl rubber

This rubber is impervious to gases and is used as a vapour barrier and for hose linings. It is highly resistant to outdoor weathering and ultraviolet radiation and is used for construction industry sealants.

Nitrile rubber

This has excellent resistance to oils and solvents and can be readily bonded to metals. It is used for petrol and fuel oil hoses, hose linings and aircraft fuel tank linings. It is also resistant to refrigerant gases.

Polychloroprene rubber (neoprene)

This was the original synthetic rubber developed during the Second World War. It has good resistance to oxidation, ageing and weathering. It is resistant to oils, solvents, abrasion and elevated temperatures. Because of its chlorine content it is fire resistant. It is used as a flexible electrical insulator, gaskets, hoses, engine mounts, sealants, rubber cements and protective clothing.

Polyisoprene (natural rubber)

This is derived from the sap of a tree called *Hervea Brasiliensis*. It has a low hysteresis, a relatively high tensile strength and a very low glass transition temperature. Unfortunately it is readily attacked by solvents, petrol, mineral oils and ozone. It degrades (perishes) rapidly in the presence of strong sunlight. Modified by additives and vulcanisation to give it increased strength and wear-resistant properties, natural rubber is used for vehicle tyres as it has excellent anti-skid properties.

Polysulphide rubber (thiokol)

Although this rubber has low mechanical strength, its resistance to solvents and its impermeability to gases is excellent and its weathering characteristics are outstanding. It also has good bonding properties and is widely used in the construction industries as a sealant. Thiokol and polyurethane rubbers are also used as fuels for solid-fuel rockets.

Polyurethane rubber

Polyurethane can be formulated to give either plastic or elastic properties. Although it has relatively high strength and abrasion resistance, it is of little use for vehicle tyres as it has low skid resistance. However, its outstanding service life makes it suitable for cushion tyres for warehouse trucks and forklift trucks where low speeds and dry floor conditions makes its low skid resistance more acceptable. It is also used for shoe heels, painting rollers, mallet heads, oil seals, diaphragms, anti-vibration mountings, gears and pump impellers.

Rubber hydrochloride

This material is better known as 'Pliofilm' and is used to form a transparent film for the vacuum packaging of foods and DIY hardware. It is easily identified by its unusually high tensile strength and tear resistance.

Silicone rubbers

Although silicone rubber has a relatively low tensile strength, it has an exceptionally wide working temperature range of -80 to $+235\,°C$. Thus it often outperforms other rubbers which are superior at room temperature but which cannot exist at such temperature extremes. It can be used for mould linings and high-temperature seals. It is also used in space vehicles and artificial satellites.

Styrene–butadiene rubber (SBR)

A general-purpose synthetic rubber used for tyres, belts, floor tiles and latex paints. It is superior to natural rubber in respect of skid resistance, solvent resistance and weathering.

SELF-ASSESSMENT TASK 8.2

1. Discuss the main differences between natural rubber and the synthetic elastomers.

2. Select suitable elastomers for the following applications, giving reasons for your choice:

    (a) a vapour barrier
    (b) an oil-resistant hose
    (c) anti-vibration engine mounts
    (d) water pump impeller
    (e) high-temperature seal

8.5 Properties and applications of typical thermoplastics

Remember that thermoplastics is that group of polymeric materials that can be softened every time they are heated. No curing takes place during the moulding operations.

Polyethylene (polythene)

This is one of the most versatile and widely used plastic materials. It remains tough and flexible over a wide range of temperatures and has good dimensional stability. It is easily moulded and is used in a wide range of domestic goods such as buckets, bowls, food containers, bags and squeezy bottles. It is also used commercially for water piping, chemical equipment and for electrical insulation. It is resistant to most solvents and has good weathering properties. Unfortunately it degrades when exposed to strong sunlight unless it contains a UV filter pigment such as carbon black. Polyethylene can exist in the amorphous state or with varying degrees of crystallinity.

*   *Low-density polyethylene* This has a branched molecular chain which makes the formation of crystallites difficult. The structure of the branched molecular chain is shown in Fig. 8.6(a).

- *High-density polyethylene*   This has a simple linear chain which leads itself to the close packing essential for high density and for the orderly structure required to form crystallites. The structure of a simple linear chain of high-density polyethylene is shown in Fig. 8.6(b).

Fig. 8.6   *Polymer chains for polyethylene: (a) polymer chain for low-density (branched) polyethylene; (b) polymer chain for high-density (linear) polyethylene*

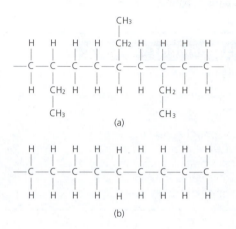

(a)

(b)

Typical properties for polyethylene are:

	Low density	*High density*
Crystallinity (%)	60	95
Density (kg/m^3)	920	950
Tensile strength (MPa)	11.0	31.0
Elongation (%)	100–600	50–800
Impact value (J)	no fracture	5–15
T_g (°C)	−120	−120
T_m (°C)	115	138
Maximum service temperature (°C)	85	125

Polypropylene

This is a tough, rigid, lightweight material with similar properties to polythene but with better heat resistance. It has good mechanical properties, and good resistance to attack by acids, alkalis and salts even at high temperatures. It is widely used for chemical plant equipment and domestic hardware, and is also used for moulding hospital and laboratory equipment. It can be drawn into fibres for rope and net making, and it can be rolled into sheets for packaging. It is a good moulding material for electrical insulators. Typical properties for polypropylene are:

Crystallinity (%)	60
Density (kg/m^3)	900
Tensile strength (MPa)	30–35
Elongation (%)	50–600
Impact value (J)	1–10
T_g (°C)	−25
T_m (°C)	176
Maximum service temperature (°C)	150

Polystyrene

This is a tough dense plastic which is hard and rigid, and has good dimensional stability. It can be moulded to give a high surface gloss and is used for such articles as domestic hollowware and refrigerator trays. Although it has good mechanical properties even at low temperatures, it tends to be brittle because of its high glass transition temperature. Being an amorphous material it has no clearly defined melting temperature.

Unfortunately it is attacked by petrol and other organic solvents. Since it can be foamed, rigid foamed polystyrene is used for heat and sound insulation blocks in building and refrigeration. It is also used for ceiling tiles, moulded packaging, and for buoyancy aids. Polystyrene has an aromatic ring as a side branch to the monomer, and this is shown in Fig. 8.7. Thus a long-chain polymer made from such monomers is too bulky to pack closely together to form crystallites and the crystallinity is zero.

Fig. 8.7 *Segment of the polymer chain for polystyrene*

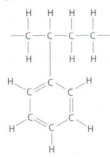

Typical properties of amorphous, low-impact, general-purpose polystyrene are:

Crystallinity (%)	0
Density (kg/m^3)	1070
Tensile strength (MPa)	28–53
Elongation (%)	1–35
Impact value (J)	0.25–2.5
T_g (°C)	100
T_m (°C)	not applicable
Maximum service temperature (°C)	65–85

(*Note*: for the properties and uses of 'high-impact' polystyrene, see 'ABS'.)

Polyvinyl chloride (PVC)

Unplasticised polyvinyl chloride is hard and tough but it can be moulded into a variety of hard-wearing hollowware articles such as buckets and plant pots. It is also used for builders' hardware such as rain guttering, downpipes and drainpipes. To form and mould rigid PVC it has to be heated above its T_g of 87 °C to make it soft and flexible. Upon cooling below its T_g, it becomes rigid again. When plasticised it becomes flexible and rubbery. In this condition it can be used for waterproof clothing, hose pipes, electric cable insulation and chemical tank linings. It offers good resistance to attack by water, acids, alkalis and most common solvents. Unfortunately it hardens and becomes brittle with age. Typical properties for PVC are:

	Unplasticised	*Plasticised*
Crystallinity (%)	0	0
Density (kg/m³)	1400	1300
Tensile strength (MPa)	49	7–25
Elongation (%)	10–130	240–380
Impact value (J)	1.5–18.0	not applicable
T_g (°C)	87	87
T_m (°C)	not applicable	not applicable
Maximum service temperature (°C)	70	60–105

Polytetrafluoroethylene (PTFE)

This is one of the most versatile and important plastics despite its relatively high cost. It is suitable for the manufacture of tough mouldings and as a non-stick, anti-friction coating (Teflon). It does not burn, neither is it attacked by any known reagent or solvent. It is a good electrical insulator and it has the lowest coefficient of friction of any known solid. It is widely used for bearings, fuel hoses, gaskets and tapes, and as a non-stick coating for cooking utensils. It is also used as a lining for chemical equipment because of its resistance to chemical attack. Typical properties for PTFE are:

Crystallinity (%)	90
Density (kg/m³)	2170
Tensile strength (MPa)	17–25
Elongation (%)	200–600
Impact value (J)	3–5
T_g (°C)	−126
T_m (°C)	327
Maximum service temperature (°C)	260

Polymethyl methacrylate (Perspex)

This is a relatively lightweight material with excellent optical properties. It is also strong and rigid with excellent electrical insulating properties. Unfortunately it is easily scratched and it softens in boiling water. Also it is attacked by petrol and many organic solvents. It is used for such diverse purposes as lenses, dentures, aircraft windows, sinks, baths, lighting fittings and diffusers, advertising displays and display lighting. Typical properties for 'Perspex' are:

Crystallinity (%)	0
Density (kg/m³)	1180
Tensile strength (MPa)	50–70
Elongation (%)	3–8
mpact value (J)	0.5–0.7
T_g (°C)	0
T_m (°C)	not applicable
Maximum service temperature (°C)	95

Acrylonitrile–butadiene–styrene (ABS)

This is a tough, strong material with high impact resistance, hence it is often referred to as 'high-impact' polystyrene. It is widely used for moulding television and radio set cabinets, motor vehicle radiator grills and panels, battery cases, crash helmets, water pump and refrigerator parts. It has high dimensional stability and remains tough at subzero temperatures. It also resists attack by acids, alkalis and most petroleum derivatives. High-impact polystyrene is made by copolymerising low-impact polystyrene with about 5 per cent butadiene rubber. An even tougher variant can be produced by copolymerising the high-impact polystyrene a stage further with acrylonitrile, as shown in Fig. 8.8, to form acrylonitrile–butadiene–styrene (ABS).

Fig. 8.8 *Constituent monomers of the copolymer ABS (monomers present in varying proportions): (a) acrylonitrile; (b) butadiene; (c) styrene*

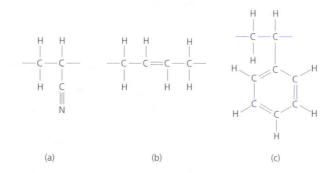

| | (a) | (b) | (c) |

Typical properties for ABS are:

Crystallinity (%)	0
Density (kg/m³)	1100
Tensile strength (MPa)	30–35
Elongation (%)	10–40
Impact value (J)	7–12
T_g (°C)	−55
T_m (°C)	not applicable
Maximum service temperature (°C)	100

Polyamides (nylon)

Polyamides are that group of polymeric materials known as 'nylons' and were the first of the 'engineering' or high-strength thermoplastics. Polyamides are produced by the reaction of diamine with an organic acid. One of the most commonly used nylons is produced by reacting hexamethylenediamine with adipic acid to give hexamethylenedipamide (commonly abbrieviated 'polyamide'). This is referred to as nylon 6/6 ('66'), and this notation indicates that there are six carbon atoms in the amine and six in the acid segments. Other grades of nylon are 6/10, 6/12, 13/13, etc. The molecular structure of nylon '66' is shown in Fig. 8.9. This is just one segment from the molecular chain. Since the chain is linear, it gives rise to crystalline structures.

Fig. 8.9 *Segment from nylon '66' polymer chain*

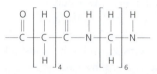

The nylon range of materials are strong, tough and flexible. They have good resistance to abrasion, but their dimensional stability and electrical insulation properties are affected by the fact that nylon absorbs water, even from the atmosphere. This absorbed moisture acts as a plasticiser, lowering the stiffness, strength and hardness. The effect of moisture absorption on the strength of nylon '66' is shown in Fig. 8.10.

Fig. 8.10 *Effect of moisture content on the strength of nylon '66'*

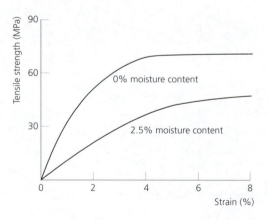

The nylons show good resistance to most common solvents but, unfortunately, they do not weather well and degrade rapidly when used out of doors. Nylons can be moulded into such components as gears, valves, bearings, cams and surgical equipment. When used as bearings, the ability of nylon to absorb moisture from the atmosphere renders the need for a lubricant unnecessary. This is particularly useful in office machinery and food-processing equipment. Nylons are also extruded and drawn into fibres for bristles, textiles, fishing lines, climbing ropes and rigging for small boats. Nylon ropes have a high strength/

weight ratio, but tend to fray easily and have little resilience to shock loads. For this reason 'Terylene' ropes are preferred by rock climbers, mountaineers and small boat sailors. Typical properties of nylon '66' are:

Crystallinity (%)	variable
Density (kg/m^3)	1100
Tensile strength (MPa)	50–85
Elongation (%)	60–300
Impact value (J)	1.5–15.0
T_g (°C)	50
T_m (°C)	265
Maximum service temperature (°C)	120

Some further typical 'nylons' and their properties are listed in Table 8.2.

Table 8.2 Properties of typical nylons

Material	Properties			
	Density (kg/m^3)	Tensile strength (MPa)	Elongation (%)	Maximum service temperature (°C)
Nylon 6	1100	70–90	60–300	120
Nylon 6/10	1100	60	80–230	120
Nylon 11	1100	50	75–300	120

Polyesters

These can be thermoplastic or thermosetting, and an example of the former type is polyethylene teraphthalate ('Terylene'). (The thermosetting polyesters will be considered in Section 8.5.) Thermoplastic polyesters are usually produced in the form of a fibre or a film. They have good dimensional stability and are resistant to most organic solvents, although they are attacked by strong acids and alkalis. They provide excellent insulation and are used as the dielectric in capacitors. Polyester films and fibres are used in such applications as textiles, loudspeaker cones, recording tapes, draughting materials, photographic films and 'papers' for special purposes, and electrical insulating tapes. They are also now preferred to nylon for climbing and rigging ropes as they have more 'give' when subjected to shock loads and do not fray so easily. Typical properties of the fibre are:

Crystallinity (%)	60
Density (kg/m^3)	1350
Tensile strength (MPa)	over 175
Elongation (%)	60–110
Impact value (J)	1.0
T_g (°C)	70
T_m (°C)	267
Maximum service temperature (°C)	69

Polyacetals

These materials are very strong and stiff with good creep resistance and fatigue endurance. They are beginning to replace metals in a number of applications because of their high crystallinity and clearly defined, high melting temperature. They are good electrical insulators and are resistant to alkalis and most common solvents.

However, they are attacked by mineral acids and degrade in strong sunlight, so that they are not recommended for outdoor use. Typical applications are: water pump impellers and housings for domestic appliances, extractor fan components, moulded instrument panels for road vehicles, plumbing fittings, bearings, cams, gears, hinges, window catches and door-lock components. Typical properties for polyformaldehyde are:

Crystallinity (%)	70–90
Density (kg/m^3)	1410
Tensile strength (MPa)	50–70
Elongation (%)	15–75
Impact value (J)	0.5–2.0
T_g (°C)	−73
T_m (°C)	180
Maximum service temperature (°C)	105

Typical properties for polyoxymethylene are:

Crystallinity (%)	70–90
Density (kg/m^3)	1410
Tensile strength (MPa)	60–70
Elongation (%)	15–70
Impact value (J)	0.5–2.0
T_g (°C)	−76
T_m (°C)	180
Maximum service temperature (°C)	120

Polycarbonates

These materials have good impact strength, good heat resistance, good dimensional stability, good electrical insulation properties and good optical properties which are superior to Perspex. Further, polycarbonates are extremely tough and have a much higher scratch resistance than any other transparent plastic. Polycarbonates are resistant to petroleum products and most solvents. Typical applications are:

- Electrical insulators.
- Capacitor dielectrics.
- Lightweight, 'unbreakable' spectacle lenses.
- Aircraft components and windows, and vehicle components.

Typical properties for polycarbonates are:

Crystallinity (%)	0
Density (kg/m^3)	1200
Tensile strength (MPa)	60–70
Elongation (%)	60–100

Impact value (J)	10–20
T_g (°C)	150
T_m (°C)	not applicable
Maximum service temperature (°C)	30

Cellulose acetate

Unlike cellulose nitrate (celluloid), which is dangerously flammable, *cellulose acetate* is a tough, durable material that is not particulary flammable. Unfortunately, cellulose acetate is highly water absorbent and this adversely affects its dimensional stability and electrical insulation properties; it is unsuitable for use out of doors. The water absorption problem can be controlled to some extent by the addition of butyrate.

All cellulosics are resistant to most household chemicals but are attacked by acids, alkalis and alcohol. All cellulosics soften in boiling water. Cellulose acetate is used for lampshades and diffusers, machine guards, spectacle frames, screwdriver and other small tool handles, photographic (safety) film base, toys and many small components. It is easily moulded and takes a sharp and detailed impression. Typical properties are:

Crystallinity (%)	0
Density (kg/m^3)	1280
Tensile strength (MPa)	24–65
Elongation (%)	5–55
Impact value (J)	0.7–7.0
T_g (°C)	120
T_m (°C)	not applicable
Maximum service temperature (°C)	70

8.6 Properties and applications of typical thermosetting plastics

The thermoplastics described in Section 8.4 soften every time they are heated above their glass transition temperatures and they can then be reshaped. Remember that *thermosetting plastics* (thermosets) undergo chemical change (curing) during moulding and can never again be softened by heating (see Section 2.15.1 and 2.17). Thermosets are stronger, more rigid and more brittle than thermoplastics.

Phenol–formaldehyde

Phenolic resins are never used by themselves but in conjunction with fillers and other additives which reduce the inherent brittleness, improve the mechanical and electrical properties, make them more amenable to moulding and enhance their natural brown colour (see Section 8.1). The properties of phenolic mouldings varies widely depending upon the additives used. Typical applications are:

- Electrical insulators.
- Electrical plugs and sockets (industrial).

- Handles, knobs, vehicle ignition system mouldings.
- Clutch and brake linings.

Phenolic resins are also used in the manufacture of laminates such as 'Tufnol', (see Section 9.2). The dark colour of phenolic resins precludes their use for many domestic electrical applications. Typical properties of a wood-flour filled phenolic moulding powder are:

Density (kg/m^3)	1350
Tensile strength (MPa)	35–55
Elongation (%)	0–1
Impact value (J)	0.5–1.5
Maximum service temperature (°C)	75

Urea–formaldehyde

The basic resin is hard, brittle, rigid and scratch resistant. Like phenol–formaldehyde it is never used by itself but in conjunction with fillers and other additives (see Section 8.1). It is resistant to most solvents and household detergents. It has good electrical insulation properties and, being virtually colourless, it can be coloured by the addition of pigments to suit any decorative requirements. For this reason it is widely used for domestic electrical equipment (plugs, sockets and switches) and most domestic utensils and toys. It is also used as a binder for foundry core-sands and for shell moulding. Typical properties for urea–formaldehyde with a cellulose filler are:

Density (kg/m^3)	1500
Tensile strength (MPa)	50–75
Elongation (%)	0–1
Impact value (J)	0.3–0.5
Maximum service temperature (°C)	75

Melamine–formaldehyde

This material is similar to urea–formaldehyde, but is more resistant to heat and is less water absorbent. This not only improves its electrical properties but makes it suitable for tableware. Typical properties for melamine–formaldehyde with a cellulose filler are:

Density (kg/m^3)	1500
Tensile strength (MPa)	56–80
Elongation (%)	0–0.7
Impact value (J)	0.2–0.5
Maximum service temperature (°C)	100

Epoxides

Epoxy resins are used for bonding glass fibre fillers, as described in Section 9.3.2. They are resistant to water and most reagents and have excellent electrical insulation properties, therefore epoxides are also widely used as a casting material for small components and as a 'potting' material for sealing electrical equipment such as transformers and chokes. Their use as the basis of high-strength adhesives will be considered in *Engineering Materials*, Volume 2. Typical properties of an unfilled resin casting are:

Density (kg/m^3)	1150
Tensile strength (MPa)	35–50
Elongation (%)	5–10
Impact value (J)	0.5–1.5
Maximum service temperature (°C)	200

Polyesters (unsaturated)

The mechanical properties of unsaturated polyesters vary quite considerably according to the polymer used. These materials have good impact resistance and surface hardness. They are also resistant to water and most common solvents but are attacked by acids and alkalis. Their low water absorption and resistance to weathering make them an excellent binder for use with glass fibre reinforcement (see Section 9.3) for mouldings ranging from domestic baths to boat hulls for pleasure craft and small commercial and naval craft. Typical properties of the unfilled (not reinforced) resin are:

Density (kg/m^3)	1120
Tensile strength (MPa)	55
Elongation (%)	2
Impact value (J)	0.5–1.0
Maximum service temperature (°C)	200

Polyesters (alkyds)

These materials have good heat resistance and excellent electrical insulation properties. They have good dimensional stability and are unaffected by water and most organic solvents. This makes them suitable for mouldings for high-voltage insulators in television sets, and for mouldings for the electrical equipment of road vehicles and aircraft. Alkyd resins are also used as the basis for the paint systems used on cars and domestic appliances. Typical properties are:

Density (kg/m^3)	2000
Tensile strength (MPa)	25
Elongation (%)	0
Impact value (J)	0.25–0.5
Maximum service temperature (°C)	150

Comparing the properties for the thermosets with those for the thermoplastics shows that the main advantage of the former over the latter lies in their superior abrasion resistance (hardness), rigidity and high maximum service temperature. However, their impact value is lower and they tend to be brittle and break more easily. Because curing takes place in the mould, the moulding time cycle takes longer and processing is more costly than for thermoplastic materials.

1. Explain what is meant by the term 'thermoplastic'.

2. Explain what is meant by 'crystallinity' in thermoplastics and show how it affects the properties of thermoplastic materials.

3. Explain what is meant by the term 'thermosetting plastic'.

4. Group the following polymeric materials as thermoplastics, elastomers or thermosetting plastics: epoxide potting compound, polyethylene, nylon, neoprene, polyisoprene, polypropylene, thiokol, urea–formaldehyde, polytetrafluoroethylene.

We now have to consider reinforced plastic materials such as the laminates that are available in sheets, sections and rounds. We also have to consider glass fibre reinforced mouldings. Since these are 'composite' materials, they are discussed in Chapter 9.

EXERCISES

8.1 Select a suitable polymeric material for the following applications giving reasons for your choice:
(a) flexible electric cable insulation
(b) non-stick anti-friction coating
(c) high-strength, lightweight ropes for mountaineering
(d) low friction, oil-less bushes for food-processing machinery
(e) heat insulating panels

8.2 (a) Explain what is meant by an 'elastomer' and how it differs from other polymeric materials.
(b) State typical applications for the following elastomers giving reasons for your choice:
  (i) styrene–butadiene–rubber (SBR)
  (ii) polyisoprene (natural) rubber
  (iii) polychloroprene (neoprene) rubber
  (iv) silicone rubber
  (v) acrylic rubber

8.3 Sate the reasons for the inclusion of the following additives in a thermosetting moulding powder:
(a) filler
(b) pigment
(c) catalyst
(d) accelerator
(e) mould release agent

8.4 State typical applications for the following filler materials:
  (a) glass fibre
  (b) wood flour
  (c) calcium carbonate
  (d) aluminium powder
  (e) shredded paper
  (f) mica granules

8.5 With the aid of a diagram explain how a typical thermosetting moulding powder is cured (polymerised) in the mould by the condensation of water molecules. Describe the precautions which have to be taken in the design of a mould to compensate for the curing process.

8.6 Contrast and compare the general properties of polymeric materials with those of the more common metals.

8.7 Explain why the following additives are used in association with polymeric materials:
  (a) plasticisers
  (b) stabilisers
  (c) antistatic agents
  (d) antioxidants

8.8 Compare the properties of the following thermoplastic materials and state a typical application for each:
  (a) polyethylene
  (b) polystyrene
  (c) unplasticised polyvinyl chloride
  (d) polymethyl methacrylate
  (e) cellulose acetate

8.9 Compare the properties of the following thermosets and state a typical application for each:
  (a) phenol–formaldehyde
  (b) urea–formaldehyde
  (c) epoxy resin
  (d) alkyd resin

8.10 Explain why the popular name 'plastic' is inappropriate for most polymeric materials.

9 Composite materials

The topic areas covered in this chapter are:

- Laminate materials.
- Fibre-reinforced materials.
- Particle-hardened and reinforced materials.
- Sintered materials including 'cermets'.
- 'Whiskers' and dispersion hardening.

9.1 Introduction

In Chapter 3 you were introduced to alloys. You were told that alloys of metals and metals with non-metals could only occur if all the component materials were *miscible* – that is, soluble in each other in the molten state. Composite materials can be made up from materials that are not soluble in each other. Composite materials are not alloys.

In its simplest form a composite material consists of two dissimilar materials in which one material forms a matrix to bond together the other (reinforcement) material. The matrix and reinforcement are chosen so that their mechanical properties complement each other, whilst their deficiencies are neutralised. For example, in a GRP moulding, the polyester resin is the matrix that binds together the glass fibre reinforcement. Some examples of composite materials were introduced in Section 1.6.

In a 'composite', the *reinforcement* material, in the form of rods, strands, fibres or particles, is bonded together with the other *matrix* material(s). For example, the fibres may have some of the highest moduli and greatest strengths available in tension, but little resistance to bending and compressive forces. On the other hand, the matrix can be chosen to have high resistance to bending and compressive forces. Used together these two different types of material produce a composite with high tensile and compressive strengths and a high resistance to bending.

9.2 Lamination

In Fig. 1.22 you were introduced to the composite structure of natural wood. This figure showed how the tubular cellulose fibres forming the 'reinforcement' are bonded together by the relatively weak lignin 'matrix'.

Brittle materials such as concrete and ceramics are strong in compression but weak in tension since they are susceptible to crack propagation, as shown in Fig. 9.1(a). The use of composite materials overcomes this problem by preventing the crack from 'running'. Two ways in which a crack can be prevented from running are:

- The use of lamination, as shown in Fig. 9.1(b).
- The use of reinforcing fibres which hold the brittle matrix in compression, so an external tensile load cannot open up any surface cracks and discontinuities (Fig. 9.1(c)).

Fig. 9.1 *Principles of reinforced composite materials: (a) crack propagation in a non-reinforced ceramic material; (b) behaviour of a laminated composite when in tension; (c) fibre reinforcement*

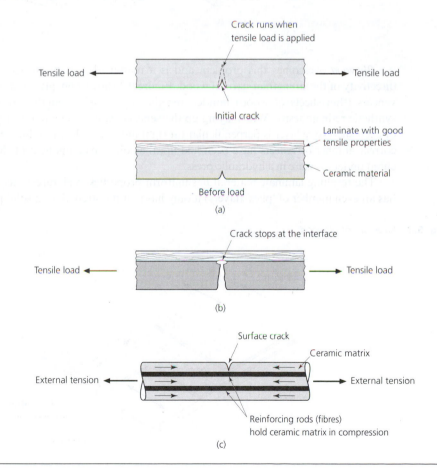

9.2.1 *Plywood*

Wood, like all materials reinforced by parallel fibres, has highly directional strength characteristics. Figure 9.2(a) shows a plank being bent in a plane *perpendicular* to the lay of the grain. Loading the plank in this way exploits its greatest strength characteristics. Figure 9.2(b) shows what would happen if a piece of wood is loaded so that bending occurs *parallel* to the lay of the grain. The wood breaks easily since the lignin bond is relatively weak compared with the natural cellulose reinforcement fibres.

Fig. 9.2 *Effect of the direction of grain on the strength of wood*

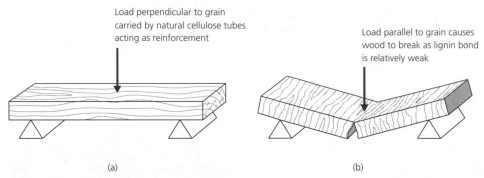

Load perpendicular to grain carried by natural cellulose tubes acting as reinforcement

Load parallel to grain causes wood to break as lignin bond is relatively weak

(a)

(b)

Plywood overcomes this problem and is a man-made composite which exploits the directivity of the strength of natural wood. Figure 9.3 shows how plywood is built up from veneers (thin sheets of wood) bonded together by a high-strength and water-resistant synthetic resin adhesive. When laying up the veneers, care is taken to ensure that the grain of each successive layer is perpendicular (at right angles) to the preceding layer. When the correct number of veneers have been laid up, the adhesive component of the composite is cured under pressure in a hydraulic press.

The resulting laminate material has uniform properties in all directions. However, if it has an even number of 'plies' (layers) it only has half the strength of a solid piece of timber

Fig. 9.3 *Structure of plywood*

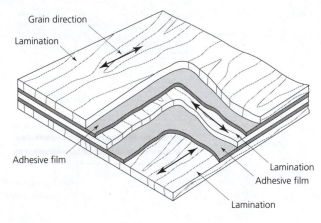

Grain direction

Lamination

Adhesive film

Lamination
Adhesive film

Lamination

when the load is applied uniaxially, as shown in Fig. 9.4. In practice an odd number of plies are used so that the grain in the outer laminations lie in the same direction.

Fig. 9.4 *Uniaxial loading of (a) solid timber (100% support for uniaxial load) and (b) plywood (50% support)*

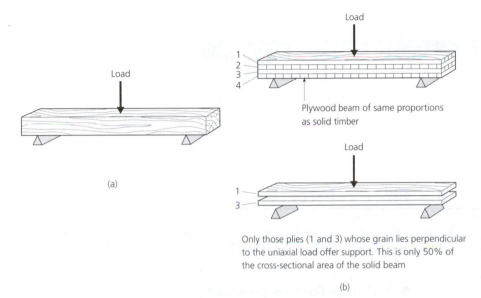

Plywood beam of same proportions as solid timber

Only those plies (1 and 3) whose grain lies perpendicular to the uniaxial load offer support. This is only 50% of the cross-sectional area of the solid beam

(b)

9.2.2 *Laminated plastic (Tufnol)*

Fibrous material such as paper, woven cotton and woollen cloth, woven glass fibre cloth, or woven asbestos cloth (not now recommended as it is a potential health hazard) can be used to reinforce phenolic and epoxy resins. Sheets of these fibrous reinforcement materials are impregnated with the resin and they are then laid up between highly polished metal plates in hydraulic presses, the thickness of the finished sheet being determined by the number of layers of impregnated reinforcement in the laminate. Each layer of reinforcement is rotated through 90 degrees of arc so as to ensure uniformity of mechanical properties. Although paper and cloth do not have the same uniaxial strength characteristics of the wood veneers of plywood, nevertheless cloth varies in strength significantly between the warp and the weft. Try tearing a thin piece of cloth (e.g. an old handkerchief) in two directions at right angles. It will tear more easily one way than the other. The laminates are then heated under pressure until they become solid sheets, rods or tubes.

Laminated composites can be machined dry with ordinary engineers' tools using a low value of rake angle and fairly high cutting speeds. However, the material is rather abrasive and carbide tooling can be used to advantage, particularly when machining this material on a production basis. Although not toxic, the dust and fumes generated during machining are unpleasant and adequate extraction should be provided. Care must be taken in the design and use of plastic laminates because of the 'grain' of the material which makes it behave in a similar manner to plywood.

'Tufnol' composites are widely used for making bearings, gears and other engineering products which have to operate in hostile environments or where adequate lubrication is often not possible, as in food processing and office machinery. In the electronic industry, copper clad sheet tufnol is used for making printed circuit boards. It is also used for making insulators for heavy-duty electrical equipment in the electrical power engineering industry.

SELF-ASSESSMENT TASK 9.1

1. Explain briefly what is meant by a 'composite' material.

2. Discuss the differences between composites and alloys.

3. Explain how lamination is used to improve the properties of some materials.

9.3 Fibre reinforcement

This method of reinforcement can range from the glass fibres used in GRP plastic mouldings to the thick steel rods used to reinforce concrete.

9.3.1 *Reinforced concrete*

Concrete itself is a particle-reinforced material. It consists of a mortar made from a hydraulic cement and sand, reinforced with an *aggregate* of chipped stones as shown in Fig. 9.5. The stones are crushed so that they have a rough texture and sharp corners that will *key* into and bond with the mortar matrix. This basic concrete has a very high compressive strength but is very weak in tension. To improve its performance overall, metal reinforcing rods are added, as shown in Fig. 9.6

As well as the amount of reinforcement, the positioning of the reinforcement is also important. Since concrete is strong in compression but weak in tension, it would seem sensible to insert the reinforcement near the tension side of the beam, as shown in Fig. 9.6. The tension side is the side of the beam that is on the outside of the 'bend' when the beam 'gives' under the effect of an externally applied load.

Fig. 9.5 *Structure of concrete*

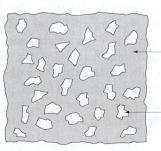

Cement and sand mortar acting as a matrix

'Aggregate' of stone chippings acting as a particle reinforcement

Fig. 9.6 *Simple reinforced concrete beam*

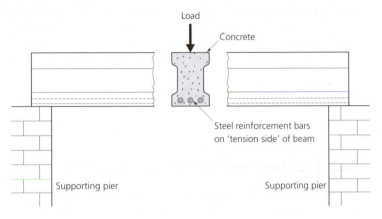

However, concrete beams often have to be craned into position because of their weight and this causes reverse stressing leading to fracture, as shown in Fig. 9.7. For this reason, a small amount of additional reinforcement is introduced on what is normally the compression side of the beam. This is also shown in Fig. 9.7.

Fig. 9.7 *If a simple reinforced concrete beam is lifted by a crane the stresses are reversed due to the beam's own weight; the reinforcement designed to support the applied load cannot support this reversed stress: the beam fractures and the reinforcing bars bend – beams likely to be craned into position have additional reinforcement as shown*

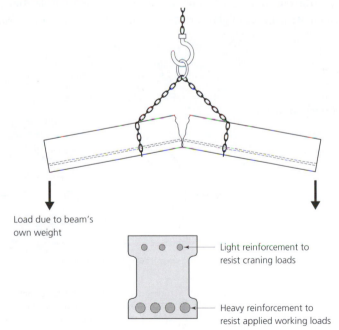

Unfortunately the concrete is more rigid than the reinforcement. The steel reinforcement can 'give' without cracking, but the concrete cannot. Therefore surface cracking can still occur. Although catastrophic failure is not likely to result directly from such surface cracking, the penetration of moisture can cause corrosion of the reinforcement and weakening of the structure. This is overcome by prestressing the reinforcement so that the concrete is always in compression under normal load conditions. Concrete and its reinforcement are dealt with more fully in *Engineering Materials*, Volume 2).

9.3.2 *Glass-reinforced plastic (GRP)*

This important composite material is produced when a plastic material, usually a polyester resin, is reinforced with glass fibre in strand or mat form. The resin is used to provide shape, colour and finish, whilst the glass fibres, which are laid in all directions, impart mechanical strength.

The amount of reinforcement which can be used depends upon the orientation of the reinforcement. With long strands laid up parallel to each other the reinforcement area fraction can be as high as 0.9. The reinforcement area fraction is the cross-sectional area of reinforcement divided by the total cross-sectional area, as shown in Fig. 9.8. With woven strand fabrics the reinforcement area fraction can be as high as 0.75; with chopped strand mat the reinforcement area fraction is substantially reduced but a figure of 0.5 should be considered the minimum for satisfactory reinforcement. Obviously the lower the reinforcement area fraction, the weaker the composite produced. Tables 9.1 and 9.2 lists the properties of some typical glass fibres. Example 9.1 shows how the 'reinforcement area fraction', tensile modulus and tensile strength of a composite can be calculated. Types of glass are dealt with more fully in *Engineering Materials*, Volume 2.

Fig. 9.8 *Reinforced composite*

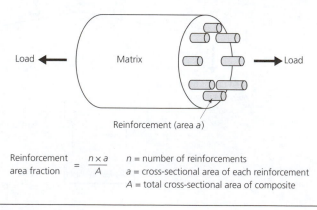

$$\text{Reinforcement area fraction} = \frac{n \times a}{A}$$

n = number of reinforcements
a = cross-sectional area of each reinforcement
A = total cross-sectional area of composite

Table 9.1 *Properties of typical glass fibres*

Steel wire is included for comparison

Material	Relative density	Tensile strength (GPa)	Tensile modulus (GPa)	Specific strength (GPa)*	Specific modulus (GPa)[†]
E-Glass[‡]	2.55	3.5	74	1.4	29
S-Glass	2.50	4.5	88	1.8	35
Steel wire	7.74	4.2	200	0.54	26

* Specific strength = (tensile strength)/(relative density)
[†] Specific modulus = (tensile modulus)/(relative density)
[‡] Typical composition for E-glass

Silicon dioxide	52–56%	Boron oxide	8–13%
Calcium oxide	16–25%	Sodium and potassium oxides	1%
Aluminium oxide	12–16 %	Magnesium oxide	0–6%

Table 9.2 *Properties of GRP composites*

Material	Reinforcement (weight %)	Tensile strength (MPa)	Tensile modulus (GPa)
Chopped strand mat	10–45	45–180	15–15
Plain weave cloth	45–65	250–375	10–20
Long fibres (uniaxially loaded)	55–80	500–1200	25–50

EXAMPLE 9.1

Given the following data for a fibre-glass-reinforced polyester component of rectangular cross-section in which the strands lie parallel to the direction of loading, calculate:
(a) The matrix area fraction and the reinforcement area fraction.
(b) The tensile modulus for the composite.
(c) The tensile strength of the composite.

Average fibre diameter	*0.005 mm*
Number of fibres per strand	*204*
Number of strands	*51,470*
Tensile modulus for polyester	*4 GPa*
Tensile modulus for glass fibre	*75 GPa*
Tensile strength of polyester	*50 MPa*
Tensile strength of glass fibre	*1500 Mpa*
Cross-sectional area of the component	*300 mm^2*

(a)   area of one filament of glass $= \dfrac{0.005^2 \times \pi}{4} = 0.000\,02\,\text{mm}^2$

∴   area of one basic strand $= 0.000\,02 \times 204 = 0.004\,08\,\text{mm}^2$

∴   total reinforcement area $= 0.004\,08 \times 51\,470 = 210\,\text{mm}^2$

$$\textbf{reinforcement area fraction} = \frac{\text{total reinforcement area}}{\text{total component area}}$$

$$= \frac{210\,\text{mm}^2}{300\,\text{mm}^2}$$

$$= \mathbf{0.7}$$

$$\textbf{matrix area} = (\text{total area}) - (\text{reinforcement area})$$

$$= 300\,\text{mm}^2 - 210\,\text{mm}^2$$

$$= 90\,\text{mm}^2$$

$$\textbf{matrix area fraction} = \frac{\text{total matrix area}}{\text{total component area}}$$

$$= \frac{90\,\text{mm}^2}{300\,\text{mm}^2}$$

$$= \mathbf{0.3}$$

(b)   **modulus of composite** $=$ (modulus of matrix × modulus area fraction)
$+$ (modulus of reinforcement × reinforcement area fraction)

$$= (4\,\text{GPa} \times 0.3) + (75\,\text{GPa} \times 0.7)$$

$$= \mathbf{53.7\,GPa}$$

(c)   **strength of composite** $=$ (tensile strength of matrix × matrix AF)
$+$ (tensile strength of reinforcement × reinforcement AF)

$$= (50\,\text{MPa} \times 0.3) + (1500\,\text{MPa} \times 0.7)$$

$$= 15\,\text{MPa} + 1050\,\text{MPa}$$

$$= \mathbf{1065\,MPa}$$

Example 9.1 indicates the importance of the reinforcement in any composite material when the contribution of the reinforcement is compared with the contribution of the matrix. If the calculation is repeated for a range of reinforcement area fractions, the curves shown in Fig. 9.9 will be obtained. Figure 9.9(a) plots the tensile modulus against reinforcement area fraction, and Fig. 9.9(b) plots the tensile strength of the composite against the reinforcement area fraction.

When a GRP moulding is correctly 'laid up', the thick, viscous resin completely surrounds and adheres to the glass fibres of the 'chopped strand mat' or the woven fibre glass cloth. The resin sets to give a hard, rigid structure. To shorten the setting time a catalyst (accelerator) is often added to the resin immediately before moulding. To improve the appearance of the 'rough side' of the moulding (the side not in contact with the polished surface of the mould), a surfacing tissue is added made up from finer strands of mat, as shown in Fig. 9.10. Glass-reinforced plastic is widely used for products that require high strength to weight ratios such as boat hulls, car bodies, gliders, etc. It has the advantages of being resistant to most environments and it can be used to form more complex shapes than is possible in wood or sheet metal.

Fig. 9.9 *Effect of reinforcement area on a typical GRP composite*

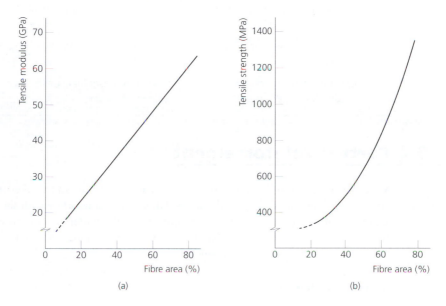

(a)   (b)

Fig. 9.10 *Composition of a GRP lay up*

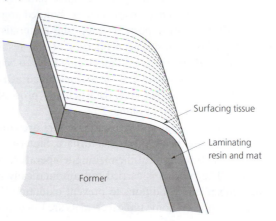

Surfacing tissue

Laminating
resin and mat

Former

9.3.3 *Carbon fibres*

These have a higher elastic modulus and lower density than glass fibres, and can be used to reinforce composite materials having a higher strength to weight ratio. Carbon fibres are used as a reinforcement in polymeric materials for a wide range of lightweight, high-strength applications. For example, carbon fibres are used in the manufactures of gas turbine fan blades, racing car body panels, high-performance tennis racket frames and high-performance golf club shafts.

SELF-ASSESSMENT TASK 9.2

1. Briefly explain the principles of fibre reinforcement.
2. Explain the importance of 'area fraction' and how this influences the properties of a fibre-reinforced composite.
3. Explain why the reinforcing rods in concrete are found mainly on the tension side of a cast beam.
4. List the main ingredients of a GRP moulding and briefly explain how a moulding is made. You may use sketches to illustrate your answer.

9.4 Particle reinforcement

We have already seen that high-conductivity copper is softer and weaker than tough pitch copper and how the oxide particles in the latter material interfere with the flow of the slip bands, so increasing the hardness and strength of the material. This is, in fact, a simple form of particle reinforcement. Similarly, the stone chippings in concrete also provide a primitive form of particle reinforcement.

9.4.1 *Cermets*

These are materials which combine the ductility and toughness of a metal matrix with the hardness and compressive strength of ceramic particles to provide *particle reinforcement*. Certain metal oxides and carbides can be bonded together and 'sintered' into a metal powder matrix to form important composite materials called *cermets*. This is short for *ceramic* reinforced *metals*. The properties of the resulting composite materials enable extremely hard and abrasion-resistant cutting tools to be produced. Such tools enable hard, tough and high-strength materials to be machined at production rates and with surface finishes not previously possible. (Particulate processes are dealt with in detail in *Manufacturing Technology*, Volume 2.)

The mixture of metal and ceramic powders are compressed in dies to form 'powder compacts'. These are then *sintered* in a furnace. Sintering consists of heating the compact to between 60 and 90 per cent of the melting temperature of the majority material present in the compact. The process is carried out automatically in a conveyor type furnace under reducing atmospheric conditions to prevent oxidation. The particles form 'weld necks' at their points of contact. Some typical cermets are listed in Table 9.3.

Table 9.3 *Typical cermets*

Type	Ceramic particles	Metal matrix	Applications
Borides	Titanium boride Molybdenum boride Chromium boride	Cobalt/nickel Chromium/nickel Nickel	Mostly cutting tool tips
Carbides	Tungsten carbide Titanium carbide Molybdenum carbide Silicon carbide	Cobalt Cobalt or tungsten Cobalt Cobalt	Mostly cutting tool tips and abrasives
Oxides	Aluminium oxide Magnesium oxide Chromium oxide	Cobalt/chromium Cobalt/nickel Chromium	Disposable tool tips; refractory sintered components, e.g. spark plug bodies, rocket and jet engine parts

Although not particle reinforcement, it is worth mentioning at this point that particle processes can be used to form composite alloys between metals that are partially or totally immiscible (won't form the solutions necessary for the formation of true alloys). Figure 9.11(a) shows the formation of a cermet cutting tool material by powder particulate processing, whilst Fig. 9.11(b) shows how an alloy system can be made by dispersion between materials that are normally immiscible.

Fig. 9.11 *Sintered particle systems: (a) metal–metal compound; (b) alloy formed by dispersion*

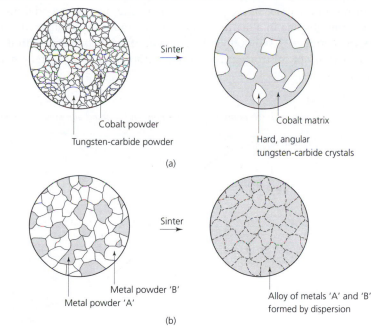

Sinter

Cobalt powder

Tungsten-carbide powder

Cobalt matrix

Hard, angular tungsten-carbide crystals

(a)

Sinter

Metal powder 'B'

Metal powder 'A'

Alloy of metals 'A' and 'B' formed by dispersion

(b)

9.4.2 'Whiskers' and dispersion hardening

'Whiskers' are high aspect ratio (hair like) crystals grown with very high strengths. They may have a diameter of only 0.5 to 2.0 microns, yet have a length of up to 20 millimetres. The properties of some whiskers are shown in Table 9.4. The boron and carbon whiskers are used to harden and reinforce polymeric materials, whilst whiskers of alumina are used to harden and reinforce the metal nickel to produce materials that retain their strength at high temperatures.

Table 9.4 *Properties of some reinforcement 'whiskers'*

Material	Relative density	Tensile strength (GPa)	Tensile modulus (GPa)	Specific strength* (GPa)	Specific modulus† (GPa)
Alumina	3.96	21.0	430	5.3	110
Boron carbide	2.52	14.0	490	5.6	190
Carbon (graphite)	1.66	20.0	710	12.0	430

* Specific strength = (tensile strength)/(relative density)
† Specific modulus = (tensile modulus)/(relative density)

Alumina (aluminium oxide) can also be used in spheroidal particle form to increase the strength of composites rather than their hardness. Aluminium metal is ground into a fine powder in an oxygen-rich atmosphere. This causes aluminium oxide (alumina) to form on the surface of the particles. After sintering, the material produced consists of 6 per cent alumina in a matrix of aluminium, and is called sintered aluminium powder or SAP. Figure 9.12 shows how the strength of SAP compares with duralumin. Although the SAP is not so strong at room temperature, it retains its strength at very much higher temperatures.

Fig. 9.12 *Comparative effects of temperature on duralumin and sintered aluminium powder*

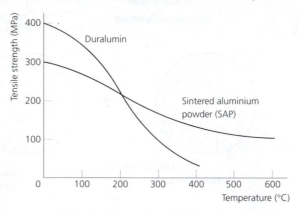

9.4.3 *Rubber*

Naturally occurring rubber (latex) comes from the sap of the tree called *Hervea Basiliensis*. This material is of little use to engineers since it becomes tacky at elevated temperatures, has a very low tensile strength at room temperature, and is attacked by oils and quickly perishes in the presence of the ultraviolet rays of sunlight.

When compounded with other substances and vulcanised, it changes to become a useful and versatile material. The reinforcing filler material is *carbon black*. The latex and the carbon black are mixed together with small amounts of chemicals such as sulphur which speed up the vulcanising process and extend the working life of the product by improving its ageing characteristics. Although rubber is a naturally occurring elastomer, this term is mainly used for the synthetic rubbers that are an important sector of the polymer industry. Typical products made from natural rubber are vehicle tyres, inner tubes, footwear, respirator masks and gloves. Reclaimed rubber is widely used for lightly stressed applications where strength is not of overall importance. The use of carbon black to improve the properties of the rubber is, again, an example of a particle-reinforced composite material.

SELF-ASSESSMENT TASK 9.3

1. Explain what is meant by particle reinforcement.

2. Explain what is meant by the term 'cermet'.

3. Briefly discuss the principle of dispersion hardening.

4. Explain why an SAP is preferable to duralumin when operating at temperatures approaching 600 °C.

EXERCISES

9.1   Explain how lamination can be used to control the spread of cracks in brittle materials.

9.2   Describe how the laminated plastic called 'Tufnol' is manufactured, and list **four** typical applications, giving reasons for your choice.

9.3   (a)   Explain why plywoods normally have an odd number of plies.
      (b)   Explain how the properties of plywood compare with solid timber of the same thickness.

9.4   Select a suitable glass fibre and a suitable resin bond for moulding a small boat hull. Describe the stages of laying-up such a moulding.

9.5   (a)   Explain how concrete can be reinforced with steel rods and wire mesh.
      (b)   Describe what is meant by the prestressing of reinforcement and why it is required.

9.6   Given the data listed below, calculate:
   (a)   the matrix area fraction
   (b)   the reinforcement area fraction
   (c)   the tensile modulus for the composite
   (d)   the tensile strength for the composite

Average fibre diameter	0.005 mm
Number of fibres per strand	200
Number of strands	4800
Tensile modulus of polyester resin	3.8 GPa
Tensile modulus of glass fibre	70 GPa
Strength of polyester resin	50 MPa
Strength of glass fibre	1450 MPa
Diameter of component	25 mm

9.7   Explain what is meant by 'particle reinforcement' and state how the particles increase the strength and hardness of the composite.

9.8   Discuss the advantages and limitations of 'cermets' as cutting tool materials.

9.9   Explain how sintered metallic materials can be produced that cannot exist as true alloys.

9.10   (a)   Explain what is meant by the term 'whisker'.
   (b)   Explain how whiskers can be used to reinforce materials.

10 Shaping and joining materials

The topic areas covered in this chapter are:

- Casting, shrinkage and machining allowance, casting defects.
- Plastic deformation.
- Cold working of metals and the heat treatment of cold-worked metals.
- Hot working of metals.
- Hot- and cold-working processes.
- Moulding polymeric materials.
- Compression joints (mechanical); compression joints (thermal).
- Soft soldering, hard soldering, brazing, solders, spelters and fluxes.
- Fusion welding, oxyacetylene welding, manual metallic-arc welding.
- Plastic welding.
- Adhesive bonding.

10.1 Casting

Long before mankind had the knowledge and tools to forge metals to shape, small metal objects were being cast. Casting is still one of the most versatile and widely used processes for shaping metal. The metal is formed directly from the molten state by pouring it into a mould and allowing it to cool and solidify.

- The mould must be made from a material with a higher melting point than that of the molten metal from which the casting is to be made.
- The mould contains a cavity in the form of the finished product into which the molten metal is poured.
- The form of this cavity is determined by ramming sand round a wooden or GRP pattern.
- The pattern is the same shape as the finished casting but *slightly larger* to allow for *shrinkage*.
- After ramming, the mould is opened so that the pattern can be removed from the cavity.
- The mould is then reassembled, ready for pouring.

Let's now see how a simple sand mould is made. One of the most widely used materials for making moulds is moulding sand, and sand moulds are very widely used for most high-melting point materials such as cast iron. The moulds can be made individually by hand where small quantities are required, or by semi- and fully automated processes where they are required for quantity production. Figure 10.1 shows a section through a typical two-part sand mould.

Fig. 10.1 *Two-part sand mould: (a) half pattern in position; (b) complete mould, pattern removed*

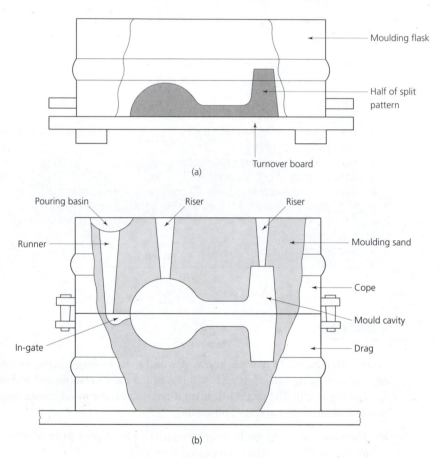

In this example, a *split pattern* is being used. Half the pattern is placed on the *turnover board* as shown in Fig. 10.1(a). The bottom half of a *moulding box* (called the 'drag or flask') is placed around the half pattern. It is also supported on the turnover board. A *parting powder* is sprinkled over the pattern and board to stop the sand sticking to them. The special *moulding sand* is then rammed into the moulding box. This is a very skilled operation requiring a lot of experience. If the sand is rammed too hard it will not vent properly. If it is not rammed hard enough the mould will collapse as the metal is poured in. The sand is smoothed off level with the top of the moulding box.

The turnover board and competed drag are turned over onto a flat surface and the turnover board is removed. The second half of the pattern is placed in position. Dowel pegs protruding from the second half of the pattern locate in holes in the first half of the pattern to ensure alignment.

The top part of the mould box (called the 'cope') is placed in position. It is located by taper pegs in the lugs at the ends of the moulding boxes. Parting powder is again applied, and the sand is rammed in as previously. This time tapered pegs are positioned in the sand to form the runners and risers.

When the mould is complete the cope and drag are carefully separated and the two halves of the pattern are removed. The pegs are also removed. Whilst the mould is open, the in-gate is cut in the sand using a special trowel and the pouring basin is cut in the cope, also with a trowel. Any loose sand is gently blown away and the cope and drag are reassembled ready for pouring. The finished mould is shown in Fig. 10.1(b).

The molten metal is poured from a ladle into the runner and the air displaced from the cavity by the molten metal escapes through the risers. There must be a riser above each high point of the cavity to prevent air locks. Pouring continues until the molten metal is seen at the top of each riser. This ensures that the mould cavity is full. It also provides surplus metal which can be drawn back into the mould as shrinkage takes place during cooling. This avoids shrinkage cavities occurring in the casting.

The mould must also contain vents. These are fine holes made with a wire after the mould is complete. The holes stop just short of the mould cavity. The purpose of the vents is to release steam and other gases which are generated when the hot metal comes into contact with the moist moulding sand. (Moisture is required in moulding sand so that it will bind together and keep its shape.) If the mould was not vented, then the release of steam and gases would cause bubbles to collect in the casting, causing 'blow holes' and 'porosity'. When the metal has solidified to form the casting the mould is broken open and the casting is removed. The runners and risers are cut off and the casting is ready for machining. The runners and risers are not wasted, they can be melted down and re-used.

10.2 Shrinkage and machining allowance

It has already been stated that the pattern has to be made oversize to allow for shrinkage of the metal as it cools. This is called the *shrinkage allowance*. Table 10.1 lists some common metals and the magnitude of the shrinkage allowance. The pattern maker does not have to calculate this allowance but uses a special rule which is already engraved oversize. This is called a *contraction rule*. A different rule or scale has to be used for each type of metal being cast. In addition to shrinkage allowance, the pattern must also be made oversize wherever the casting is to be machined – that is, a *machining allowance* has to be superimposed on top of the shrinkage allowance. Not only must sufficient additional metal be provided to ensure that the casting 'cleans up', but sufficient metal must be provided to ensure that the tip of the cutting tool operates well below the hard and abrasive skin at the surface of the casting. This was shown in Fig. 2.17.

Table 10.1 *Shrinkage allowance*	
Material	Shrinkage allowance
Aluminium	21.3 mm/m
Brass	16.0 mm/m
Cast iron	10.5 mm/m
Steel	16.0 mm/m

10.3 Casting defects

The mechanism by which metals solidify and the effect this has upon their grain structures and properties has already been discussed in Sections 2.11 and 2.12. The practical defects which occur in the foundry can be listed as follows.

10.3.1 *Blow holes*

These are smooth round holes with a shiny surface usually occurring just below the surface of the casting. Because they are not normally visible until the casting is machined, their presence can mean scrapping a casting on which costly machining has already been carried out. Blow holes are caused by steam and gases being trapped in the mould (Section 10.1). This may result from inadequate venting, incorrectly placed risers, excessive moisture in the sand or excessive ramming reducing the *permeability* of the sand. Permeability is the ability of the sand to allow entrapped gases to escape between the individual sand particles.

10.3.2 *Porosity*

This is also caused by inadequate venting, but in this instance the trapped gases do not form large bubbles near the surface of the casting. Instead, a mass of pinpoint bubbles are spread throughout the casting rendering it porous and 'spongy'. Inadequate 'degassing' of the metal immediately prior to pouring is a frequent cause of porosity. Apart from being a source of weakness, porosity renders castings useless where pressure tightness is required, as in fluid valve bodies and pipe fittings.

10.3.3 *Scabs*

These are blemishes on the surface of the casting resulting from sand breaking away from the wall of the mould cavity. This may be due to lack of cohesiveness in the sand resulting from too low a clay content or from inadequate ramming. Too rapid pouring can also result in the scouring away of the walls of the mould cavity.

10.3.4 *Uneven wall thickness*

This can occur in hollow castings and may be due to two causes, either individually or in combination. Figure 10.2 shows a mould with the core in position. The core is located in the 'prints' left by suitable projections on the pattern.

- *Misplaced cores* These are caused by the moulder not assembling the core correctly in the mould. Ill-fitting or inadequate core prints can also allow the core to move.

- *Displaced cores* These are caused by the buoyancy of the sand in the molten metal. The core will tend to float upwards. Although trapped in the core prints, the core may bow upwards if it is long and slender and left unsupported. This can be prevented by the use of *chaplets*, as shown in Fig. 10.3, and *core irons*. The latter are metal reinforcements around which the core may be built.

Fig. 10.2 *Use of cores*

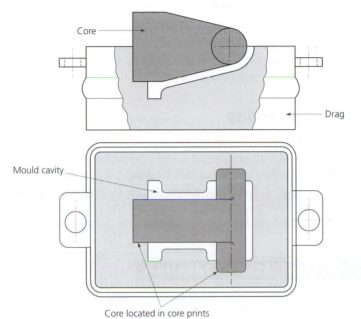

Fig. 10.3 *Use of chaplets to support a core*

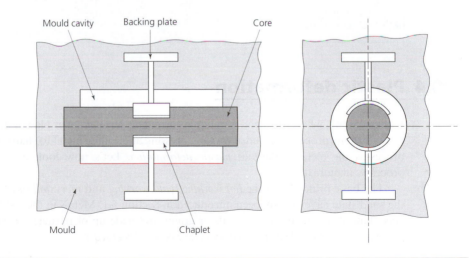

10.3.5 *Fins*

These are caused by badly fitting mould parts or badly fitting cores which do not fit into the core prints snugly and allow a thin layer of metal to escape and solidify leaving a projection along the joint line. Fins can be removed by 'fettling' after casting, but this is an extra and unnecessary expense and detracts from the appearance of the casting.

10.3.6 *Cold shuts*

These usually result from casting intricate components with thin sections from metal which is lacking in fluidity or is at too low a temperature. Consequently, sections of the mould may not fill completely or the metal may flow too sluggishly and at too low a temperature to unite when separate streams meet.

10.3.7 *Drawing*

This results from lack of risers, or their incorrect positioning, so that thick sections are not adequately fed with extra metal as they cool and contract. Further, since thick sections cool slowly and solidify last, metal may be drawn from them to feed other parts of the casting. This will result in holes and hollows in the casting which are not only unsightly but are sources of weakness and the casting may not clean up when machined. Sufficient and correctly placed risers feed metal back into the casting as it solidifies and prevents drawing.

SELF-ASSESSMENT TASK 10.1

1. Explain briefly how metals are cast to shape and why the pattern has to be made larger in size than the finished casting.

2. Explain the meaning of the following terms related to casting:

   (a) blow holes
   (b) porosity
   (c) scabs
   (d) cold shuts

10.4 Plastic deformation

Unlike casting, where the metal is in liquid form, *flow forming* takes place in the *solid state*. Flow-forming processes included forging, pressing and rolling metal to shape. During these processes the metal undergoes *plastic deformation*. Let's now look at some of these processes in more detail.

The basic principles of the *hot working, cold working* and *recrystallisation* of metals in the solid state have already been introduced in Section 5.1. Metals in the solid state have a crystalline structure and the crystals or grains are made up of particles arranged in strict geometric patterns called space lattices (see Section 2.8 *et seq.*).

It is due to the strict geometric symmetry of these space lattices that plastic deformation can occur in the solid state. When plastic deformation occurs during hot working or cold working, planes of atoms slip past each other, as shown in Fig. 10.4(a). These planes of movement are called *slip planes*. Usually, slip planes lie between, and parallel to, the planes of greatest atomic density, as shown in Fig. 10.4(b). When a ductile or a malleable crystalline material is subjected to an applied force of sufficient magnitude, movement of the lattice structure can occur along the *slip planes* within a crystal, as shown in Fig. 10.5.

Fig. 10.4 *Slip planes: (a) slip between planes of atoms under an applied load; (b) some possible orientations*

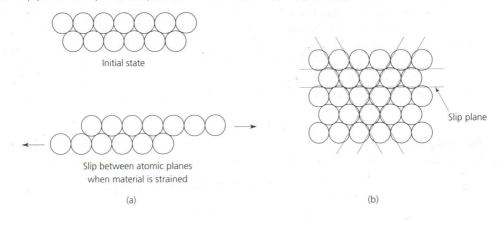

Fig. 10.5 *Formation of slip bands during plastic deformation: (a) before the application of a force; (b) subsequent movement along slip planes*

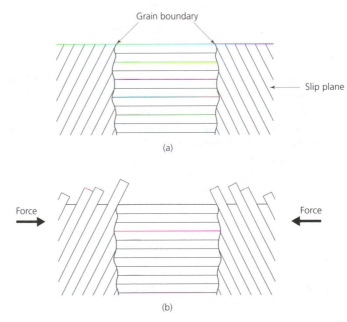

Movement does not occur in all the slip planes available in the material, but only in those planes which are at a suitable angle to the applied force. Slip can only occur where the grain is not constrained, and slip cannot cross grain boundaries. Thus slip can only occur within a grain; the bigger the grain, the greater the number of slip planes available, and the greater the amount of slip which can take place.

This is borne out in practice, since fine-grain materials are generally less ductile and malleable than the same materials after processing by heat treatment to increase the grain size (see Fig. 10.6). Thus a material which has been fully annealed can undergo greater plastic flow than when the same material has only been normalised. Some caution must be observed in developing this argument. Increased grain growth usually improves malleability, but this may not always be true for ductility. Ductility is plastic flow caused by the application of a tensile load. Therefore, although extreme grain growth may improve the plastic properties of the material, it will reduce the ability of that material to withstand the tensile loading which is causing the plastic flow. Further, the grain growth encouraged during annealing has to be controlled to provide a compromise between plasticity and strength where ductile deformation is required.

Fig. 10.6 *'Carpet' analogy*

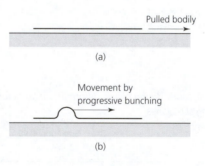

Initially, it was thought that slip occurred as the result of one block of atoms moving simultaneously over another, rather like sliding a carpet bodily along a floor, as shown in Fig. 10.6(a). However, calculations on the strength of metals proved that this was not true, and that slip occurs through a system of *dislocations* rather like moving a carpet a little at a time, by bunching it, as shown in Fig. 10.6(b), and moving the 'bunch' along progressively. Similarly, slip can be assumed to take place progressively by the movement of a dislocation along the slip planes.

This dislocation is a misfit in the geometry of the crystal lattice and it separates the region that has been subjected to slip from regions where slip has not occurred. In this manner, the dislocation can travel across the crystal with little applied force (far less that would be required to cause the slip of a block of atoms simultaneously). When the dislocation has travelled across the crystal, the whole plane has slipped one atomic distance (approximately 4×10^{-10} mm). This is obviously a very small distance, and a measurable plastic strain involves the movement of a large number of dislocations. The mechanism of a dislocation is shown in Fig. 10.7.

Fig. 10.7 *Principle of dislocation: (a) unstressed crystal lattice; (b) elastic deformation – lattice is distorted but slip has not yet occurred; (c) plastic deformation – dislocation commences at ⊥; (d) and (e) dislocation moves across lattice; (f) deformation complete – dislocation moves out of the crystal to form a slip band*

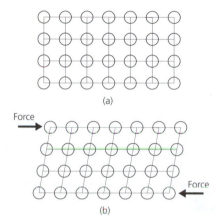

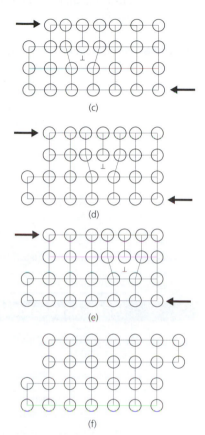

- Figure 10.7(a) shows the basic crystal lattice in the unstressed condition.
- Figure 10.7(b) shows elastic deformation where the lattice is distorted but not yet dislocated.
- Figure 10.7(c) shows the start of plastic deformation at the point marked ⊥ resulting from an increase in the applied force.
- Figures 10.7(d) and (e) show the dislocation moving progressively across the crystal.
- Figure 10.7(f) shows the dislocation complete.

Another mechanism by which deformation can take place is twinning. The principle of deformation by twinning is shown in Fig. 10.8(a). Unlike slip, where all the atoms in a block move the same distance, in twinning each successive plane of atoms in a block moves a different distance. When twinning is complete, the deformation of the crystal lattice will result in one half of the twin becoming the mirror image of the other half, as shown in Fig. 10.8(b). Like slip, twinning proceeds by a series of dislocations. The force required to cause twinning is generally greater than the force to cause slip. Twinning is not as common as block slip and occurs mainly when metals are shock loaded at low temperatures (e.g. cold-heading rivets).

Fig. 10.8 *Principle of twinning: (a) crystal lattice prior to deformation; (b) crystal lattice after deformation*

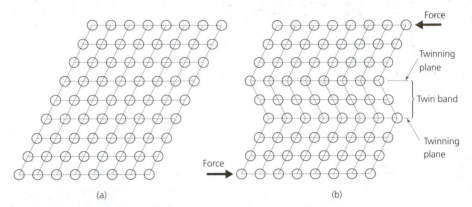

(a)   (b)

SELF-ASSESSMENT TASK 10.2

1. State the property exploited when flow-forming metals, giving the reason for your choice.

2. Compare the differences between forming metals by

    (a) cold working
    (b) hot working

3. With reference to the flow forming of metals, explain what is meant by the terms:

    (a) slip planes
    (b) slip bands
    (c) dislocation

10.5 Cold working

Figure 10.9(a) shows the effect of drawing a metal rod through a die so as to reduce its diameter and increase its length. As it passes through the die, the metal undergoes severe plastic deformation and the equi-axed crystals, typical of a metal in the annealed condition, become elongated and distorted. In cold working this distortion occurs below the temperature of recrystallisation (see Section 5.1) and the crystals remain in the distorted condition. This affects the properties of the metal as follows.

Fig. 10.9 *Some effects of cold working metals: (a) effects of cold working on crystal structure; (b) distorted slip planes in a cold-worked metal; (c) mechanism of crack propagation; (d) effect of cold working on electrical conductivity*

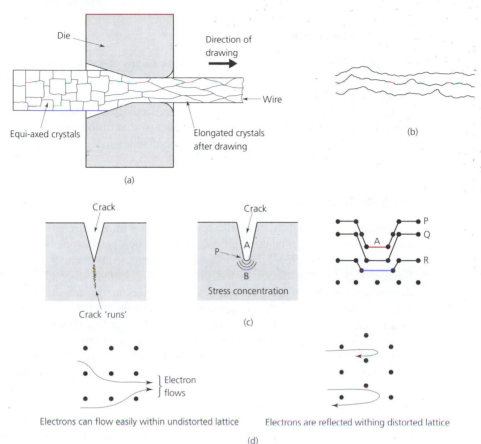

10.5.1 *Strength*

The tensile strength of the metal is increased. This is because metals only fracture by movement of the lattice along the slip planes. In the case of metals which have undergone cold working, the slip planes have become distorted ('rough') as shown in Fig. 10.9(b), and movement between adjacent, distorted slip planes is like the movement between two sheets of sandpaper. In this case, the planes cannot easily slip over one another so fracture becomes more difficult and the metal becomes stronger.

10.5.2 *Hardness*

Softness in a metal is the result of slip taking place easily with the application of only low values of applied force. Therefore, the metal becomes harder if slip becomes more difficult and requires a greater applied force to make it occur. It follows that the distortion of the crystals during cold working which made slip more difficult and the metal stronger, also makes the metal harder. Hence, hardness and strength are interrelated (see Section 11.11).

10.5.3 *Elasticity*

Reducing the ease with which slip can take place results in greater distortion of the crystal lattice before dislocation can occur. This raises the yield point of the metal and results in a longer elastic range (see Section 11.3). Under these conditions any deformation will be elastic, and plastic deformation cannot occur until the applied load is sufficient to overcome the 'roughness' between the slip planes so that dislocation can occur. Thus, not only is the elasticity of the material increased, but its ductility and malleability are correspondingly reduced.

10.5.4 *Toughness*

Distortion of the crystal structure also results in loss of impact strength (toughness). Again this is because slip is difficult to produce when friction between the slip planes has been increased by cold working. Lack of toughness (or increased brittleness) is aggravated by the behaviour of the slip planes in the presence of a surface discontinuity. If a surface discontinuity (such as a sharp corner) or a crack is present, slip cannot take place so easily. This is because it is difficult for the essential dislocations to flow around the apex of the discontinuity, resulting in the discontinuity or crack spreading (running) and the material showing a loss of toughness and a corresponding increase in brittleness. This is why impact (toughness) test specimens are usually notched. Figure 10.9(c) shows the mechanism of crack propagation in a brittle metal. It makes no difference whether the brittleness has been imparted by crystal distortion during cold working or whether the brittleness has been imparted by heat treatment (quench hardening).

At the base of the discontinuity the stress will be large when the metal is in tension; that is, there will be a stress concentration at the point P. At the base of the discontinuity, the plane P tries to slip over the plane Q and prevent the discontinuity from running and spreading. If slip cannot easily take place, then the stress at P will cause the intermolecular bonds at Q to break and transfer the stress to the plane R, and so on. Thus the crack spreads and the metal is said to be brittle.

10.5.5 *Machinability*

This is improved by cold working since the metal becomes stiffer, resulting in a cleaner shear at the point of cutting and a correspondingly improved surface finish.

10.5.6 *Electrical conductivity*

Cold-worked metals have a lower electrical conductivity than annealed metals. This is because cold working distorts the whole crystal lattice and makes it more difficult for electron flow to occur, as shown in Fig. 10.9(d).

10.6 The heat treatment of cold-worked metals

If the working of the metal is excessive it will work harden to such an extent that its increased brittleness will result in the metal breaking under reduced load conditions. This is precisely what happens during a tensile test to destruction (see Section 11.2). The properties imparted to the metal by cold working will remain until the structure of the metal is changed by heat treatment.

The heat-treatment process will depend upon the composition of the material and the severity of the cold working it has received. The process annealing of plain carbon steels has been described in Sections 5.3 and 5.4, and the solution treatment of the heat-treatable non-ferrous alloys was described in Section 7.5. These processes are not appropriate to pure metals such as aluminium and copper and non-ferrous alloys such as brass, which do not respond to solution and precipitation treatments. Let's now look at some of the heat-treatment processes appropriate to cold-worked metals

10.6.1 *Recovery*

The minimum heat treatment for a cold-worked metal is simple stress relief. This occurs at quite low temperatures, particularly for non-ferrous metals and alloys. There is no change in grain structure, but individual atoms move within the crystal lattice. This effect is called *recovery* and occurs at a temperature of about one-third the melting point of the metal when calculated on the *Kelvin* temperature scale (see Example 10.1).

EXAMPLE 10.1

Estimate the recovery temperature for a mild steel pressing if its melting temperature is taken as 1520 °C.

$$\text{melting point for steel (given)} = 1520\,°C$$
$$\text{or} \qquad\qquad\qquad = (1520 + 273)\,K$$
$$= 1793\,K$$

$$\therefore \quad \text{recovery temperature} = 1793 \times 0.33$$
$$= 598\,K\ (\text{approx.})$$
$$\text{or} \qquad\qquad\qquad = (598 - 273)\,°C$$
$$= \mathbf{325\,°C}$$

Treatment at the recovery temperature does not adversely affect the hardness and strength resulting from cold working, and may even enhance these properties in some instances as shown in Fig. 10.10. Stress relief annealing of 70/30 brass after it has been severely cold worked prevents 'season cracking' (see Section 13.10.3) occurring later in its service life.

Fig. 10.10 *Effect of heat treatment on cold-worked 70/30 brass*

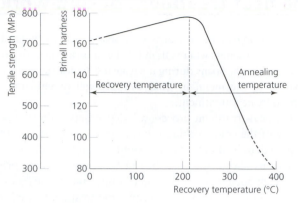

10.6.2 *Recrystallisation*

We considered the recrystallisation of metals earlier in Section 5.1. If cold-worked metal is to undergo further plastic deformation, then it must be heat treated to restore the grain structure to the annealed condition by recrystallisation. The recrystallisation temperature depends largely upon the degree of cold working which the material has undergone prior to heat treatment. The more severe the cold working, the lower will be the recrystallisation temperature. The recrystallisation temperature is generally lower for pure metals than it is for alloys. For example, it was stated in Chapter 7 that the addition of 0.5 per cent of the metal arsenic to copper produced an alloy which would sustain its strength and hardness at elevated temperatures. This is due to the presence of arsenic in the alloy raising the recrystallisation temperature from about 200 °C for pure copper to 500 °C for the arsenical copper alloy.

Just as the recovery temperature can be approximated as one-third the melting temperature as calculated on the Kelvin scale, so the recrystallisation temperature can be approximated as one-half the melting temperature of the metal or alloy as calculated on the Kelvin scale. Some metals, such as copper, recrystallise at a slightly lower temperature since they can sustain very severe cold working and crystal deformation prior to heat treatment (see Example 10.2).

EXAMPLE 10.2

Estimate the recrystallisation temperature for a mild steel pressing if its melting temperature is taken as 1520 °C.

melting point for steel (given) = 1520 °C
or   = (1520 + 273) K
                            = 1793 K

$\therefore$ recrystallisation temperature = 1793 × 0.5
                            = 896.5 K
or   = (896.5 − 273) °C
                            = **623.5 °C**

Reference back to Section 5.3 for subcritically annealing mild steel shows that this approximation gives an acceptable recrystallisation temperature. Some metals, such as lead and tin, recrystallise at temperatures below room temperature and it is virtually impossible to cold work such metals under normal working conditions.

10.6.3 *Grain growth*

Finally, the effect of crystal structure on grain growth must be considered. Figure 10.11 shows the relationship between grain size after recrystallisation annealing and the amount of deformation received prior to annealing. If the metal is only lightly worked (5 to 10 per cent reduction) the crystals will be, correspondingly, only slightly distorted with few stress points. Therefore, upon annealing, few nuclei will be created and these will have room to grow into large, coarse grains. This coarse-grain structure leads to loss of strength and the formation of an 'orange peel' surface. Therefore, if the metal is only lightly worked, heat treatment should be restricted to 'recovery' treatment.

Recrystallisation annealing should be reserved for severely cold-worked metals if a fine grain is to be achieved. It must be remembered that, for most metals and alloys – other than the ferrous metals, the heat-treatable aluminium alloys and some aluminium bronzes – cold working is the only way in which the metals may be hardened and that the subsequent recrystallisation is the only way in which grain refinement can be achieved.

Fig. 10.11 *Relationship between grain size and deformation*

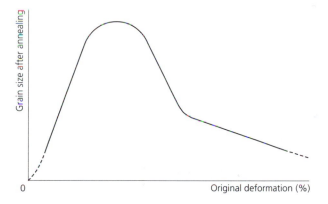

10.7 Hot working

This has already been described as the forming of metals by plastic flow above the temperature of recrystallisation. If the temperature of the metal being worked is sufficiently high, recrystallisation takes place as quickly as the crystals become deformed and the metals can be heavily worked with ease without risk of cracking. As the temperature falls during processing, recrystallisation occurs more slowly. Not only is more force required to achieve plastic deformation, but there is an increased risk of surface cracks appearing.

A hot-working process should be matched to the heating and cooling cycle of the component so that the process is completed at a temperature sufficiently above its recrystallisation temperature to avoid cracking, but not so far above the recrystallisation temperature that grain growth occurs. Care must be taken not to raise the temperature of the component too high initially so that, in the case of non-ferrous metals, they are melted or, in the case of ferrous metals, they become 'burnt' – that is, oxidation of the grain boundaries occurs so that the material is severely weakened and the component must be scrapped. It may be necessary to reheat the component during sustained hot working.

If a hot-worked component such as a forging is cut in half and etched so that its grain structure becomes visible, then it is apparent that after hot working the grain flow follows the profile of the component, as shown in Fig. 10.12. This figure compares a gear blank machined from a bar with one that has been forged to shape. Metals break more easily along the grain flow than across it (just like wood). So you will see that in Fig. 10.12(a) the teeth cut in forged gear blank will be much stronger than those cut in the blank which has been machined from a bar, as shown in Fig.10.12(b).

Fig. 10.12 *Grain orientation: (a) machined from bar (b) machined from forging*

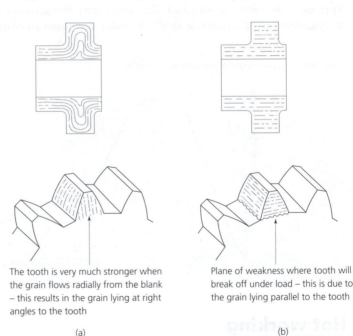

The tooth is very much stronger when the grain flows radially from the blank – this results in the grain lying at right angles to the tooth

(a)

Plane of weakness where tooth will break off under load – this is due to the grain lying parallel to the tooth

(b)

Grain orientation in the stock material has an important influence on subsequent processing. For example, Fig. 10.13 shows the effect of grain flow on a simple bracket bent from cold-rolled strip. It can be seen that when the direction of bending is parallel to the lay of the grain, the component is more liable to crack than when the direction of bending is perpendicular to the lay of the grain.

Fig. 10.13 *Effect of grain flow on subsequent processing*

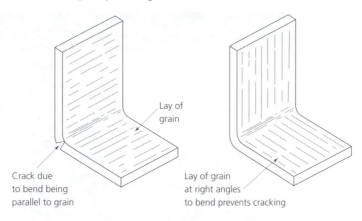

Lay of grain

Crack due to bend being parallel to grain

Lay of grain at right angles to bend prevents cracking

SELF-ASSESSMENT TASK 10.3

1. Explain how cold working can affect the following material properties:

    (a) hardness
    (b) elasticity
    (c) toughness
    (d) electrical conductivity

2. Explain why metals have to be heat treated before and during severe cold working.

3. Explain why the heat treatment process discussed in task 2 is not required before and during hot working.

10.8 Some hot-working processes

10.8.1 *Hot forging*

The basic forging operations of drawing down, upsetting, piercing and drifting, and swaging are not only performed by the blacksmith on the anvil using hand tools, but these techniques can be applied to larger components by substituting pneumatic and steam hammers for the blacksmith's hand tools. Hot forging under a steam hammer is shown in Fig. 10.14.

Where a large number of the same component is to be produced, closed-die forging is used. The die cavity is the shape of the finished component and a section through such a die is shown in Fig. 10.15. A set of dies for use with a drop-hammer are shown in Fig. 10.16.

For very large forgings hydraulic presses are used, as their slow but steady squeeze ensures that grain flow occurs right to the centre of the work, whereas the sharp blow of the hammer only produces surface flow in any but small components. Figure 10.17 shows the temperature ranges for the hot working of some typical metals.

Fig. 10.14 *Forging – steam hammer*

Fig. 10.15 *Closed-forging die (source: B & S Massey Ltd)*

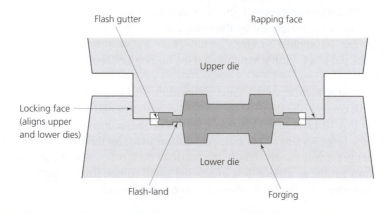

Fig. 10.16 *Drop-forging dies (source: B & S Massey Ltd)*

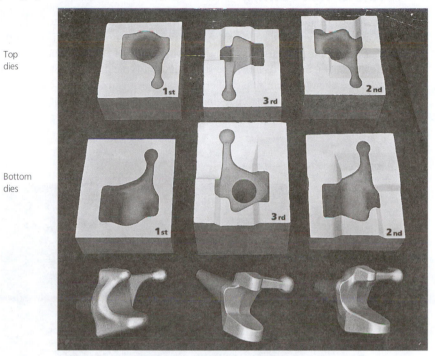

Top
dies

Bottom
dies

Fig. 10.17 *Hot-working temperatures*

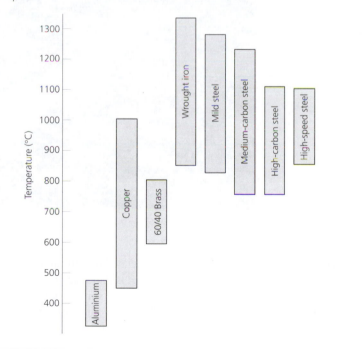

10.8.2 *Hot-rolling*

Figure 10.18 shows an ingot being hot rolled into a slab of steel. The white-hot slab is shown just leaving the rolls and it is supported on a motorised roller bed. The slab passes backwards and forwards between the work rolls of the mill and is gradually reduced in thickness. (Figure 2.20 shows the structure of a typical cast ingot prior to hot rolling.) The structure changes from fine chill crystals at the surface to coarse equi-axed crystals at the centre, resulting from the changing rate of cooling. In addition there will be a physical discontinuity in the centre of the ingot called the pipe.

Fig. 10.18 *Hot rolling (source: Davey-Loewy Ltd)*

These changes of structure and the associated discontinuities adversely affect the properties of the ingot. The top or neck of the ingot is removed and, with it, most of the pipe and any slag and impurities which have floated to the surface of the molten metal. The structure of the ingot is then rectified by the first hot-rolling process. The pipe and other internal discontinuities are pressure welded by the heat of the ingot and the pressure of the rolls to make the metal homogeneous. The crystal structure is broken down and, since this is a hot-working process, recrystallisation occurs simultaneously to give a refined and uniform grain structure, as shown in Fig. 10.19(a). The effect of recrystallisation on the strength and ductility of a metal after hot rolling is shown in Fig. 10.19(b).

Fig. 10.19 *Hot rolling: (a) effect on structure; (b) effect on properties of grain size after rolling*

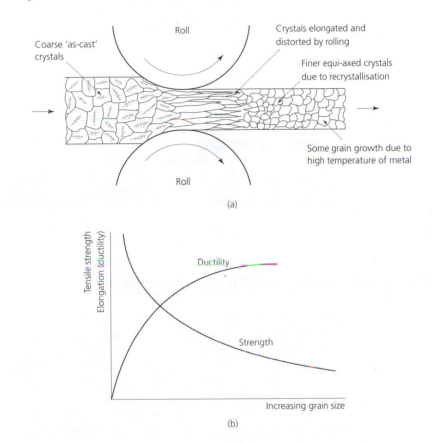

10.8.3 *Hot extrusion*

The principle of hot extrusion is shown in Fig. 10.20. A hydraulic ram squeezes a billet of metal – heated above its temperature of recrystallisation – through a die just like tooth-paste being squeezed out of the end of a tooth-paste tube. The hole in the die is shaped to produce the required section, and long lengths of material are produced by this process. Materials that are commonly extruded are: copper, brass alloys, aluminium and aluminium alloys. After hot extrusion the sections are often finished by cold drawing to improve the surface finish, dimensional accuracy and stiffness.

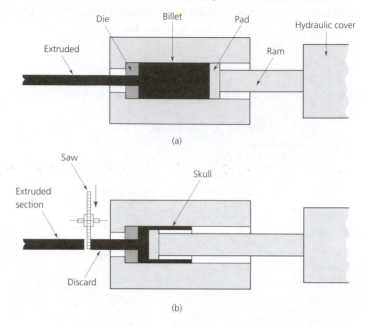

(a)

(b)

10.9 Some cold-working processes

Cold-working processes are essentially finishing processes. The forces required to produce quite modest reductions in cross-sectional area are very much higher than those for hot working, so that the amount of reduction is kept to a minimum. However, the finish and dimensional accuracy produced by cold working is much superior to that produced by hot working. Since cold working results in some work hardening of the metal, there is a corresponding improvement in its mechanical properties.

The metal to be cold worked is usually broken down by hot working so that there is only a finishing allowance left. The oxide film (scale) on the surface of the hot-worked metal is removed by pickling the metal in acid, after which it is passed through a neutralising bath and oiled to prevent rusting and passed to the cold-working process.

10.9.1 *Cold-rolling*

Figure 10.21 shows a typical cold strip rolling mill. The part-processed strip is unwound from the decoiler situated to the extreme left of the figure. This strip is straightened and flattened in the pinch rolls and leveller and passed to the mill rolls themselves for reduction in thickness. The mill shown is a four-high, single-stand reversing mill with the lower rolls hydraulically loaded. (In earlier mills the top rolls were screwed down mechanically.) After passing through the mill the strip, now reduced in thickness, is recoiled on the right-hand coiler. After this first pass the mill rolls are reversed and the strip is returned through the mill for further reduction and is finally recoiled on the left-hand, middle coiler.

Fig. 10.21 *Cold rolling: the typical line shown consists of coil storage, coil car, undriven decoiler with snubber roll, spade opener with debender rolls, four-high reversing hydraulic mill, reversing coilers, coil car and storage station; separate high- and low-pressure hydraulic packs provide the mill-loading system and the operations of the ancillary equipment; a high-capacity soluble oil system supplies strip lubrication and roll cooling (source: Sir James Farmer Norton & Co. (International) Ltd)*

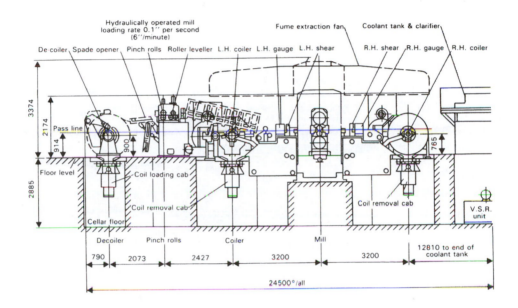

10.9.2 *Tube drawing*

In this process a pickled and oiled hot-drawn tube is further reduced and finished by drawing it through a die on a draw-bench, as shown in Fig. 10.22(a). (The initial hot-drawing is similar, but the raw material is a hollow billet heated to above its temperature of recrystallisation.) In order to control the wall thickness and internal finish of the tube, a 'plug' mandrel may be used, as shown in Fig. 10.22(b). The floating mandrel or 'plug' is drawn forward with the tube but cannot pass through the die. This technique is used for long, thin-walled tubes but is of limited accuracy. An alternative technique is to use a fixed mandrel, as shown in Fig. 10.22(c). Obviously there are limitations to the length of tube which can be drawn by this latter technique. However, it is widely used for thick-walled tubes and where greater accuracy is required.

Fig. 10.22 *Tube drawing: (a) simple draw bench; (b) use of a plug; (c) use of a mandrel*

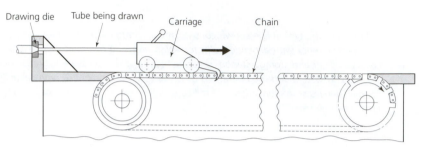

(a)

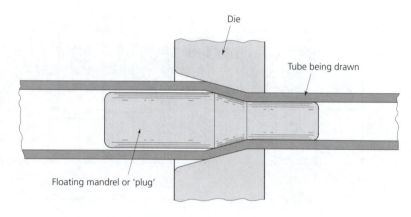

(b)

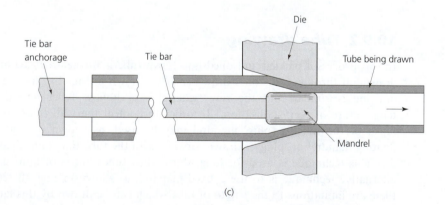

(c)

10.9.3 *Wire drawing*

This process is shown in Fig. 10.23(a). It is similar in principle to tube drawing, but since a longer length of material is involved, the wire is pulled through the die by a capstan or 'bull-block'. It may be coiled up on the capstan or passed to a separate coiler after taking only one or two turns round the capstan. Fine wire, as used for electrical conductors, is produced on multiple head machines, as shown in Fig. 10.23(b).

As the wire becomes thinner it becomes progressively longer. Thus each successive capstan has to run faster than the preceding one. The speed is controlled by the pull of the wire on the tension arm, which is coupled to the motor speed control. If the tension on the arm increases, the capstan is slightly slowed, but if the tension on the arm decreases, the capstan motor in the next stage is speeded up. If tension on the arm is increased, the capstan motor of the next stage is slowed down.

Fig. 10.23 *Wire drawing: (a) single die; (b) multiple die*

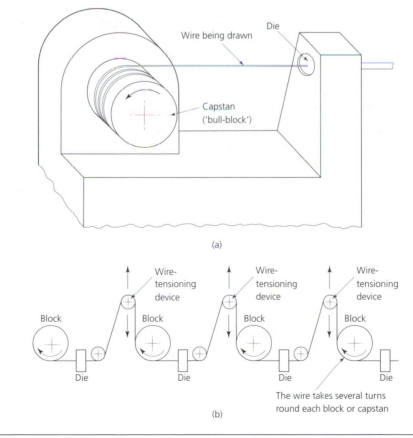

(a)

(b)

10.9.4 *Impact extrusion*

This process differs fundamentally from the hot-extrusion process described in Section 10.8. In impact extrusion, which is a cold-working process, a slug of metal is struck by a punch and made to flow up between the punch and the die, as shown in Fig. 10.24. Tooth-paste tubes are made from aluminium 'slugs' by this process. At one time only very soft and malleable metals could be cold impact extruded (e.g. aluminium, tin, lead). However, modern techniques allow even alloy steels to be processed in this manner and the impact extrusion and 'warm forging' of such materials are discussed in *Engineering Materials*, Volume 2.

Fig. 10.24 *Impact extrusion*

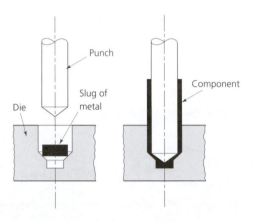

10.9.5 *Rivet heading*

This process is shown in Fig. 10.25. Here, the preformed head of the rivet is supported by a hold-up or 'dolly', whilst the opposite end of the rivet is headed by a pneumatic hammer. The shape of the rivet head being formed is controlled by the rivet 'snap' which is fitted to the hammer. In the example shown, both ends of the rivet will be finished with a rounded or *snap head*.

Fig. 10.25 *Cold heading rivets*

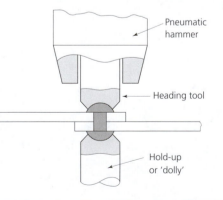

The advantages and limitations of hot-working, cold-working and casting processes are compared in Tables 10.2, 10.3 and 10.4 respectively.

Table 10.2 *Hot-working processes*

Advantages	Limitations
Low cost.	Poor surface finish – rough and scaly.
Grain refinement from cast structure.	Due to shrinkage on cooling the dimensional accuracy of hot-worked components is of a low order.
Materials are left in the fully annealed condition and are suitable for cold working (heading, bending, etc.).	Due to distortion on cooling and to the processes involved, hot working generally leads to geometrical inaccuracy.
Scale gives some protection against corrosion during storage.	Fully annealed condition of the material coupled with a relatively coarse grain leads to a poor finish when machined.
Availability as sections (girders) and forgings as well as the more usual bars, rods, sheets and strip and butt-welded tube.	Low strength and rigidity for metal considered.
	Damage to tooling from abrasive scale on metal surface.

Table 10.3 *Cold-working processes*

Advantages	Limitations
Good surface finish.	Higher cost than for hot-worked materials. It is only a finishing process for material previously hot worked. Therefore, the processing cost is added to the hot-worked cost.
Relatively high dimensional accuracy.	
Relatively high geometrical accuracy.	
Work hardening caused during the cold-working processes: (a) Increases strength and rigidity (b) Improves the machining characteristics of the metal so that a good finish is more easily achieved.	Materials lack ductility due to work hardening and are less suitable for bending, etc.
	Clean surface is easily corroded.
	Availability limited to rods and bars also sheets and strip, solid drawn tubes.

Table 10.4 *Casting processes (gravity, sand only)*

Advantages	Limitations
Virtually no limit to the shape and complication of the component to be cast.	Strength and ductility low, as structure is un-refined.
Virtually no limit to the size of the casting.	Quality is uncertain, as local differences of structure and mechanical defects such as blow holes cannot be controlled or corrected.
Low cost, as no expensive machines and tools are required as in forging.	Low accuracy due to shrinkage.
	Poor surface finish.
Scrap metal can be reclaimed in the melting furnace. (Wrought and machined components have to be made from relatively expensive pre-processed materials.)	Component must be designed without sudden changes of section, so that molten metal flows easily and cooling cracks and warping will not occur.
	Not all metals are suitable for casting. The best metals have a low shrinkage, a short freezing range and high fusibility (melt at relatively low temperatures), and have a high fluidity when molten.

SELF-ASSESSMENT TASK 10.4

1. Describe **two** cold-working processes of your choice.

2. Describe **two** hot-working processes of your choice.

3. When hot working metals, explain why the metal must neither be heated to too high a temperature at the start, nor allowed to cool to too low a temperature during processing.

10.10 Moulding polymeric materials

10.10.1 *Thermo-setting plastics*

These are usually moulded in dies fitted into hydraulic presses. A typical upstroke press is shown in Fig. 10.26. The platens are provided with electrical or steam heating elements so that the moulds fastened to them can be heated to the curing temperature of the thermosetting plastic being processed. Modern presses are automatic in operation and this ensures the constant moulding conditions at each stroke essential if mouldings of consistent quality are to be produced. The three factors which require to be preset to ensure correct polymerisation (curing) of the moulding powder are:

- The moulding pressure.
- The time for which the mould is closed and kept under pressure.
- The temperature of the mould.

Fig. 10.26 *Moulding press*

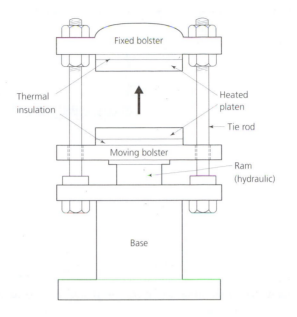

Figure 10.27 shows a section through a simple flash-type mould. It is suitable for shallow components in which the thickness is not critical. To ensure complete filling of the mould a slight excess of moulding powder is placed in the mould cavity. The excess powder is forced out into the flash-gutter as the mould is closed and must be subsequently trimmed off when the moulding is cold. The flash-land holds back the flow of the excess material and ensures complete filling of the mould.

Allowance must be made for thermal shrinkage of the moulding as it cools, and also for shrinkage due to chemical changes resulting in loss of water as steam during the curing reaction. The steam generated during polymerisation (curing) must be allowed to escape from the mould. This is only one of many different moulding techniques available for processing thermosetting plastics.

The moulding material may be fed into the mould in the form of powder or granules containing the thermosetting resin together with its additives and fillers, or compacted into a preformed shape to ensure uniform filling of the mould cavity. Correct loading of the mould is critical, insufficient moulding material resulting in voids and porosity through the cavity not being properly filled. A slight excess of material is to be preferred as it ensures uniform filling of the mould, any surplus being allowed for in the flash. Excessive overcharging of the mould must be avoided as this can result in damage to the mould.

The moulding material can be loaded either cold or preheated. Preheating reduces the curing time and reduces erosion of the mould since the material is in a less abrasive condition. As mentioned above, volatile gases are released during curing and these must be allowed to escape through mould clearances and vents machined into the dies.

Fig. 10.27 *Flash mould*

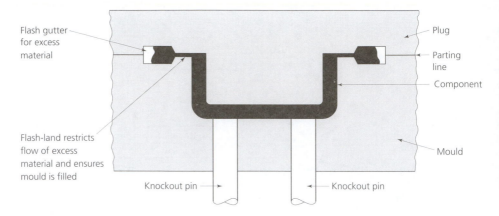

Flash gutter
for excess
material

Flash-land restricts
flow of excess
material and ensures
mould is filled

Knockout pin

Plug

Parting
line

Component

Mould

Knockout pin

A release agent (lubricant) must be sprayed into the mould cavity to prevent the plasticised resin from sticking to the walls of the mould. Remember that most resins are also high-strength adhesives (see Section 10.18). The correct curing time and temperature, which is critical, generally has to be arrived at by trial and error based upon experience of previous, similar mouldings.

An *overcured* moulding has a dull or blistered surface, and will be crazed and brittle with internal cracks and poor mechanical properties. *Undercuring* may produce a moulding with a correct appearance, but with poor mechanical properties. Moisture in the moulding material may also cause blisters and porosity. The moisture from damp moulding material must not be confused with the water vapour produced during the curing process.

10.10.2 *Thermoplastics*

The moulding of thermoplastic materials requires quite different techniques to those described above as no curing takes place in the mould. The most common process for moulding thermoplastic materials is *injection moulding*. In this process a measured amount of thermoplastic material is heated until it becomes a viscous fluid. It is then injected into the mould under high pressure. Since no curing takes place in the mould, it can be opened as soon as the moulding has cooled sufficiently to become self-supporting. Injection-moulding machines are generally arranged with the mould parting line vertical and the axis of injection horizontal, as shown in Fig. 10.28.

Like so many processes that are simple in principle, the practice is fraught with difficulties. For instance, heating the plastic until it is a viscous fluid is not easy. If care is not taken the surface becomes degraded by the heat before the inner mass of the plastic has reached moulding temperature. Again, the injected plastic material will not necessarily fill the mould cavity; there may be blisters, voids, sinks, shorts, and some cavities in a multi-impression mould may not receive any moulding material at all. Ejection of the completed moulding is also difficult if distortion is to be avoided and,

Fig. 10.28 *Injection moulding*

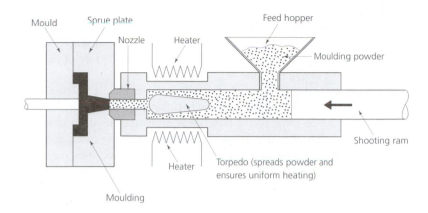

since thermoplastic materials are adhesive, they will stick to the mould whenever the opportunity presents itself despite the use of a release agent. The following principle variables need to be controlled.

- The quantity of plastic material that is injected into the mould.
- The injection pressure.
- The injection speed.
- The temperature of the plastic moulding material.
- The temperature of the mould. Too cold and the plastic may solidify before the mould is full; too hot and the moulding may be soft and lacking in rigidity when the mould is opened.
- The injection plunger forward-time (i.e. the time the plastic material is maintained under pressure in the mould as it cools and becomes rigid). This reduces the effect of shrinkage and ensures that the mould cavities are kept filled.
- The mould closed time.
- The mould clamping force.
- The mould open time.

The temperature of the plastic during the moulding cycle is extremely critical as this controls its viscosity. Too high a temperature leads to degradation and poor mechanical properties, whilst too low a temperature leaves the plastic too viscous so that it will not fill the mould properly.

As soon as the mould is filled, the pressure is increased and 'packing' of the mould commences. This ensures that the moulding has high density and good mechanical properties; it also prevents sinks and shorts occurring due to shrinkage of the plastic as it cools.

Unlike thermosetting plastics where any trimmings such as the flash and the cull are scrap and cannot be re-used, any trimmings from thermoplastic mouldings can be recycled and are not wasted.

10.10.3 *Extrusion*

The extrusion of thermoplastic materials is, in principle, a continuous injection-moulding process. Any thermoplastic material can be extruded to produce lengths of uniform cross-section such as rods, tubes, sections and filaments. To produce a continuous flow of plastic material through the die, a screw conveyor is used in place of the ram and cylinder of the injection-moulding machine.

The general arrangement of an extrusion-moulding machine is shown in Fig. 10.29. Note that screw conveyor type feed is also used for injection moulding for very large components, as some of the large body mouldings used on modern cars are beyond the volumetric capacity of the type of injection-moulding machine previously shown in Fig. 10.28. This latter type of machine is still the most widely used for small mouldings since it is quicker in operation.

Fig. 10.29 *Extrusion moulding*

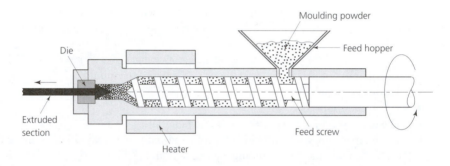

So far in this chapter we have seen how the properties of various materials are exploited in the forming and shaping processes described. It's now time to turn our attention to the joining of materials. Again we see how the various joining processes exploit the properties of the materials involved.

SELF-ASSESSMENT TASK 10.5

1. Explain why moulds for thermosetting plastics have to be vented.

2. State what precautions have to be taken in the design and use of plastic moulds to ensure complete filling.

3. (a) State why molds have to be sprayed with a release agent and lubricant before each moulding is made.
   (b) Explain briefly why the moulding of thermoplastics is a quicker process than the moulding of thermosets.

10.11 Compression joints

10.11.1 *Mechanical*

Compression joints rely upon the *elasticity* of components being joined to secure one component to another. No additional fastenings, such as bolts or rivets, are required. Basically, a mechanical compression joint consists of an oversize peg being forced into an undersize hole so that the peg is compressed and the hole is stretched. In a lightly compressed joint, friction will be generated between the two components as they try to spring back to their original dimensions. This friction, alone, secures the components together. Where the size difference is greater, the one component may actually 'bite' into the other giving some degree of positive locking. Figure 10.30 shows a typical pressed or 'staked' compression joint.

Fig. 10.30 *Mechanical compression joint: (a) the bush becomes compressed as it is inserted in the bush plate and the bush plate expands; (b) the bush plate springs back on the bush and grips it – the bush plate is in tension and the bush is in compression*

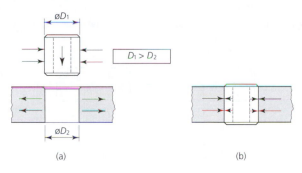

(a)   (b)

You have to choose the materials for compression joints with care. The outer component is subjected to considerable tensile stress and the inner component is subjected to considerable compression stress. Therefore steel is suitable for both the inner and outer components as it is equally strong in tension and compression. Alternatively, you could use cast iron for the inner component because it is strong in compression. However, you should not use cast iron for the outer component since it is weak in tension and would crack.

10.11.2 *Thermal (hot shrunk)*

Instead of forming a compression joint by mechanical pressing, the *thermal expansion* of the outer component is exploited. You may remember that when metals are heated they expand. So, if the outer component is heated sufficiently it expands to a size where it can be slipped easily over the inner component. As the outer component cools it shrinks back to its original size and forms a compression joint on the inner component as shown in Fig. 10.31(a). Heating must be uniform to prevent distortion and the temperature has to be closely controlled.

- Too low a temperature results in insufficient expansion for the components to be assembled.
- Too high a temperature can result in changes in the properties and structure of the material used for the outer (heated) component.

Fig. 10.31 *Thermal compression joint: (a) hot shrink; (b) cold expansion*

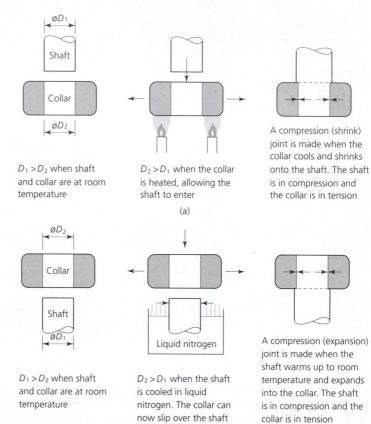

$D_1 > D_2$ when shaft and collar are at room temperature

$D_2 > D_1$ when the collar is heated, allowing the shaft to enter

A compression (shrink) joint is made when the collar cools and shrinks onto the shaft. The shaft is in compression and the collar is in tension

(a)

$D_1 > D_2$ when shaft and collar are at room temperature

$D_2 > D_1$ when the shaft is cooled in liquid nitrogen. The collar can now slip over the shaft

A compression (expansion) joint is made when the shaft warms up to room temperature and expands into the collar. The shaft is in compression and the collar is in tension

(b)

10.11.3 *Thermal (cold expansion)*

In a cold-expansion joint, the inner component is cooled down until it will slip easily into the hole in the outer component. As it warms up to room temperature, it expands and forms a compression joint with the outer component, as shown in Fig. 10.31(b). This technique requires the use of such coolants as solid carbon dioxide (dry ice) or liquid nitrogen. These are best used under carefully controlled workshop conditions as such low temperatures are potentially dangerous and special equipment is required in their use. The appropriate codes of practice and safety regulations must be rigidly adhered to.

Cooling has the advantage that it does not affect the physical properties of the material, whereas heating may do so. For example, heating a starter ring gear to slip it over a flywheel of a car engine may soften the teeth of the gear. Cooling the flywheel will not affect its properties once it has returned to room temperature. The precautions concerning the choice of materials for thermal compression joints are the same as those for mechanical compression joints.

10.12 Soft soldering

The process of soft soldering exploits the fusibility (low melting temperature) of tin–lead alloys. The molten 'solder', as the alloy is called, *bonds* to an unmelted parent metal by the application of heat and a suitable flux. The solder does not just 'stick' to the surface of the parent metal. This bonding is the result of solder molecules migrating into the surface of the parent metal to form a permanent amalgam. The *parent metal* is the metal from which the components being joined are manufactured. Thus, the solder must have a lower melting temperature than the parent metal and it must also be capable of reacting together with the parent metal to form a bond. The composition of some soft solders, together with their melting temperature range and some typical applications, have already been listed in Tables 7.13 and 7.14.

The bonding action of the solder cannot take place unless the two surfaces to be joined are chemically as well as physically clean. The surfaces of the joint should be degreased and then scoured with 'steelwool' to make them physically clean. They are then chemically cleaned and prepared for tinning by the action of a *flux*. The purpose of a flux is to:

- Remove the oxide film from the surfaces to be soldered.
- Prevent the oxide film from reforming during the soldering process.
- 'Wet' the surfaces being joined so that the molten solder will run out into an even film.
- Allow itself to be easily displaced by the molten solder so that a metal to metal contact is achieved.

The action of the flux is shown in Fig. 10.32. The fluxes used for soft soldering fall into two categories: *active fluxes* and *passive (inactive) fluxes*.

10.12.1 *Active fluxes*

Active fluxes such as Baker's fluid (acidified zinc chloride solution) quickly dissolve the oxide film and prevent it reforming. They also etch the surface to be soldered, ensuring good wetting and bonding. Unfortunately all active fluxes leave a corrosive residue which has to be washed off immediately after soldering and the joint has to be treated with a rust inhibitor.

10.12.2 *Passive fluxes*

Passive fluxes, such as resin, are used for those applications where it is not possible to remove any corrosive residue by washing – for example, electrical connections. Unfortunately passive fluxes do not remove oxide films to any appreciable extent, they only prevent them from reforming during the soldering process. Therefore the initial mechanical scouring of the joint faces has to be very thorough.

Fig. 10.32 *The soldering flux is displaced by molten solder, which is said to 'wet' when it leaves a continuous permanent film on the surface of the parent metal instead of rolling over it: A – flux solution lying above oxidised metal surface; B – boiling flux solution removing the film of oxide (e.g. as chloride); C – bare metal in contact with fused flux; D – liquid solder displacing fused flux; E – tin reacting with the base metal to form compound; F – solder solidifying*

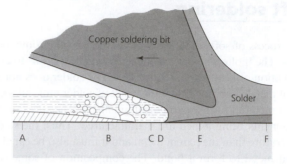

10.13 Hard soldering (brazing)

Hard soldering is the general term used for silver soldering and brazing, and it can be defined as:

> *A process of joining metals in which a molten filler metal is drawn by capillary attraction into the space between closely adjacent surfaces of the parts to be joined.*

As in soft soldering only the filler metal (solder) becomes molten and the parent metal remains solid throughout the process. Therefore, like soft solders, hard solders also have a melting temperature range below that of the parent metal. However, this melting temperature range (generally above 500 °C) is well above that of soft solder and a soldering iron cannot be used to heat the joint and load the solder into it.

The heat required for the process of hard soldering is provided by some form of gas blow-pipe. Natural gas from the mains supply or bottled gas (propane) may be used. Some typical hand torches for flame brazing are shown in Fig. 10.33(a) and a typical set up for flame brazing is shown in Fig. 10.33(b).

A hard-soldered joint is much stronger than a soft-soldered joint. You require special fluxes for hard-soldering processes and the flux must match the filler alloy being used. Such fluxes are supplied by the manufacturers of hard solders and brazing spelters and you should follow their instructions carefully. Hard-soldering fluxes are usually supplied as powders and have to be mixed into a paste with water before applying to the joint. The success of all hard-soldering processes depends upon the following conditions:

- Selection of a suitable filler alloy which has a melting range appreciably lower than the parent metals being joined.

- Thorough cleanliness of the surfaces to be joined by hard soldering.
- Complete removal of the oxide film from the joint surfaces before and during hard soldering by means of a suitable flux.
- Complete 'wetting' of the joint surfaces by the molten filler alloy. When a surface is 'wetted' by a liquid, a continuous film of the liquid remains on that surface after draining. This condition is essential for hard soldering and the flux, having removed the oxide film, must completely wet the joint surfaces. This 'wetting' action by the flux assists the spreading and feeding of the molten filler alloy into the joint by capillary attraction. This ensures a completely filled joint.
- Since the molten filler alloy is drawn into the joint by capillary attraction, the space between the joint surfaces must be kept to a minimum and it must also be kept constant. Any local increase in the gap can present a barrier to the feeding of the filler alloy. This will prevent the joint from being uniformly filled, resulting in serious loss of strength.
- Melting the filler alloy alone is not sufficient to produce a sound joint. The parent metal must itself be raised to the brazing temperature so that the filler alloy melts on coming into contact with the joint surfaces even after the flame has been withdrawn.

Fig. 10.33 *Flame brazing: (a) typical hand torches used for hard soldering and brazing; (b) fire bricks or other suitable insulating material is packed around the component to form a brazing hearth which contains and reflects the torch's heat*

Oxy-acetylene, the most versatile of all the hand torches

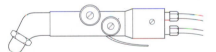

Small compressed-air torch for precision brazing

Large compressed air-gas torch for general brazing

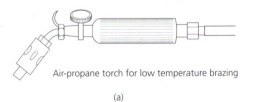

Air-propane torch for low temperature brazing

(a)

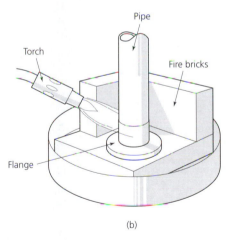

(b)

Unlike welding, dissimilar metals and alloys may be joined by hard soldering. For example: copper to brass, steel to brass, mild steel to malleable cast iron, etc. The groups of filler material most widely used for hard soldering are summarised in Table 10.5.

Table 10.5 Composition of hard solders and brazing spelters

Category	BS 1845 type	Composition (%)						Approximate melting range (°C)
		Ag	Cu	Z	Cd	P		
Silver solders	3*	49–51	14–16	15–17	18–20	—		620–640
	4†	60–62	27.5–29.5	9–11	—	—		690–735
	5‡	42–44	36–38	18.5–20.5	—	—		700–775
Self-fluxing brazing alloys containing phosphorus	6	13–15	Balance	—	—	4–6		625–780
	7	—	Balance	—	—	7–7.5		705–800
Brazing spelters (brass alloys)	8	—	49–51	Balance	—	—		860–870
	9	—	53–55	Balance	—	—		870–880
	10	—	59–61	Balance	—	—		885–890

* Type 3 is extremely fluid at brazing temperatures and is, therefore, ideal when brazing dissimilar metals. It is a low melting range alloy.
† Type 4 has high electrical conductivity and is, therefore, very suitable for joining electrical conductors. It is the most expensive because of its high silver content.
‡ Type 5 is a general-purpose silver solder which can be employed at much higher brazing temperatures. It is the strongest of the silver solders.

Silver solders are expensive material since they contain the precious metal silver. However, they produce strong and ductile joints and are used for the finest work as the melting temperature range is sufficiently low not to affect the parent metal, and a very neat joint can be made.

10.13.1 *Brazing alloys containing phosphorus*

Brazing alloys containing phosphorus are usually referred to as 'self-fluxing' alloys. These alloys contain silver, phosphorus and copper. They are cheaper and stronger than the silver solders, but they can only be used to braze copper and copper alloy components in air. No separate flux is needed. The phosphorus content reacts with the oxygen in the air to form a compound which acts as a flux. These filler alloys must not be used for brazing nickel, nickel alloys or copper-based alloys containing more than 10 per cent nickel. Nor must they be used for brazing ferrous metals and alloys.

10.13.2 *Brazing spelters*

Brazing spelters are brass alloys and are the oldest filler alloys used. It is from the use of these brass alloys that the process called 'brazing' gets its name. These *spelters* make the strongest joints but they also have the highest melting temperatures. They are mainly used to braze copper, steel and malleable cast iron components.

10.14 Fusion welding

In the soldering and brazing processes previously described, joints are made by a thin film of metal that has a lower melting point and inferior strength than the metals being joined. In *fusion welding* any additional material added to the joint has a similar composition and strength to the metals being joined. Figure 10.34 shows the principle of fusion welding, where not only the filler metal but also the edges of the components being joined are melted. The molten metals fuse together and, when solid, form a homogeneous joint whose strength is equal to the metals being joined. In this chapter we are only going to consider the welding of plain carbon steels.

Fig. 10.34 *Fusion welding: (a) before – a single 'V' butt requires extra metal; (b) after – the edges of the 'V' are melted and fused together with the molten filler metal*

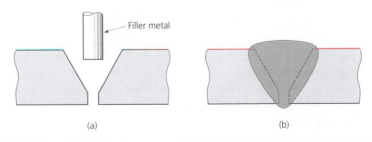

Filler metal

(a)   (b)

10.14.1 *Oxyacetylene welding*

In this process the heat source is a mixture of oxygen and acetylene burning to produce a flame whose temperature can reach 3250 °C, and this is above the melting point of most metals. Figure 10.35 shows a typical set of welding equipment. The welding gases form a highly flammable and even explosive mixture, so this equipment must only be used by a suitably qualified person or a trainee under the direct instruction of such a person. Figure 10.36 shows the two basic techniques for fusion welding using an oxyacetylene torch. No flux is required when welding ferrous metals as the products of combustion from the burnt gases protect the molten weld pool from atmospheric oxygen.

Fig. 10.35 *Oxyacetylene welding equipment*

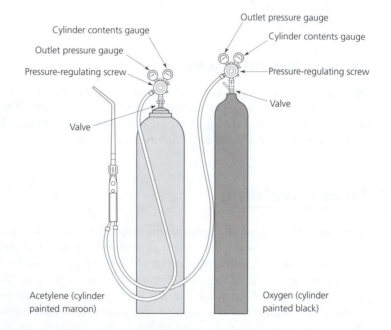

10.14.2 *Metallic arc welding*

This is a fusion-welding process where the heat energy required to melt the edges of the components being joined and also the filler rod is supplied by an *electric arc*. The arc is the name given to the prolonged spark struck between two electrodes. In this process the filler rod forms one electrode and the work forms the other electrode. The filler rod/electrode is coated with a flux which melts and shields the joint from atmospheric oxygen at the very high temperatures involved. (Average arc temperature is about 6000 °C.) The flux also stabilises the arc and prevents the rod from short circuiting against the sides of the joint when welding thick metal. Figure 10.37 compares the principles of gas and metallic arc welding.

Fig. 10.36 *Oxyacetylene welding techniques: (a) the leftward method of welding – this is the easiest technique for a right-handed operator; it is used for sheet metal; (b) the rightward method of welding – this method is used for thicker plate as it gives better penetration*

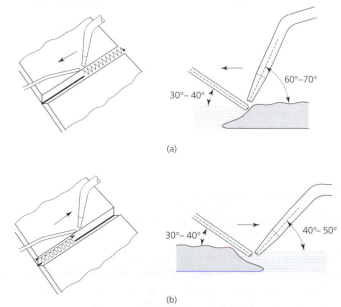

Fig. 10.37 *Comparison of (a) oxyacetylene welding and (b) manual metallic arc welding*

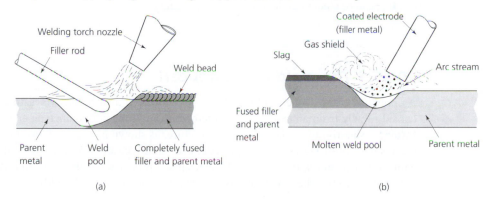

A transformer is used to reduce the mains voltage to a low-voltage, heavy current supply which is not only safe but suitable for arc welding. As with gas welding, arc-welding equipment must not be used by untrained persons except under the closest supervision. The dangers with arc welding arise from the very high temperatures and very heavy electric currents involved. Also high voltages are present in the primary circuit (supply side) of the transformer, and these can lead to accidents involving electrocution. Figure 10.38 shows the general arrangement of a metallic arc-welding installation. For both gas and electric arc welding, suitable eye protection equipment must be used.

Fig. 10.38 *Manual metallic arc-welding circuit diagram*

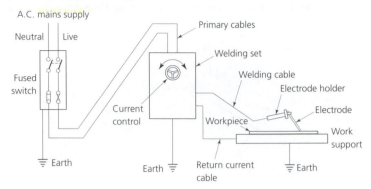

10.15 Effect of welding on the structure and properties of materials

The structures in a welded joint range from the wrought structures of the parent metal to the cast structures of the weld itself, all of which will have been subjected to heat treatment by the high temperatures involved in the process. The *weld deposit* will possess a typical cast structure with all its inherent defects. The *heat-affected zone* of the parent metal will exhibit the effects of *heat treatment*. The unaffected regions, where the temperature has not been so high, will retain the original wrought structure of the parent metal. Therefore the effects of welding can be studied under the following headings:

- The weld-metal deposit.
- The heat-affected zone.

10.15.1 *The weld-metal deposit*

As previously stated, the weld metal can be considered as a miniature casting which has cooled rapidly from an extremely high temperature. Long columnar type crystals may be formed giving rise to a relatively weak structure, as shown in Fig. 10.39(a). In a multi-run weld each deposit *normalises* the preceding run and considerable grain refinement occurs with a consequent improvement in the mechanical properties of the joint. In this case, only the top run exhibits the coarse 'as-cast' structure as shown in Fig. 10.39(b).

Non-metallic inclusions

The formation of oxide and nitride inclusions due to atmospheric contamination is reduced by the blanket of burnt gases (products of combustion) in the case of gas welding, and by the use of a flux when electric arc welding. Modern flux-coated electrodes usually provide good quality weld deposits free from harmful inclusions. In the argon arc-welding process the metal is deposited under a shroud of the inert gas argon. This prevents oxidation and the formation of nitrides, so no flux is necessary. Further, since no flux is required there will be no slag inclusions. In multi-run welds using coated electrodes the slag must be removed between each run.

Fig. 10.39 *Weld metal deposit structure: (a) large single-run weld; (b) metallic arc weld*

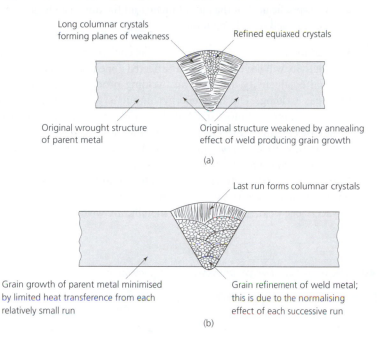

Long columnar crystals forming planes of weakness

Refined equiaxed crystals

Original wrought structure of parent metal

Original structure weakened by annealing effect of weld producing grain growth

(a)

Last run forms columnar crystals

Grain growth of parent metal minimised by limited heat transference from each relatively small run

Grain refinement of weld metal; this is due to the normalising effect of each successive run

(b)

Gas porosity

The chief cause of gas porosity is the presence of hydrogen in the weld metal or the formation of steam from the reaction of hydrogen with any oxide present in the molten parent metal. In addition, hydrogen is present in the welding flame when gas welding and in the flux coatings of electrodes when arc welding.

Weld-metal cracking

Welded joints that are prepared under restraint are liable to intercrystalline cracking in the weld deposit due to contractional strains set up during the cooling of the metal. Such cracking, usually known as 'hot cracking', is largely related to the grain size and the presence of grain boundary impurities. At high temperatures, the grain boundaries are more able to accommodate shrinkage strains than the grains themselves. A coarse grain deposit with large columnar crystals possesses a relatively small grain boundary area and is, therefore, more susceptible to hot cracking.

10.15.2 *The heat-affected zone*

The heat-affected zone of the parent metal is difficult to define. It will depend upon such factors as:

- The temperature of the weld pool.
- The time taken to complete the weld.

- The thermal conductivity of the parent metal.
- The specific heat of the parent metal and the dimensions of the parent metal.
- The method of welding used.

Welded joints produced in metals such as copper and aluminium that have a high thermal conductivity will have a wider heat-affected zone than a plain carbon steel that has a lower thermal conductivity. Metallic arc welding produces a more concentrated heating effect than gas welding. The heat energy output is greater with arc welding, so the welding process can proceed more quickly. Therefore the heat-affected zone when arc welding will be narrower than that when gas welding the same materials.

The heat-affected zone in mild steel plate can exhibit various structures These range from an overheated structure for those parts adjacent to the weld pool and, therefore, heated to well above the upper critical temperature, to those parts whose temperature has hardly risen above room temperature. These are shown in Fig. 10.40 for both a single-run oxyacetylene weld, and a single-run metallic arc weld.

Fig. 10.40   *Macrostructure of single-run welds in mild steel: (a) oxyacetylene weld; (b) metallic arc weld*

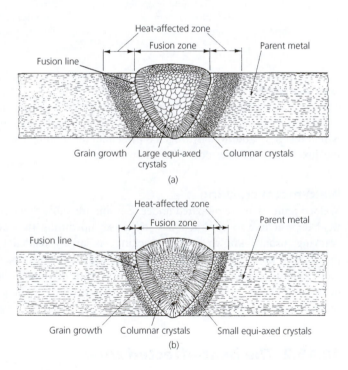

- The molten metal of the weld pool will be made up from the deposited metal and the molten parent metal.
- There will be a *fusion line* between the molten metal and the unmolten metal.

- The *heat-affected zone* will extend from the fusion line to the parent metal where the temperature has not be sufficiently raised to change the original wrought structure.
- Adjacent to the fusion line is a zone of coarse grains. Here the metal has been heated nearly to its *melting point* and considerable *grain growth* will have occurred.
- Progressing away from the weld, the grains become smaller, and the zone where very fine grains appear is called the *refined zone*. Here the metal has been heated sufficiently for *recrystallisation* to occur but cooling has been sufficiently rapid to prevent any growth.
- Beyond the refined zone is the transition zone. In this zone some of the metal will have recrystallised and some will not. There will be a mixed structure.
- Finally we come to the unaffected zone where the parent metal has not been sufficiently heated for any structural changes to occur and the original wrought structure will be in evidence. The temperature zones are summarised in Table 10.6.

The properties of the material will change with these changes in structure. The coarser grains will show greater ductility and softness but reduced strength. The finer crystals will show less ductility but greater hardness and strength. These effects become more apparent as the carbon content of the steel increases.

Table 10.6 *Crystal structure of the metal*

Effect of temperature gradient

Temperature zones	Remarks
Fusion zone	Temperature reaches melting point. The cooling rate is in the order of 350–400 °C/min, which is the maximum quenching range. The weld is less hard than the adjacent area of the parent metal because of loss of useful elements (carbon, silicon and manganese).
Overheated zone	The temperature reaches 1100–1500 °C. Cooling is extremely rapid in the order of 200–300 °C/min. Some grain coarsening occurs.
Annealed zone	Here the temperature reaches slightly higher than 900 °C. The parent metal has a refined normalised grain structure. The change is not complete because the cooling rate is still high, in the order of 170–200 °C/min.
Transformation zone	The temperature here is between 720 and 910 °C. These are the upper and lower critical temperatures between which the iron in steel transforms from a body-centred-cubic to a face-centred-cubic structure. The parent metal tends to recrystallise.

10.16 Spot welding

This is a resistance welding process widely used in the sheet metal industry. The joint is produced by making a series of spot welds side by side at regular intervals. Apart from ensuring that the joint faces are clean and free from corrosion, no special joint preparation is required.

The temperature of the metal components to be joined is raised locally by the passage of a heavy electric current at low voltage through the components, as shown in Fig. 10.41.

The components are gripped between the *copper electrodes*. When the current flows through the components their *resistance* to the current causes local heating. Sufficient heat is generated to raise the metal to its welding temperature at this spot. The current is then switched off and the pressure exerted by the electrodes is increased to form a weld. No filler metal is added and the process is very rapid. The cycle of events is controlled automatically by a programmable logic controller (PLC).

The effect of welding temperature on a spot weld is shown in Fig. 10.42. If the welding temperature is too high, the columnar crystals will meet at the centre of the spot weld. This will form a plane of weakness which may lead to intercrystalline cracking. If the temperature is correct, equi-axed grains will form in the centre of the weld before the columnar crystals can meet. The importance of time and temperature control when spot welding should now be apparent.

Fig. 10.41 *Principles of resistance welding: (a) schematic diagram of electric spot-welding machine; (b) spot welding*

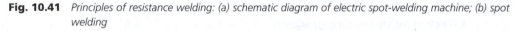

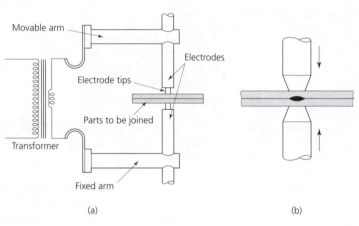

Fig. 10.42 *Structure of spot welds: (a) correct welding temperature; (b) temperature too high*

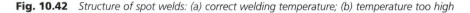

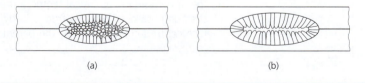

10.17 Welding plastic materials

Heat-welding techniques can only be used to join *thermoplastic materials,* since only these plastic materials soften upon being heated. A distinction must be made between sealing and welding. The term *sealing* is reserved for the thermal joining of thin films and foils (plastic bags containing foods such as crisps). The term *welding* is reserved for joining relatively heavy gauge (thick) sheet plastic components.

The low thermal conductivity and softening temperatures of plastic materials necessitates the use of a low welding temperature. This enables the heat to penetrate the body of the plastic before the surface is overheated. Heat is normally applied to the joint using a hot-air gun. Two types of gun are shown in Fig. 10.43. Oxidation can weaken the joint, so where the joint is critical hot nitrogen gas is used instead of hot air.

Fig. 10.43 *Plastic-welding guns: (a) electrically heated hot air plastic-welding gun; (b) gas-heated hot nitrogen plastic-welding gun*

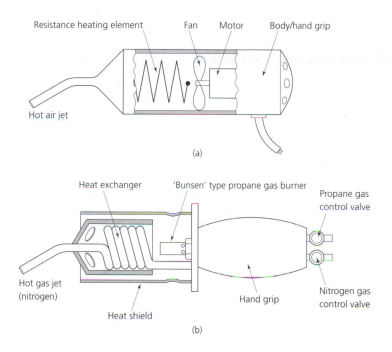

(a)

(b)

Unlike metals, which have a fairly sharply defined melting point, thermoplastics usually have a wide range of temperatures between which they start to soften and eventually degrade. The easiest plastics to weld are polyvinyl chloride (PVC) and polyethylene (PE) as they have a wide softening range.

The basic technique is to apply a jet of hot air or hot nitrogen so that the edges of the parent plastic sheet is softened. Filler material, in the form of a rod of the same material as that being welded, is added into the joint in much the same way as when gas welding metals. This is shown in Fig. 10.44. The weld 'bead' must not be removed when plastic welding as this 'bead' adds considerably to the strength of the joint.

SELF-ASSESSMENT TASK 10.6

1. (a) State the reasons for using a flux when soft soldering.
   (b) State the differences between active fluxes and passive fluxes.
2. Explain why arc-welding electrodes require a flux coating, whilst oxyacetylene welding rods do not.
3. (a) State the causes of gas porosity and intercrystalline cracking when welding and suggest ways of preventing these faults.
   (b) Explain what is meant by 'weld decay' and state how this problem can be overcome.
4. Discuss the essential difference between the fusion welding of metals and the heat welding of plastic sheet.

10.18 Adhesive bonding

Naturally occurring adhesives fall into two categories:

- *Glues* These are made from the bones, hooves and horns of animals and the bones of fishes. Derivatives of milk and blood are also used. Glues were largely used for joining wood and were used in the furniture and toy manufacturing industries. They have now been almost totally replaced with modern high-strength synthetic adhesives, but they are still used where their non-toxic and non-narcotic properties are important.
- *Gums* These are still made from vegetable matter, resins and rubbers being extracted from trees and starches from the by-products of flour milling. Since they are non-toxic they are used for such low strength applications as stamp adhesives and envelope flap adhesives which have to be licked.

Fig. 10.44 *Plastic heat-welding technique*

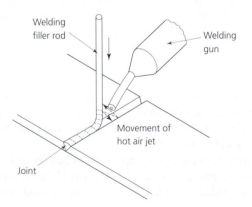

Modern high-strength synthetic adhesives have been developed by the plastics industry. Figure 10.45 lists some typical adhesives and the adherend materials for which they can be used, whilst Table 10.7 lists some of the more important advantages and limitations of such adhesives compared with the jointing processes discussed earlier in this chapter.

Fig. 10.45 *Selection of adhesives; in general, any two adherends may be bonded together if the chart shows that they are compatible with the same adhesive*

Adherends	Natural: Animal glues	Starch	Dextrine	Casein	Elastomers: Acrylonitrile butadiene	Polychloroprene	Polyurethane	Silicone rubber	Polybutadiene	Natural rubber	Butyl	Thermoplastics: Cellulose nitrate	Polyvinyl alcohol	Polyvinyl acetate	Polyarylate	Silicone resin	Cyanoacrylate	Thermosets: Phenolic formaldehyde	Urea formaldehyde	Resorcinol formaldehyde	Melamine formaldehyde	Polyesters (unsaturated)	Epoxy resins	Polyimides	Phenolic-vinyl formal	Phenolic-polyvinylacetal	Phenolic nitrile	Phenolic epoxy	Inorganic: Sodium silicate
Metals					•	•					•				•	•	•					•	•	•			•	•	
Glass, Ceramics	•				•		•						•	•	•		•						•					•	•
Wood	•		•	•						•			•	•				•	•	•	•		•						
Paper	•	•	•	•	•								•	•	•														•
Leather	•				•	•					•		•	•															
Textiles, Felt	•				•					•			•	•															
Elastomers																													
Polychloroprene (Neoprene)					•																								
Nitrile					•												•												
Natural					•		•										•									•			
Silicone							•																						
Butyl					•																								
Polyurethane					•	•	•				•																		
Thermoplastics																													
Polyvinyl chloride (flexible)					•	•	•																						
Polyvinyl chloride (rigid)					•	•	•																•						
Cellulose acetate						•						•		•															
Cellulose nitrate						•						•		•															
Ethyl cellulose												•		•													•		
Polyethylene (film)						•		•						•									•				•		
Polyethylene (rigid)																							•				•		
Polypropylene (film)					•		•							•									•				•		
Polypropylene (rigid)																							•				•		
Polycarbonate						•																	•						
Fluorocarbons									•							•		•					•						
Polystyrene						•											•						•						
Polyamides (nylon)						•											•	•					•						
Polyformaldehyde (acetals)						•											•					•	•						
Methyl pentane						•											•												
Thermosets																													
Epoxy														•						•			•						
Phenolic						•								•					•	•			•				•		
Polyester																						•	•						
Melamine					•	•																	•						
Polyethylene terephthalate					•	•											•						•						
Diallyl phthalate					•																	•	•						
Polyimide																							•	•					

Table 10.7 *Advantages and limitations of bonded joints*	
Advantages	Limitations
The ability to join dissimilar materials, and materials of widely different thicknesses.	The bonding process is more complex than mechanical and thermal processes, i.e. the need for surface preparation, temperature and humidity control of the working atmosphere, ventilation and health problems caused by the adhesives and their solvents. The length of time that the assembly must be jigged up whilst setting (curing) takes place.
The ability to join components of difficult shape that would restrict the application of welding or riveting equipment.	
Smooth finish to the joint which will be free from voids and protrusions such as weld beads, rivet and bolt heads, etc.	
Uniform distribution of stress over entire area of joint. This reduces the chances of the joint failing in fatigue.	Inspection of the joint is difficult.
	Joint design is more critical than for many mechanical and thermal processes.
Elastic properties of many adhesives allow for flexibility in the joint and give it vibration damping characteristics.	
The ability to electrically insulate the adherends and prevent corrosion due to galvanic action between dissimilar metals.	Incompatibility with the adherends. The adhesive itself may corrode the materials it is joining.
The join may be sealed against moisture and gases.	Degradation of the joint when subject to high and low temperatures, chemical atmospheres, etc.
Heat-sensitive materials can be joined.	Creep under sustained loads.

Let's now look at the successful joining of materials using adhesives in rather more detail. Figure 10.46(a) shows a typical bonded joint and explains the terminology used for the various features of the joint. The strength of the joint depends upon the two following factors:

- *Adhesion* This is the ability of the bonding material (adhesive) to stick (adhere) to the materials being joined (adherends). There are two ways in which the bond can occur. These are shown in Fig. 10.46(b) and (c).
- *Cohesion* This is the ability of the adhesive film itself to resist the applied forces acting on the joint.

Figure 10.46(d) shows the three ways in which a bonded joint may fail under load. These failures can be prevented by careful joint design and their correct selection of the adhesive.

Careful preparation of the joint surfaces is essential for a sound joint. The surfaces must be chemically and physically clean. Also the atmospheric temperature and humidity of the working environment must be closely controlled. No matter how effective the adhesive, and how carefully you apply it, the joint will be a failure if it is not correctly designed. It is bad practice to apply adhesive to a joint which was originally proportioned for bolting, riveting or welding. You must design the joint to exploit the special properties of adhesives. Some typical adhesive joint designs are shown in Fig. 10.47. Most adhesives are relatively strong in tension and shear but weak in cleavage and peel. These terms are explained in Fig. 10.48.

Any adhesive must 'wet' the joint surfaces thoroughly or voids will occur and the actual bonded area will be substantially less than the design area. This will result in a weak joint. Figure 10.49 shows the effect of wetting on the adhesive film.

Fig. 10.46 *The bonded joint: (a) elements of the bonded joint; (b) simple cemented joint in which the adhesive penetrates the pores of the adherends (occurs with rough or porous surfaces); (c) adhesive and adherends react together chemically so that an intermolecular bond is formed; (d) adhesive and cohesive failure*

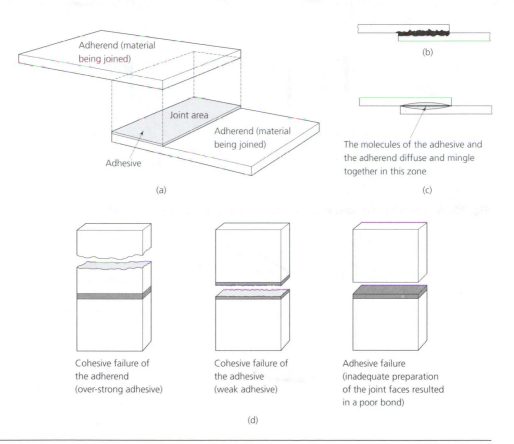

Fig. 10.47 *Suitable joints for bonding*

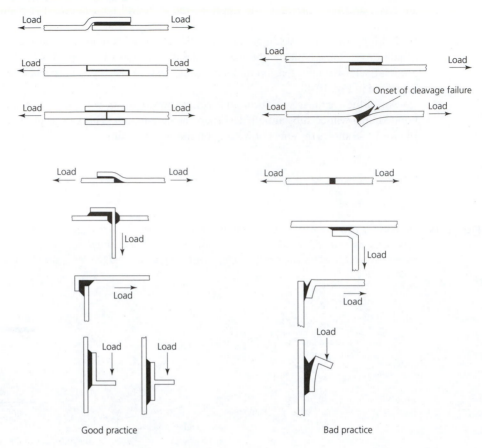

Good practice

Bad practice

Fig. 10.48 *Stressing of bonded joints: (a) tension; (b) cleavage; (c) shear; (d) peel*

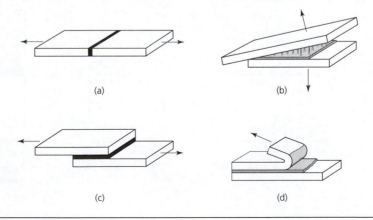

(a)

(b)

(c)

(d)

Fig. 10.49 *Wetting capacity of an adhesive: (a) an adhesive with a poor wetting action does not spread evenly over the joint area, which reduces the effective area and weakens the joint; (b) an adhesive with a good wetting action will flow evenly over the entire joint area, which ensures a sound joint of maximum strength*

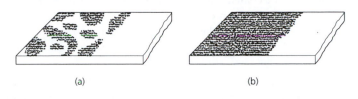

(a)   (b)

10.19 **Thermoplastic adhesives**

Thermoplastic materials are those plastic materials, like polystyrene washing-up bowls, which soften when heated and harden again when cooled. The adhesives derived from this group of synthetic materials may be applied in two ways.

- *Heat activated*   This is where the adhesive is softened by heating until fluid enough to spread freely over the joint surfaces. These are brought into contact immediately, whilst the adhesive is still soft, and pressure is applied until the adhesive has cooled to room temperature and set.
- *Solvent activated*   This is where the adhesive is softened by a volatile solvent. The dissolved adhesive is applied to the joint and a bond is achieved by the solvent evaporating. The cellulose adhesive (balsa cement) used by aero-modellers is an example of a solvent-activated adhesive.

Because the evaporation is essential to the setting of the adhesive, a sound bond is almost impossible to achieve at the centre of a large joint area when joining in impervious materials. This is shown in Fig. 10.50.

Fig. 10.50 *Solvent-activated adhesive fault. Joints made between non-porous adherends (such as metal or plastic) with solvent-activated adhesives may fail due to lack of evaporation of the solvent. This solvent around the edge of the joint sets off, forming a seal and preventing further evaporation of the solvent. This reduces the effective area of the joint and reduces its strength*

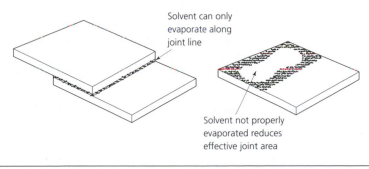

Solvent can only
evaporate along
joint line

Solvent not properly
evaporated reduces
effective joint area

10.20 Impact adhesives

These are solvent-based adhesives which are spread separately on the joint faces and left to dry. Because the coated joint faces are open to the atmosphere, evaporation of the solvent is rapid and total. When dry, the coated joint faces are brought into contact whereupon they instantly bond together by intermolecular attraction. This enables non-absorbent materials to be successfully joined over large contact areas. A typical joint is shown in Fig. 10.51.

Fig. 10.51 *The use of an impact adhesive: (a) adhesive is spread thinly and even on both joint surfaces and left to dry by evaporation (avoiding the problem in Fig. 10.49); (b) when dry, the surfaces are brought into contact – they form an immediate intermolecular bond*

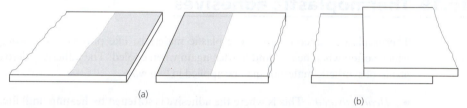

(a)   (b)

Thermoplastic adhesives are based upon synthetic materials such as polyamides, vinyl and acrylic polymers and cellulose derivatives. They are also based upon naturally occurring materials such as resins, shellac, mineral waxes and rubber. Such adhesives are not as strong as the thermosetting adhesives but, being generally more flexible, are more suitable for joining non-rigid materials. Unfortunately, they are heat sensitive and lose their strength rapidly as the temperature rises. For example, natural glues become liquid at the temperature of boiling water.

10.21 Thermosetting adhesives

Thermosetting plastic materials change chemically when heated (cured) and can never again be softened. This makes them less heat sensitive than thermoplastic adhesives. The heat necessary to cure the adhesive can be applied externally, as when phenolic resins are used, or the heat may be generated internally by chemical reaction (addition of a chemical hardener), as when epoxy and polystyrene resins are used.

Since the setting process is a chemical reaction and is not dependent on the evaporation of a solvent, the area of the joint does not affect the setting process. Thermosetting adhesives are very strong and are used to make structural joints between high strength materials such as metals. Joints in the body shells of motor cars and stressed components in aircraft are increasingly dependent upon high-strength adhesives in place of welding and riveting.

- The stresses are more uniformly transmitted from one component of the joint to the other.
- The joints are sealed against corrosion.

- Dissimilar materials can be joined without the risk of electrochemical attack.
- The relatively low curing temperatures, compared with welding temperatures, do not adversely affect the structure and properties of the materials being joined.

Unfortunately thermosetting materials tend to be rigid when cured, so they are unsuitable for joining flexible material or structures, or for making joints which are subjected to high levels of vibration.

10.22 Safety in the use of adhesives

One of the great advantages of naturally occurring glues and gums is that they are neither toxic nor narcotic. Further, they are not particularly flammable. Therefore they are widely used for labelling and packaging foodstuffs, and for the adhesives on stamps and envelope flaps, which have to be licked.

Unfortunately most synthetic adhesives and their solvents, hardeners, catalysts, etc., are toxic and narcotic to some degree. Also their solvents are invariably highly flammable. Therefore these adhesives, together with their solvents, hardeners and catalysts, must be stored and used only in *well-ventilated conditions* and the working area must be declared a *no-smoking zone*. The health hazards associated with these materials range from dermatitis and sensitisation of your skin, to permanent damage to your brain, liver, kidneys and other internal organs if inhaled or accidentally swallowed.

All safety regulations concerning these materials must be rigidly adhered to (if you'll excuse the pun) and any protective clothing provided must be worn.

10.22.1 *Precautions*

- Use only in well-ventilated areas.
- Wear protective clothing appropriate to the process, no matter however inconvenient.
- If you don't wear gloves, protect your hands with a barrier cream.
- After use, wash thoroughly in soap and water; do not use solvents except under medical supervision.
- Do not smoke in the presence of solvents. Not only are they highly flammable but, when the vapours are drawn in through a cigarette or a pipe, some of the vapours change chemically into highly poisonous gases.

SELF-ASSESSMENT TASK 10.7

1. State the essential differences between glues, gums and synthetic adhesives and suggest an appropriate use for each.

2. Explain the difference between heat-activated and solvent-activated adhesives.

3. Explain what is meant by an 'impact' adhesive and suggest an appropriate application.

4. Discuss the precautions that should be taken when working with solvent-activated adhesives.

10.1 Draw a section through a typical sand mould for a hollow casting and describe the essential parts of the mould and the technique for producing the mould.

10.2 describe the following defects which may be found in castings:
(a) blow holes
(b) porosity
(c) scabs
(d) fins
(e) cold shuts
(f) uneven wall thickness
(g) drawing

10.3 Discuss the advantages and limitations of shaping material by:
(a) machining from the solid
(b) forging to shape

10.4 Describe **three** hot-working processes which exploit the property of malleability in a metal.

10.5 (a) Describe **two** cold-working processes which exploit the property of ductility in a metal.
(b) Describe **two** cold-working processes which exploit the property of malleability in a metal.

10.6 With the aid of sketches, explain the principles of injection moulding polymeric materials and list the criteria essential for the production of a polymeric material and list the criteria essential for the production of a successful moulding. State the group of polymeric materials which are usually injection moulded.

10.7 Draw a section through a simple mould suitable for thermosetting polymeric materials and explain the importance of the following features:
(a) ejector
(b) vents
(c) flash gutter
(d) flash land
(e) draught

10.8 (a) Estimate the recovery temperature for cold-worked sheet copper if its melting temperature is taken as 1080 °C.
(b) Estimate the recrystallisation temperature for cold-worked aluminium sheet if its melting temperature is taken as 660 °C.

10.9 Describe the essential differences between the processes of soft soldering, hard soldering and fusion welding.

10.10 Explain what is meant by the *heat-affected zone* of a welded joint, and discuss the changes in structure that are likely to be found in this zone.

10.11 Explain how a compression joint exploits the properties of the component materials, and state the properties required.

10.12 Explain how thermoplastic sheet materials can be joined by:
(a) heat welding
(b) solvent welding

10.13 Discuss the advantages and limitations of joining metals by adhesive bonding as an alternative to mechanical and welded joining techniques.

10.14 Discuss the main causes of failure in an adhesive-bonded joint.

10.15 Discuss the differences between an impact adhesive and a solvent-activated adhesive. State where each should be used.

11 Materials testing (destructive)

The topic areas covered in this chapter are:

- Properties of materials.
- Tensile test, proof stress, interpretation of test results, effect of grain size and structure, testing polymeric materials.
- Impact testing, Izod test, Charpy test, interpretation of test results, effect of processing on toughness.
- Hardness testing, Brinell test, Vickers diamond pyramid test, Rockwell test, Shore scleroscope test, effect of processing on hardness, testing polymeric materials, comparative hardness scales.
- Ductility testing, bend tests, Erichson cupping test.

11.1 Properties of materials

Although the properties of materials were introduced in Chapter 1, these properties are so heavily influenced by the composition, processing and heat treatment to which the material has been subjected, that further consideration of material properties and the testing of those properties has been delayed until this chapter. Now that the composition, processing and heat treatment of a range of metallic and non-metallic materials widely used by the engineer have been described, you should more readily be able to understand the problems and techniques associated with the testing of material properties.

11.2 Tensile test

Strength has already been defined as the ability of a material to resist applied forces without yielding or fracturing. By convention, strength usually denotes the resistance of a material to a tensile load applied axially to a specimen; this is the principle of the *tensile test*. Figure 11.1(a) shows a popular bench-mounted tensile testing machine, whilst Fig. 11.1(b) shows a more sophisticated machine suitable for industrial and research laboratories. This latter machine is capable of performing compression, shear and bending tests as well as tensile tests.

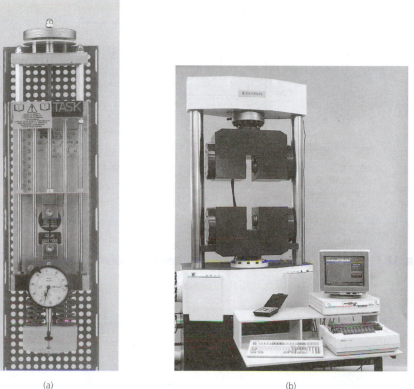

(a)   (b)

Both these machines apply a carefully controlled tensile load to a standard specimen and measure the corresponding extension of that specimen. Figure 11.2 shows some standard specimens and the direction of the applied load. These specimens are based upon British Standard BS 18. For the test results to be consistent for any given material, it is most important that the standard dimensions and profiles are adhered to.

The shoulder radii are particularly critical and small variations, or the presence of tooling marks, can cause considerable differences in the test data obtained. Flat specimens are usually machined only on their edges so that the plate or sheet surface finish, and any structural deformation at the surface caused by the rolling process, is taken into account in the test results.

Let's now look at Fig. 11.3. In this figure, the *gauge length* (L_o) is the length over which the elongation of the specimen is measured. The *minimum parallel length* (L_c) is the minimum length over which the specimen must maintain a constant cross-sectional area before the test load is applied. The lengths L_o, L_c, L_1, and the cross-sectional area (A) are all specified in BS 18.

Fig. 11.2 *Tensile test specimens: (a) turned specimen for wedge grips; (b) sheet specimen for wedge grips; (c) sheet specimens for pin-jointed grips; (d) turned specimen with screwed ends. (For specimen dimensions and details see BS 18)*

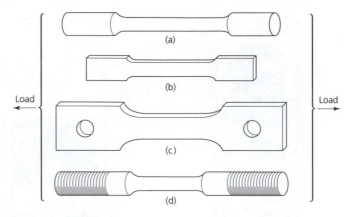

Fig. 11.3 *Proportions of tensile test specimens: (a) cylindrical; (b) flat*

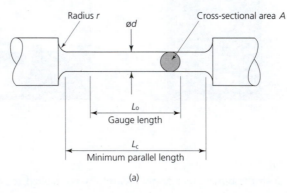

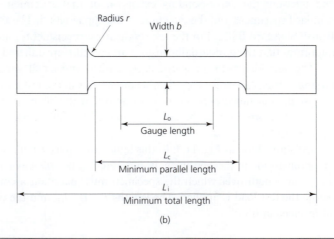

Cylindrical test specimens are proportioned so that the gauge length L_o and the cross-sectional area A maintain a constant relationship. Therefore such specimens are called *proportional test pieces*. The relationship is given by the expression:

$$L_o = 5.658\sqrt{A}$$

Since $\quad A = (\pi d^2)/4$

$$\sqrt{A} = 0.886d$$

$\therefore \quad\quad L_o = 5.658 \times 0.886d$

$$= 5.013d$$

$$= \mathbf{5d}\,(\text{approx.})$$

Therefore a specimen 5 mm diameter will have a gauge length of 25 mm.

The elongation obtained for a given force depends upon the length and area of the cross-section of the specimen or component, since:

$$\text{elongation} = \frac{\text{applied force} \times L}{E \times A}$$

where $\quad L = $ length

$A = $ cross-sectional area

$E = $ elastic modulus

Therefore if the ratio L/A is kept constant (as it is in a proportional test piece), and E remains constant for a given material, then comparisons can be made between elongation and applied force for specimens of different sizes.

11.3 Tensile test results

Let's now look at the sort of results we would get from a typical tensile test on a piece of annealed low-carbon steel. The load applied to the specimen and the corresponding extension can be plotted in the form of a graph, as shown in Fig. 11.4.

Fig. 11.4 *Load–extension curve for a low-carbon steel*

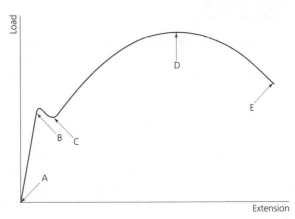

- From A to B the extension is proportional to the applied load. Also, if the applied load is removed the specimen returns to its original length. Under these relatively lightly loaded conditions the material is showing *elastic* properties.
- From B to C it can be seen from the graph that the metal suddenly extends with no increase in load. If the load is removed at this point the metal will not spring back to its original length and it is said to have taken a *permanent set*. This is the *yield point*.
- The *yield stress* is the stress at the yield point; that is, the load at B divided by the original cross-section area of the specimen. Usually, a designer works at 50 per cent of this figure to allow for a 'factor of safety' (see Section 13.1).
- From C to D extension is no longer proportional to the load, and if the load is removed little or no spring back will occur. Under these relatively greater loads the material is showing *plastic* properties.
- The point D is referred to as the 'ultimate tensile strength' when referred to load–extension graphs or the 'ultimate tensile stress' (UTS) when referred to stress–strain graphs. The ultimate tensile stress is calculated by dividing the load at D by the original cross-sectional area of the specimen. Although a useful figure for comparing the relative strengths of materials, it has little practical value since engineering equipment is not usually operated so near to the breaking point.
- From D to E the specimen appears to be stretching under reduced load conditions. In fact the specimen is thinning out (necking) so that the 'load per unit area' or stress is actually increasing. The specimen finally work hardens to such an extent that it breaks at E.
- In practice, values of load and extension are of limited use since they apply only to one particular size of specimen and it is more usual to plot the *stress–strain* curve. (An example of a stress–strain curve for a low-carbon steel is shown in Fig. 11.6(a).)
- Stress and strain are calculated as follows:

$$\text{stress} = \frac{\text{load}}{\text{area of cross-section}}$$

$$\text{strain} = \frac{\text{extension}}{\text{original length}}$$

11.4 Proof Stress

Only very ductile materials, such as fully annealed mild steel, show a clearly defined yield point. The yield point will not even appear on bright drawn low-carbon steel which has become slightly work hardened during the drawing process. In such circumstances the *proof stress* is used. The proof stress is defined as the stress that produces a specified amount of plastic strain, such as 0.1 or 0.2 per cent. Figure 11.5 shows a typical stress–strain curve for a material of relatively low ductility such as hardened and tempered medium-carbon steel.

Let's now consider the point C in Fig. 11.5. The corresponding strain is given by D and this consists of a combination of plastic and elastic components. If the stress is now gradually reduced (by reducing the load on the specimen), the strain is also reduced and the stress–strain relationship during this reduction in stress is represented by the line CB.

Fig. 11.5 *Proof stress*

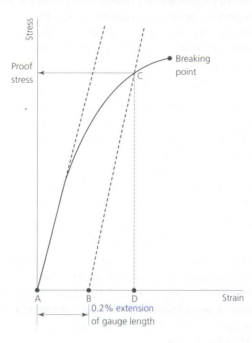

During the reduction in stress the elastic deformation is recovered so that the line CB is straight and parallel to the initial stages of the loading curve for the material; that is, the part of the loading curve where the material is showing elastic properties.

In our example, the stress at C has produced a plastic strain of 0.2 per cent as represented by AB. Thus the stress at C is referred to as *0.2 per cent proof stress*. So AB represents the plastic deformation and BD represents the elastic deformation when the specimen is stressed to the point C.

The material will have fulfilled its specification if, after the proof stress has been applied for 15 seconds and removed, the permanent set of the specimen is not greater than the specified percentage of the gauge length which, in this example, is 0.2 per cent.

11.5 The interpretation of tensile test results

The interpretation of tensile test data requires skill borne out of experience, since many factors can affect the test results – for instance, the temperature at which the test is carried out, since the tensile modulus and tensile strength decrease as the temperature rises for most metals and plastics, whereas the ductility increases as the temperature rises. The test results are also influenced by the rate at which the specimen is strained.

Figure 11.6(a) shows a typical stress–strain curve for an annealed mild steel. From such a curve we can deduce the following information.

- The material is ductile since there is a long elastic range.
- The material is fairly rigid since the slope of the initial elastic range is steep.

- The limit of proportionality (elastic limit) occurs at about 230 MPa.
- The upper yield point occurs at about 260 MPa.
- The lower yield point occurs at about 230 MPa.
- The ultimate tensile stress (UTS) occurs at about 400 MPa.

Figure 11.6(b) shows a typical stress–strain curve for a grey cast iron. From such a curve we can deduce the following information:

- The material is brittle since there is little plastic deformation before it fractures.
- Again the material is fairly rigid since the slope of the initial elastic range is steep.
- It is difficult to determine the point at which the limit of proportionality occurs, but it is approximately 200 MPa.
- The ultimate tensile stress (UTS) is the same as the breaking stress for this sample. This indicates negligible reduction in cross-section (necking) and minimal ductility and malleability. It occurs at approximately 250 MPa.

Figure 11.6(c) shows a typical stress–strain curve for a wrought light alloy. From this curve we can deduce the following information:

- The material has a high level of ductility since it shows a long plastic range.
- The material is much less rigid than either low-carbon steel or cast iron since the slope of the initial plastic range is much less steep when plotted to the same scale.
- The limit of proportionality is almost impossible to determine, so the proof stress will be specified instead. For this sample a 0.2 per cent proof stress is approximately 500 MPa (the line AB).

Fig. 11.6 *Typical stress–strain curves: (a) annealed mild steel; (b) grey cast iron; (c) a light alloy*

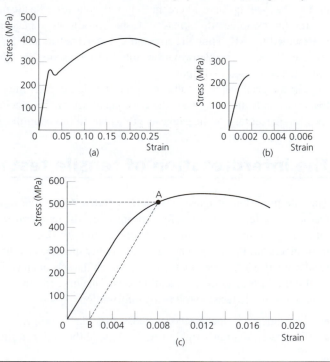

A tensile test can also yield other important facts about a material under test. For example, it can enable the elastic modulus (E) for the material to be calculated.

11.5.1 *Young's modulus of elasticity*

The physicist Robert Hooke found that within its elastic range the strain produced in a material is proportional to the stress applied. It was left to Thomas Young to quantify this law in terms of a mathematical constant for any given material.

strain $\propto$ stress

therefore $\dfrac{\text{stress}}{\text{strain}} = \text{constant } (E)$

This constant term (E) is variously known as 'Young's modulus', the 'modulus of elasticity' or the 'tensile modulus'. Thus:

$$E = \frac{\text{tensile or compressive stress}}{\text{strain}}$$

$$= \frac{(\text{force})/(\text{original cross-sectional area})}{(\text{change in length})/(\text{original length})}$$

Example 11.1 demonstrates how to calculate modulus from test data.

EXAMPLE 11.1

Calculate the modulus of elasticity for a material which produces the following data when undergoing test:

Applied load	*35.7 kN*
Cross-sectional area	*25 mm²*
Gauge length	*28 mm*
Extension	*0.2 mm*

$$E = \frac{\text{stress}}{\text{strain}}$$

where $\quad \text{stress} = \dfrac{35.7\,\text{kN}}{25\,\text{mm}^2}$

and $\quad \text{strain} = \dfrac{0.2\,\text{mm}}{28\,\text{mm}}$

therefore $\quad E = \dfrac{35.7 \times 28}{25 \times 0.2}$

$\qquad\qquad = 199.92\,\text{kN/mm}^2$

$\qquad\qquad = \mathbf{200\,GPa}$ (approx.)

This would be a typical value for a low-carbon steel.

It was stated earlier that malleability and ductility are special cases of the general property of plasticity.

- *Malleability* This refers to the extent to which a material can undergo deformation in compression before failure occurs.
- *Ductility* This refers to the extent to which a material can undergo deformation in tension before failure occurs.

All ductile materials are malleable, but not all malleable materials are ductile since they may lack the strength to withstand tensile loading.

Therefore ductility is usually expressed, for practical purposes, as the *percentage elongation* in gauge length of a standard test piece at the point of fracture when subjected to a tensile test to destruction.

$$\text{elongation } \% = \frac{\text{increase in length} \times 100}{\text{original length}}$$

The increase in length is determined by fitting the pieces of the fractured specimen together carefully and measuring the length at failure.

increase in length (elongation) = length at failure − original length

Fig. 11.7 *Elongation*

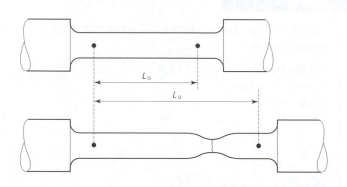

Figure 11.7 shows a specimen for a soft, ductile material before and after testing. It can be seen that the specimen does not reduce in cross-sectional area uniformly, but that severe local necking occurs prior to fracture. Since most of the plastic deformation and, therefore, most of the elongation occurs in the necked region, doubling the gauge length does not double the elongation when calculated as a percentage of gauge length. Therefore it is important to use a standard gauge length if comparability between results is to be achieved. Elongation is calculated as follows (see Example 11.2):

$$\text{elongation } \% = \frac{L_u - L_o}{L_o} \times 100$$

Calculate the percentage elongation of a 70/30 brass alloy, if the original gauge length (L_o) *is 56 mm and the length at fracture* (L_u) *is 95.2 mm.*

$$\text{elongation \%} = \frac{L_u - L_o}{L_o} \times 100$$

$$= \frac{95.2 - 56}{56} \times 100$$

$$= \mathbf{70\%}$$

Table 11.1 lists some typical test results for a range of metallic materials.

Table 11.1 *Typical tensile test results for metallic specimens*

		Tensile strength			
Material	Condition	Yield point (MPa)	0.1% PS (MPa)	UTS (MPa)	Elongation (%)
Aluminium bronze	Annealed	—	125	385	70
	Hard	—	590	775	4
Cartridge brass (70/30)	Annealed	—	77	325	70
	Hard	—	510	695	5
Commercial brass (63/37)	Annealed	—	95	340	55
	Hard	—	540	725	4
Muntz brass (60/40)	Extruded	—	110	370	40
Cupro-nickel (256 Ni)	Annealed	—	—	355	45
	Hard	—	—	600	5
Duralumin	Age hardened	—	280	400	10
Grey cast iron	—	—	—	150/250	negligible
Low-tin bronze	Annealed	—	110	340	65
	Hard	—	620	740	5
Magnesium alloy	Wrought	—	155	280	10
Medium carbon steel	Quench hardened and tempered at 550–650 °C	550	—	750	14
Low-carbon steel	Annealed	250/350	—	450/525	20/25
Stainless steel (ferritic)	Annealed	340	—	510	31
Stainless steel (austentic 18/8)	Annealed	278	—	618	50
	Cold-rolled	803	—	896	30

11.6 The effect of grain size and structure on tensile testing

The test piece should be chosen so that it reflects as closely as possible the component and the material from which the component is produced. This is relatively easy for components produced from bar stock, but not so easy for components produced from forgings as the grain flow will be influenced by the contour of the component and will not be uniform.

Castings also present problems since the properties of a specially cast test piece are unlikely to reflect those of the actual casting. This is due to the difference in size and the corresponding difference in cooling rates.

The lay of the grain in rolled bar and plate can greatly affect the tensile strength and other properties of a specimen taken from them. Figure 11.8 shows the relative grain orientation for transverse and longitudinal test pieces. The tensile test for the longitudinal test piece will show it to be stronger than the transverse test piece. This is an important factor that the designer of large fabrications must take into account.

Fig. 11.8 *Effect of grain orientation on material testing*

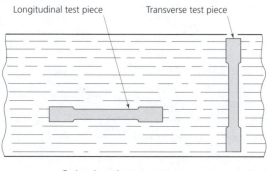

Longitudinal test piece · Transverse test piece

Grain orientation ⟶

Figure 11.9 shows the effect of processing upon the properties of a material. A low-carbon steel of high ductility, in the annealed condition, shows the classical stress–strain curve with a pronounced yield point and a long plastic deformation range. The same material, after finishing by cold drawing, no longer shows a yield point and the plastic range is noticeably reduced.

Figure 11.10 shows the effect of heat treatment upon the properties of a medium-carbon steel. In this example the results have been obtained by quench hardening a batch of identical specimens and then tempering them at different temperatures.

Figure 11.11 shows the effect of heat treatment upon the properties of a work-hardened metallic material. Stress relief (recovery) has very little effect upon the tensile strength and elongation (ductility) until the recrystallisation (annealing) temperature is reached. The metal initially shows the high tensile strength and lack of ductility associated with a severely distorted grain structure.

Fig. 11.9 *Effect of processing on properties of (a) an annealed low-carbon steel and (b) a cold-drawn low-carbon steel*

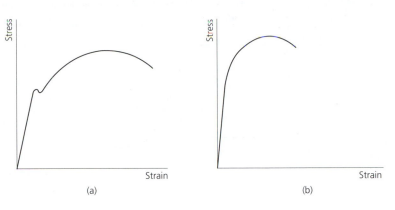

(a)   (b)

Fig. 11.10 *Effect of tempering on tensile test results*

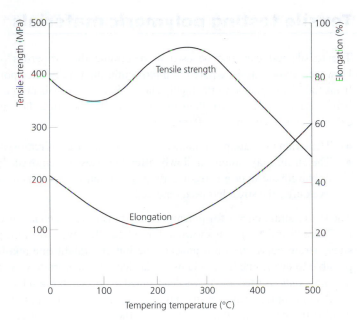

After stress relief the tensile strength rises and the ductility falls until the recrystallisation temperature is reached. During the recrystallisation range there is a marked change in properties. The tensile strength is rapidly reduced and the ductility, in terms of elongation percentage, rapidly increases.

Fig. 11.11 *Effect of temperature on cold-worked material*

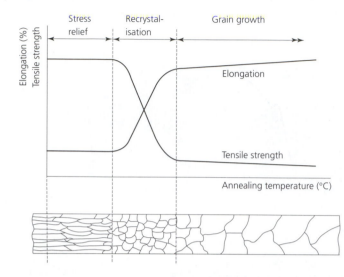

11.7 Tensile testing polymeric materials

The tensile test can also be used to determine the properties of polymeric materials. However, some care is required in interpreting the results since polymeric materials range from the highly plastic to the highly elastic. Some materials even exhibit both extremes of behaviour depending upon the strain rate and temperature. Polymeric materials can also exhibit elastic strain in two different ways:

- The strain may disappear immediately after the stress is removed.
- The strain may disappear slowly after the stress is removed. For example, a foamed polyurethane may not recover its original shape and dimensions until as much as a week after the stress has been removed.

The stress–strain curves for polymeric materials have been classified into five main groups by Carswell and Nason, as shown in Fig. 11.12. However, for some polymeric materials the stress–strain curve does not indicate the initial straight line (elastic) range and it is not possible to determine the modulus of elasticity in the normal way. For such materials, the *secant modulus* is used and this is determined as shown in Fig. 11.13.

An appropriate value of strain is specified and the corresponding stress for that strain is divided by the strain value. In Fig. 11.13 the strain is 0.2 per cent and the corresponding stress is 30 MPa. Hence:

$$\text{secant modulus} = \frac{30 \times 100}{0.2} = \textbf{15 GPa}$$

The test results obtained for polymeric materials are much more influenced by the test conditions than are the corresponding tests for metallic materials. Figure 11.14(a) shows how the temperature can affect the test results obtained. For most rigid polymers the

Fig. 11.12 *Typical stress–strain curves for polymers (after: Carswell and Nason)*

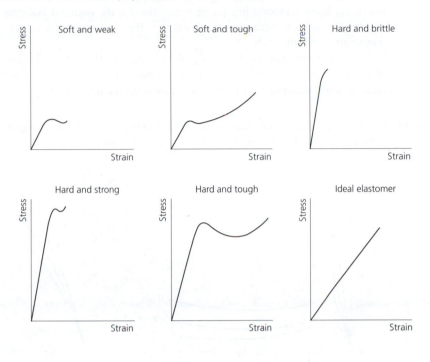

Fig. 11.13 *Secant modulus*

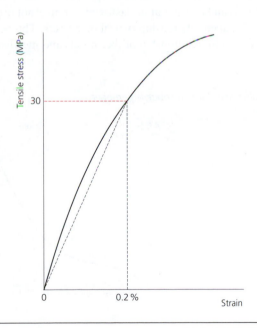

percentage elongation increases as the temperature increases, but the strength is much lower. At lower temperatures (at or below the T_g) the material becomes brittle and fails with negligible elongation. For some rubbers the elongation can actually increase at low temperatures within limits.

The strain rate can also have a significant effect on the results obtained from specimens of a given material. Figure 11.14(b) shows how the strength of the material appears to increase when straining occurs at a lower rate (over a longer time).

Fig. 11.14 *Effects of test conditions on polymers: (a) effect of temperature on tensile strengths of low-density polythene; (b) effect of strain rate on tensile strength of Perspex*

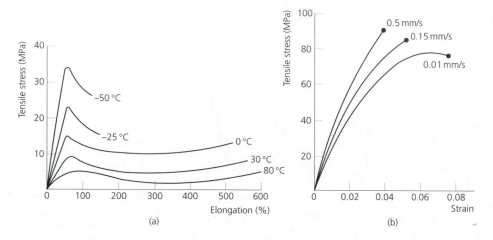

(a)

(b)

The stress–strain behaviour of a typical elastomer (polyisobutylene rubber) is shown in Fig. 11.15. It can be seen that in elastomers strain is not proportional to stress, but that they show the S-shaped relationship typical of rubbers. This is, of course, completely unlike the linear stress–strain relationship of thermosets and metals when stressed within their elastic limits.

Fig. 11.15 *Stress–strain curves for polyisobutylene rubber*

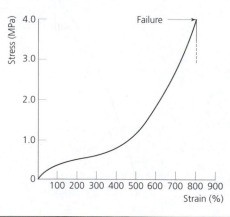

Creep is also an important factor when testing polymeric materials. It is defined as the gradual extension of a material over a period of time when subjected to a constant applied load, and this will be considered in detail in Section 13.2. Table 11.2 lists some typical tensile test results for a range of polymeric materials.

Table 11.2 *Typical tensile test results for polymeric materials*

Material	Type	Tensile strength (MPa)	Elongation (%)
Polythene	Thermoplastic		
Low density		11.0	90–650
High density		31.0	50–500
Polypropylene	Thermoplastic	30.0–35.0	50–600
Polystyrene	Thermoplastic	30.0–50.0	1–35
Polyvinyl chloride (PVC)	Thermoplastic		
Rigid		49.0	10–30
Flexible		7.0–25.0	250–350
Polytetrafluoroethylene (PTFE)	Thermoplastic	17.0–25.0	200–600
Polymethyl methacrylate (Perspex)	Thermoplastic	50.0–70.0	3–8
Polyamide (nylon '66')	Thermoplastic	50.0–87.5	60–300
Polycarbonate	Thermoplastic	60.0–70.0	60–100
Cellulose acetate	Thermoplastic	24.0–65.0	5–55
Phenol–formaldehyde	Thermosetting plastic	35–55 (wood floor filler)	1
Urea–formaldehyde	Thermosetting plastic	50–75 (cellulose filler)	1
Melamine–formaldehyde	Thermosetting plastic	55–80 (cellulose filler)	0.7
Epoxy resin rigid and unfilled	Thermosetting plastic	35–80	5–10

11.8 Impact testing

The tensile test does not tell the whole story. Figure 11.16 shows how a piece of high-carbon steel rod will bend when in the annealed condition, yet snap easily in the quench-hardened condition despite the fact that in the latter condition it will show a much higher value of tensile strength. Impact tests consist of striking a suitable specimen with a controlled blow and measuring the energy absorbed in bending or breaking the specimen. The energy value indicates the toughness of the material under test.

Figure 11.17 shows a typical impact-testing machine. This machine has a hammer which is suspended like a pendulum, a vice for holding the specimen in the correct position relative to the hammer and a dial for indicating the energy absorbed in carrying out the test in joules (J).

- If there is maximum overswing, as there would be if no specimen was placed in the vice, then zero energy absorption is indicated.
- If the hammer is stopped by the specimen with no overswing, then maximum energy absorption is indicated.
- Intermediate readings are the impact values (J) of the materials being tested (their toughness or lack of brittleness).

Fig. 11.16 *Impact loading: (a) a rod of high-carbon (1.0%) steel in the annealed (soft) condition will bend when struck with a hammer (UTS 925 MPa); (b) after hardening and lightly tempering, the same piece of steel will fracture when hit with a hammer despite its UTS having increased to 1285 MPa*

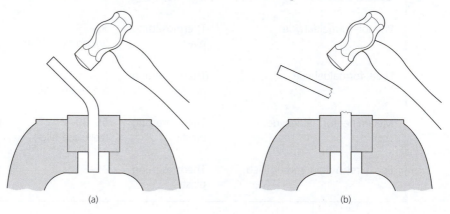

(a)   (b)

Fig. 11.17 *Typical impact-testing machine*

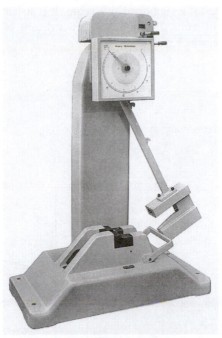

There are two standard impact tests currently in use.

• *The Izod test*   In this test a 10 mm square, notched specimen is used. The striker of the pendulum hits the specimen with a kinetic energy of 162.72 J at a velocity of 3.8 m/s. Figure 11.18 shows details of the specimen and the manner in which it is supported.

Fig. 11.18 *Izod test (all dimensions in millimetres): (a) detail of notch; (b) section of test piece (at notch); (c) position of striker*

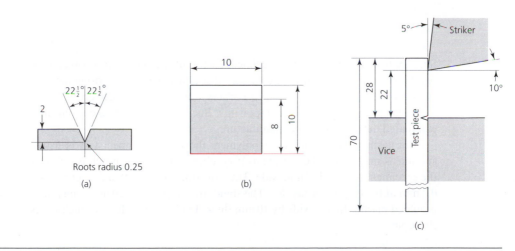

- *The Charpy test*   In the Izod test the specimen is supported as a cantilever, but in the Charpy test it is supported as a beam. It is struck with a kinetic energy of 298.3 J at a velocity of 5 m/s. Figure 11.19 shows details of the Charpy test specimen and the manner in which it is supported.

Fig. 11.19   *Charpy test (all dimensions in millimetres)*

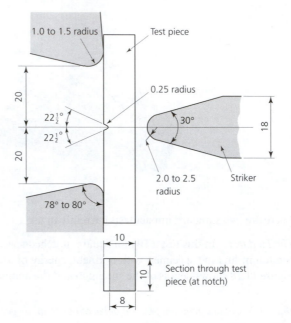

Since both tests use a notched specimen, useful information can be obtained regarding the resistance of the material to the spread of a crack which may originate from a point of stress concentration such as sharp corners, undercuts, sudden changes in section, and machining marks in stressed components. Such points of stress concentration should be eliminated during design and manufacture.

For testing the toughness of polymers the Charpy impact test is usually used. The specimens may be notched or plain. If notched, a U-shaped notch is milled in one side of the specimen, the slot is 2 mm wide, 3 mm or 5 mm deep, with a corner radius not less than 0.2 mm at the bottom of the slot. The standard test piece is 120 mm long and, in the case of moulded plastic, 15 mm wide by 10 mm thick. Different widths and thicknesses are used for sheet plastic.

11.9 The interpretation of impact tests

The results of an impact test should specify the energy used to bend or break the specimen and the particular test used, i.e. Izod or Charpy. In the case of the Charpy test it is also necessary to specify the type of notch used as this test allows for three types of notch, as shown in Fig. 11.20. A visual examination of the fractured surface after the test also provides useful information.

Fig. 11.20 *Standard Charpy notches (all dimensions in millimetres): (a) U notch; (b) keyhole notch; (c) V notch*

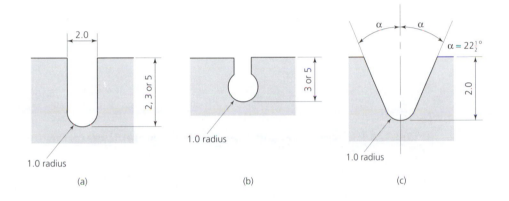

- *Brittle metals* A clean break with little deformation and little reduction in cross-sectional area at the point of fracture. The fractured surfaces will show a granular structure.
- *Ductile metals* The fracture will be rough and fibrous. In very ductile materials the fracture will not be complete – the specimen bending over and only showing slight tearing from the notch. There will also be some reduction in cross-sectional area at the point of fracture or bending.
- *Brittle polymers* A clean break showing smooth, glassy, fractured surfaces with some splintering.
- *Ductile polymers* No distinctive appearance to the fracture except for a considerable reduction in cross-sectional area and some tearing of the notch. Very ductile polymers may simply bend with some tearing at the notch, but no complete fracture will occur.

The temperature of the specimen at the time of making the test also has an important influence on the test results. Figure 11.21 shows the embrittlement of low-carbon steels at refrigerated temperatures, and hence their unsuitability for use in refrigeration plant and space vehicles. Some typical impact test results for a range of metallic materials and a range of polymers are given in Tables 11.3 and 11.4.

Whilst the energy absorbed on impact for metallic specimens is usually stated in joules (J), the impact energy absorbed by polymer specimens is often divided by the cross-sectional area of the specimen for un-notched specimens, or the area behind the notch for notched specimens.

Fig. 11.21 *Effect of test temperatures on toughness*

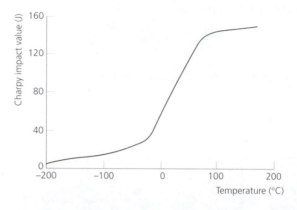

Table 11.3 *Typical impact test results for metallic materials*

Material	Condition	Charpy V (J)
Aluminium bronze	Annealed	100
	Hard	25
Cartridge brass (70/30)	Annealed	88
	Hard	20
Commercial brass (63/37)	Annealed	65
	Hard	12
Muntz brass (60/40)	Extruded	6
Cupro-nickel (25% Ni)	Annealed	157
	Hard	35
Duralumin	Age hardened	80
Grey cast iron	—	2.5
Magnesium alloy	Wrought	12
Medium-carbon steel	Quenched hardened and tempered at 550–650 °C	65
Low-carbon steel	As hot rolled	50
Stainless steel (ferritic)	Annealed	165
Stainless steel (austenitic)	Annealed	217

Table 11.4 *Typical impact test results for polymeric materials*

Material	Type	Izod value (J)
Polythene	Thermoplastic	
Low density		2–18
High density		25–30
Polypropylene	Thermoplastic	1–10
Polystyrene	Thermoplastic	0.25–2.5
Polyvinyl chloride (PVC)	Thermoplastic	
Rigid		1.5–18.0
Flexible		no fracture
Polytetrafluoroethylene (PTFE)	Thermoplastic	3.0–5.0
Polymethyl methacrylate (Perspex)	Thermoplastic	3–8
Polyamide (nylon '66')	Thermoplastic	1.5–15.0
Polycarbonate	Thermoplastic	10–20
Cellulose acetate	Thermoplastic	5–55
Phenol–formaldehyde	Thermosetting plastic	0.3–1.0 (wood floor filler)
Urea–formaldehyde	Thermosetting plastic	0.3–0.5 (cellulose filler)
Melamine–formaldehyde	Thermosetting plastic	0.2–0.5 (cellulose filler)
Epoxy resin rigid and unfilled	Thermosetting plastic	0.5–1.5

11.10 The effect of processing on toughness

Impact tests are frequently used to determine the effectiveness of annealing temperatures on the grain structure and impact strength of cold-worked, ductile metals. In the case of cold-worked low-carbon steel, the impact strength is quite low, initially, as the heavily deformed grain structure will be relatively brittle and lacking in ductility, particularly if the limit of cold working has been approached. Annealing at low temperatures has little effect as it only promotes recovery of the crystal lattice on the atomic scale and does not result in recrystallisation. In fact, during recovery there may even be a slight reduction in the impact strength.

However, at about 550–650 °C recrystallisation of low-carbon steels occurs with only slight grain growth. Annealing in this temperature range results in the impact strength increasing dramatically, as shown in Fig. 11.22, and the appearance of the fracture changes from that of a brittle material to that of a ductile material. Annealing at higher temperatures or prolonged soaking at the lower annealing temperature results in grain growth and a corresponding fall in impact strength.

Fig. 11.22 *Effect of annealing on the toughness of a low-carbon steel*

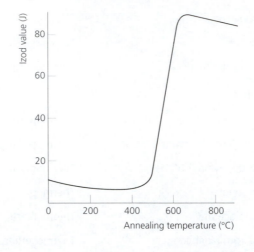

The effect of tempering on the impact value of a quench-hardened high-carbon steel is shown in Fig. 11.23. Initially, only stress relief occurs but, as the tempering temperature increases, the toughness also increases which is why cutting tools are tempered. Tempering modifies the extremely hard and brittle martensitic structure of quench-hardened plain carbon steels and causes a considerable increase in toughness with very little loss of hardness.

Fig. 11.23 *Effect of tempering on the toughness of a quench-hardened high-carbon steel*

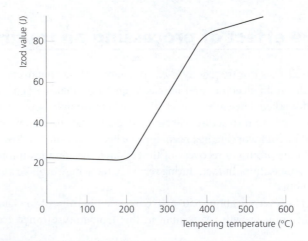

1. Name the material property that is determined by impact testing.

2. Describe the essential differences between the Izod test and the Charpy test.

3. Describe the differences in fracture appearances for:

   (a) brittle metals
   (b) ductile metals
   (c) brittle polymers

11.11 Hardness testing

Hardness has already been defined in Section 1.3 as the resistance of a material to indentation or abrasion by another hard body. It is by indentation that most hardness tests are performed. A hard indenter is pressed into the specimen by a standard load, and the magnitude of the indentation (either area or depth) is taken as a measure of hardness.

11.11.1 *The Brinell hardness test*

In this test, hardness is measured by pressing a hard steel ball into the surface of the test piece, using a known load. It is important to choose the combination of load and ball size carefully so that the indentation is free from distortion and suitable for measurement. The relationship between load P (kg) and the diameter D (mm) of the hardened ball indenter is given by the expression:

$$\frac{P}{D^2} = K$$

Where K is a constant; typical values of K are:

Ferrous metals	$K = 30$
Copper and copper alloys	$K = 10$
Aluminium and aluminium alloys	$K = 5$
Lead, tin and white-bearing metals	$K = 1$

Thus, for steel, a load of 3000 kg is required if a 10 mm diameter ball indenter is used.

Figure 11.24 shows how the Brinell hardness value is determined. The diameter of the indentation is measured in two directions at right angles and the average taken. The diameter is measured either by using a microsope scale, or by a projection screen with micrometer adjustment, similar to that used for the Vickers test as shown later in Fig. 11.26. In practice, conversion tables are used to translate the value of diameter (d) directly into hardness numbers H_B.

Fig. 11.24 *Principle of the Brinell hardness test*

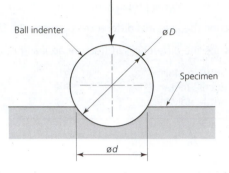

To ensure consistent results, the following precautions should be observed:

- The thickness of the specimen should be at least seven times the depth of the indentation to allow unrestricted plastic flow below the indenter.
- The edge of the indentation should be at least three times the diameter of the indentation from the edge of the test piece.
- The test is unsuitable for materials whose hardness exceeds 500 H_B, as the ball indenter tends to flatten.

Data other than hardness can be estimated from this test as follows.

- *Machinability*   With high-speed steel cutting tools, the hardness of the stock being cut should not exceed $H_B = 350$ for a reasonable tool life. Again, materials with a hardness less than $H_B = 100$ will tend to tear and leave a poor surface finish.
- *Tensile strength*   Since there is a definite relationship between strength and hardness (Sections 10.52 and 11.8), the ultimate tensile stress (UTS) of a component can be approximated as follows:

  UTS (MPa) = $H_B \times 3.54$ (for annealed plain carbon steels)

  = $H_B \times 3.25$ (for quench-hardened and tempered plain carbon steels)

  = $H_B \times 5.6$ (for ductile brass alloys)

  = $H_B \times 4.2$ (for wrought aluminium alloys).
- *Work-hardening capacity*   Materials which will cold work without work hardening unduly will pile up round the indenter, as shown in Fig. 11.25(a). Materials which work harden readily will sink around the indenter, as shown in Fig. 11.25(b).

Fig. 11.25 *Work-hardening capacity: (a) piling up; (b) sinking*

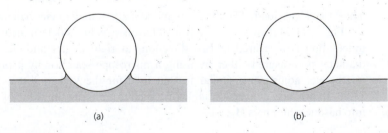

(a)   (b)

11.11.2 *The Vickers hardness test*

This test is preferable to the Brinell test where hard materials are concerned, as it uses a diamond indenter. (Diamond is the hardest material known – approximately 6000 H_B.) The diamond indenter is in the form of a square-based pyramid with an angle of 1360 between opposite faces.

Since only one type of indenter is used the load has to be varied for different hardness ranges. Standard loads are 5, 10, 20, 30, 50 and 100 kg. It is necessary to state the load when specifiying a Vickers hardness number. For example, if the hardness number is found to be 200 when using a 50 kg load, then the hardness number is written as $H_D(50) = 200$.

Figure 11.26(a) shows a universal hardness testing machine suitable for performing both Brinell and Vickers hardness tests, whilst Fig. 11.26(b) shows the measuring screen for determining the distance across the corners of the indentation. The screen can be rotated so that two readings at right angles can be taken and the average is used to determine the hardness number (H_D). This is calculated by dividing the load by the projected area of the indentation:

$$H_D = \frac{P}{D^2}$$

where D = the average diagonal (mm)

P = load (kg)

Fig. 11.26 *Hardness testing: (a) universal hardness-testing machine; (b) measuring screen showing magnified image of Vickers impression*

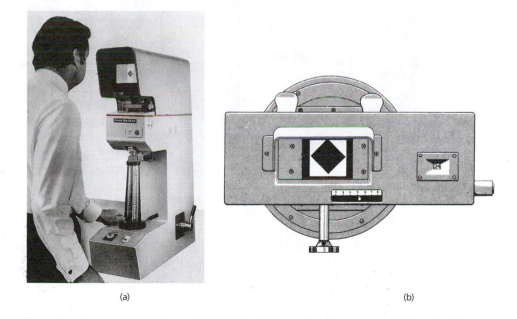

(a)   (b)

11.11.3 *The Rockwell hardness test*

Although not as reliable as the Brinell and Vickers hardness tests for laboratory purposes, the Rockwell test is widely used in industry as it is quick, simple and direct reading. Figure 11.27(a) shows a typical Rockwell hardness testing machine, whilst Fig. 11.27(b) shows the appearance of the hardness indicating scales. Universal electronic hardness testing machines are now widely used which, at the turn of a switch, can provide either Brinell, Vickers or Rockwell tests and show the hardness number as a digital readout automatically. They also give a 'hard copy' printout of the test result together with the test conditions and date. However, the mechanical testing machines described in this chapter are still widely used, and will be for some time to come.

Fig. 11.27 *Rockwell hardness test: (a) Rockwell direct reading testing machine; (b) direct reading hardness scale*

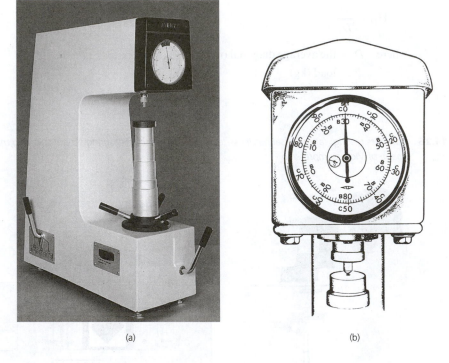

(a)   (b)

In principle the Rockwell hardness test compares the difference in depth of penetration of the indenter when using forces of two different values. That is, a minor force is first applied (to take up the backlash and pierce the skin of the component) and the scales are set to read zero. Then a major force is applied over and above the minor force and the increased depth of penetration is shown on the scales of the machine as a direct reading of hardness without the need for calculation or conversion tables.

The indenters most commonly used are a 1.6 mm diameter hard steel ball and a diamond cone with an apex angle of 120°. The minor force in each instance is 98 N. Table 11.5 gives the combinations of type of indenter and additional (major) force for the range of Rockwell scales, together with typical applications. The B and C scales are the most widely used in engineering.

Table 11.5 Rockwell hardness test conditions

Scale	Indenter	Additional force (kN)	Applications
A	Diamond cone	0.59	Steel sheet; shallow case-hardened components
B	Ball, 1.588 mm dia.	0.98	Copper alloys; aluminium alloys, and annealed low-carbon steels
C	Diamond cone	1.47	Most widely used range: hardened steels; cast irons; deep case-hardened components
D	Diamond cone	0.98	Thin but hard steel – medium depth case-hardened compounds
E	Ball, 3.175 mm dia.	0.98	Cast iron, aluminium alloys; magnesium alloys, bearing metals
F	Ball, 1.588 mm dia.	0.59	Annealed copper alloys, thin soft sheet metals
G	Ball, 1.588 mm dia.	1.47	Malleable irons; phosphor bronze; gunmetal; cupro-nickel alloys, etc
H	Ball, 3.175 mm dia.	0.59	Soft materials; high ferritic aluminium, lead, zinc
K	Ball, 3.175 mm dia.	1.47	Aluminium and magnesium alloys
L	Ball, 6.350 mm dia.	0.59	Plastics: thermoplastic
M	Ball, 6.350 mm dia.	0.98	Plastics: thermoplastic
P	Ball, 6.350 mm dia.	1.47	Plastics: thermosetting
R	Ball, 12.70 mm dia.	0.59	Very soft plastics and rubbers
S	Ball, 12.70 mm dia.	0.98	—
V	Ball, 12.70 mm dia.	1.47	—

The standard Rockwell test cannot be used for very thin sheet and foils, and for these the Rockwell superficial hardness test is used. The minor force is reduced from 98 N to 29.4 N and the major force is also reduced. Typical values are listed in Table 11.6.

Table 11.6 *Rockwell superficial hardness test conditions*

Scale	Indenter	Additional force (kN)
15-N	Diamond cone	0.14
30-N	Diamond cone	0.29
45-N	Diamond cone	0.44
15-T	Ball, 1.588 mm dia.	0.14
30-T	Ball, 1.588 mm dia.	0.29
45-T	Ball, 1.588 mm dia.	0.44

11.11.4 *Shore Sceleroscope*

In the tests previously described, the test piece must be small enough to mount in the testing machine, and hardness is measured as a function of indentation. However, the scleroscope, shown in Fig. 11.28, works on a different principle, and hardness is measured as a function of resilience. Further, since the scleroscope can be carried to the work piece, it is useful for testing large surfaces such as the slideways on machine tools. A diamond-tipped hammer of mass 2.5 g drops through a height of 250 mm. The height of the first rebound indicates the hardness on a 140-division scale.

Fig. 11.28 *Shore scleroscope*

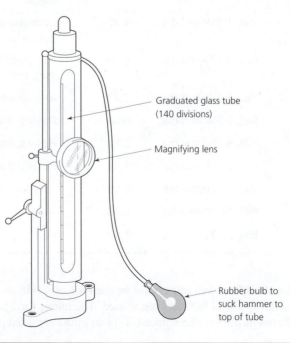

Graduated glass tube
(140 divisions)

Magnifying lens

Rubber bulb to
suck hammer to
top of tube

11.12 The effect of processing on hardness

All metals work harden to some extent when cold worked. Figure 11.29 shows the relationship between the Vickers hardness number (H_D) and the percentage reduction in thickness for rolled strip. The metals become harder and more brittle as the amount of cold working increases until a point is reached where the metal is so hard and brittle that cold working cannot be continued.

Aluminium reaches this state when a 60 per cent reduction in strip thickness is achieved in one pass through the rolls of a rolling mill. In this condition the material is said to be fully work hardened. The degree of work hardening or 'temper' of strip and sheet material is arbitrarily stated as: soft (fully annealed), $\frac{1}{4}$ hard, $\frac{1}{2}$ hard, $\frac{3}{4}$ hard, and hard (fully work hardened).

Fig. 11.29 *Effect of cold working on the hardness of various materials*

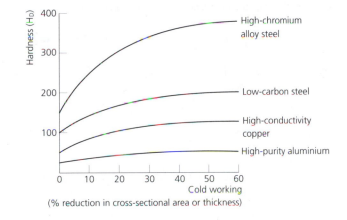

The effect of heating a work-hardened material such as a brass is shown in Fig. 11.30. Once again very little effect occurs until the temperature of recrystallisation is reached. At this temperature there is a rapid fall off in hardness, after which the decline in hardness becomes more gradual as grain growth occurs and the metal becomes fully annealed.

The effect of heating a quench-hardened plain carbon steel is more gradual, as shown in Fig. 11.31. During the tempering range of the steel no grain growth occurs, but there are structural changes. Initially, there is a change in the very hard martensite as particles of carbide precipitate out. As tempering proceeds and the temperature is increased, the structure loses its acicular martensitic appearance and spheroidal carbide particles in a matrix of ferrite can be seen under high magnification. (Spheroidising annealing was considered in Section 5.2.) These structural changes increase the toughness of the metal considerably, but with some loss of hardness.

Fig. 11.30 *Effect of heating cold-worked 70/30 brass*

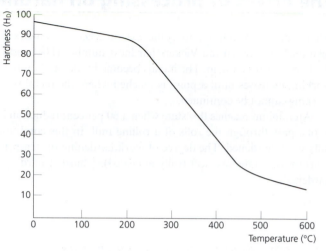

Fig. 11.31 *Effect of heating a quench-hardened 0.8% plain carbon steel*

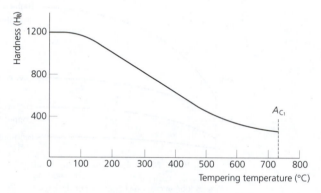

11.13 The hardness testing of polymers

Brinell and Rockwell hardness tests may be used to test polymeric materials, but the size of the indenter and the applied load have to be varied to suit the softness of the material. For example, when using the Brinell test a 10 mm diameter hardened steel ball is used with a force of 4.9 kN applied for 30 seconds. Typical results are:

Polystyrene	H_B 25
Perspex	H_B 20
Polyvinyl chloride	H_B 20
Polyethylene	H_B 2

The Rockwell test is also widely used for polymeric materials, particularly the M and the R scales (see Table 11.5). The M scale is used for the harder polymers and the R scale is used for the softer plastics and for the rubbers. Typical values are:

Melamine–formaldehyde	R_M 120
Phenol–formaldehyde	R_M 115
Urea–formaldehyde	R_M 115
Polyester (not reinforced)	R_M 70–100
Epoxy	R_M 80–110
Polymethyl methacrylate (Perspex)	R_M 80–110
Polycarbonate	R_M 75
Polystyrene	R_M 75
Nylon (66)	R_R 110
High-impact polystyrene (ABS)	R_R 105

The *Shore Durometer* is also widely used for measuring the hardness of plastics and rubbers. The type 'A' durometer test is used for soft plastics and soft rubbers, and has an indenter with a truncated cone, as shown in Fig. 11.32(a). The applied force is 8 N (newtons). The depth of penetration is measured the instant the indenter is seated on the specimen. This is because rubbers tend to 'creep' and the reading after 15 seconds is substantially lower than the initial reading. To ensure reasonable repeatability, the immediate reading is taken as the hardness number. A car tyre has a typical hardness as A70.

For hard polymers and rubbers the type 'D' durometer test is used. This has an indenter of the type shown in Fig. 11.32(b) and the applied force is 44.5 N.

Fig. 11.32 *Durometer indenters (all dimensions in millimetres): (a) type 'A'; (b) type 'D'*

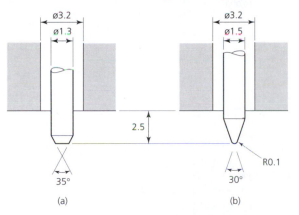

(a) (b)

A quick workshop test is the pencil test. This is in fact a scratch-resistance test. A set of pencils from 2B, B, HB, H, 2H up to 9H are carefully sharpened to a wedge and dressed on abrasive paper. Each pencil, in turn, is held at right angles to the test-piece surface and drawn across the material. The least hard pencil to leave a visible mark (indentation) on the surface of the plastic is reported as the hardness grade.

Alternatively to hardness, polymeric materials may be tested for softness (BS 2782). This test involves a ball indenter 2.38 mm diameter being pressed into the surface of the plastic by an initial force of 5.25 N applied for 5 seconds, after which an additional force of 5.25 N is applied for a further 30 seconds. The difference between the penetration depths for the two forces is measured and expressed as the softness number. This number is the actual difference in depth expressed in units of 0.01 mm. Hence, a test which gives an initial penetration of

0.1 mm and a final penetration of 0.5 mm would have a difference in penetration of 0.4 mm, or a softness number of 4. The test is performed at a temperature of $23 \pm 1\,°C$.

11.14 Comparative scales of hardness

Most hardness-testing machine manufacturers issue tables showing comparative hardness numbers for various methods of testing. These tables should be treated with caution since the tests are carried out under different conditions. For instance, the ball indenter of the Brinell test displaces the metal by plastic flow, whilst the diamond indenter of the Vickers test tends to cut its way into the test piece by shear, and the scleroscope is a dynamic test measuring hardness as a function of resilience.

Figure 11.33 shows the approximate relationship between typical harness scales. These scales are related by a horizontal line across the figure. The *Moh* scale of hardness is included for comparison and is based upon a list of naturally occurring materials. These are arranged so that each material will scratch the material appearing immediately below it. Thus the hardest material (diamond) is placed at the head of the list, and the softest material (talc) is placed at the bottom of the list. Table 11.7 lists some hardness values for a range of typical engineering materials.

Fig. 11.33 *Hardness scales*

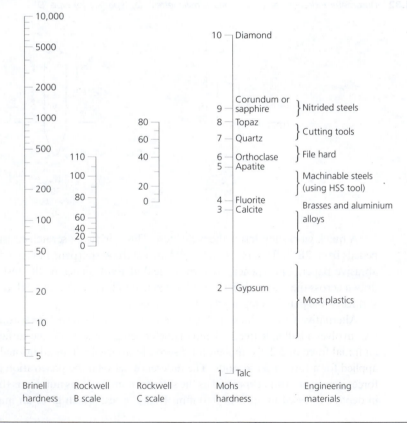

Table 11.7 *Hardness of typical engineering materials*

Material	Condition	Hardness	
Aluminium bronze	Annealed	H_D	80
	Hard	H_D	220
Cartridge brass (70/30)	Annealed	H_D	65
	Hard	H_D	185
Commercial brass (63/37)	Annealed	H_D	65
	Hard	H_D	185
Muntz brass (60/40)	Extruded	H_D	75
Cupro-nickel (25% Ni)	Annealed	H_D	80
	Hard	H_D	170
Duralumin	Age hardened	H_D	30
Grey cast iron	—	H_D	210
Low tin bronze	Annealed	H_D	60
	Hard	H_D	210
Magnesium alloy	Wrought	H_D	65
Low-carbon steel	Hot rolled	H_B	150
Medium-carbon steel	Quench hardened and tempered at 550–650 °C	H_B	250
High-carbon steel	Quench hardened and tempered at 150–300 °C	H_B	800
Stainless steel (ferritic)	Annealed	H_B	150
Stainless steel (austenitic)	Annealed	H_B	170
Stainless steel (martensitic)	Cutlery temper	H_B	534
	Spring temper	H_B	450
Polyvinyl chloride (rigid) (PVC)	—	H_{RR}	110
Polystyrene (high density)	—	H_{RM}	80
Polyamide (nylong '66')	—	H_{RR}	110
Acrylonitrite-butadiene-styrene (ABS)	—	H_{RM}	70

SELF-ASSESSMENT TASK 11.3

1. Describe the basic principle of hardness testing.

2. Describe any precautions that must be taken to avoid false readings when hardness testing.

3. Explain why it is unsafe to compare hardness scales in all but the most general terms.

11.15 Ductility testing

The percentage elongation, as determined by the tensile test, has already been discussed (Section 11.5) as a measure of ductility. Another way of assessing ductility is a simple bend test. There are several ways in which this test can be applied, as shown in Fig. 11.34. The test chosen will depend upon the ductility of the material and the severity of the test required.

Fig. 11.34 *Bend tests: (a) close bend; (b) angle bend; (c) 180°C bend*

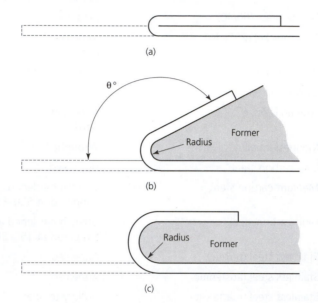

- *Close bend test*   The specimen is bent over on itself and flattened. No allowance is made for spring back, and the material is satisfactory if the test can be completed without the metal tearing or fracturing. This also applies to the following tests.
- *Angle bend test*   The material is bent over a former and the nose radius of the former and the angle of bend ($\theta°$) are fixed by specification. Again no allowance is made for spring back.
- *180° bend test*   This is a development of the angle bend test using a flat former as shown. Only the nose radius of the former is specified.
- *Reverse bend test*   This is used for very ductile materials where the preceding tests are not severe enough. The material is bent round a former of specified nose radius through 90° or 180° for a specified number of times.

In all the above tests the test piece is 10 mm wide and has rounded edges.

Another test widely used for testing the suitability of a sheet material for deep drawing in the press-shop (e.g. the cold pressing of car body panels) is the *Erichson cupping test*. This test simulates a deep-drawing operation on a sample piece of material, as shown in Fig. 11.35(a).

Fig. 11.35 *Erichson cupping test: (a) principle cupping test – the punch forces the spherical indenter into the blank, stretching it into the die – the more ductile the material, the greater will be the depth of the cup 'D' before the crack appears; (b) Erichson standard curves*

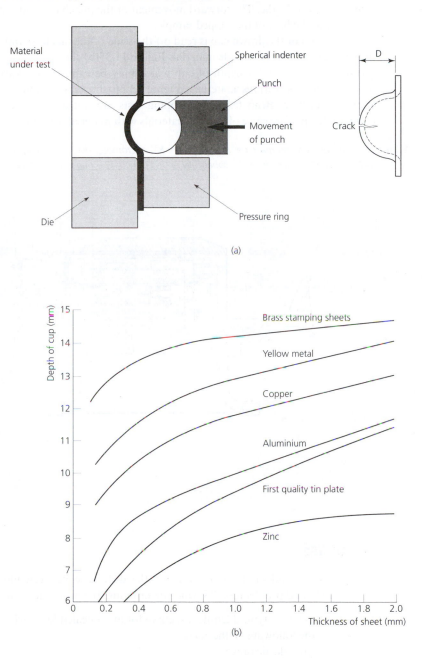

(a)

(b)

The circular, sample blank is gripped between the pressure ring (blank holder) and the die face to prevent the sample from puckering. After the pressure ring has been tightened down on the blank it is slackened back by a standard amount (0.05 mm) to give the metal freedom to draw. The spherically ended punch is then pressed into the sample blank, drawing it into the die. The forward movement of the punch continues until a crack is seen to appear at the base of the cupped sample.

The depth of the drawn cup is read off the scale of the machine and is a measure of the drawing quality of the sample material. Figure 11.35(b) shows Erichson's standard curves giving typical values for depth of cup which may be expected for various thicknesses of material and for various materials. Figure 11.36 shows a section through an Erichson cupping machine. Bend tests and cupping tests are not absolute tests of ductility, but compare the performance of sample materials with materials of known performance.

Fig. 11.36 *The Erichson cupping machine: A – specimen; B – micrometer dial; C – sliding scale; D – pressure ring lock; E – pressure ring drive; F – spherically ended punch; G – mirror to observe cup; H – depth of cup; I – die*

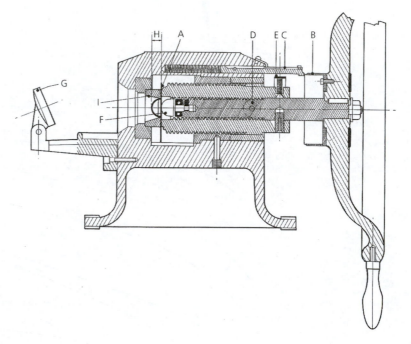

EXERCISES

11.1 With the aid of diagrams describe the difference between ultimate tensile stress (UTS) and proof stress (PS) when determining the properties of a ductile metal.

11.2 (a) Sketch a typical tensile test curve for an annealed low-carbon steel and indicate the following on the curve:
(i) elastic range
(ii) elastic limit

(iii) yield point
   (iv) plastic range
   (v) breaking point of the specimen
(b) Describe how the ductility of a material can be calculated from a tensile test in terms of elongation percentage.

11.3 (a) With the aid of sketches explain how the secant modulus for polymeric materials is determined.
   (b) Explain why creep is such an important factor when testing polymeric materials and distinguish between creep and instantaneous elongation.

11.4 (a) With the aid of sketches show the essential differences between the Charpy and Izod impact tests, paying particular attention to the proportions of the specimens, the notch profile, method of support and criteria for determining the impact number.
   (b) Describe how the results of an impact test can be interpreted to indicate the toughness of a material, its susceptibility to crack propagation, and how the appearance of the fractured surfaces differ between ductile and brittle materials.

11.5 With the aid of sketches describe:
   (a) how the Brinell hardness test is performed
   (b) the relationship between load and the diameter of the indenter and how the hardness number is derived
   (c) any precautions which should be taken to ensure that the test result is valid
   (d) how properties other than hardness may be derived by calculation and visual inspection of the test results

11.6 Compare and contrast the Brinell hardness test, the Vickers hardness test and the Rockwell hardness test. Pay particular attention to the essential differences between the tests, suitable applications and the dangers inherent in using hardness conversion tables to compare the results of these tests.

11.7 Calculate the modulus of elasticity for a material, given the following data:

Applied load	7.2 kN
Cross-sectional diameter	10 mm
Gauge length	35 mm
Extension	0.5 mm

11.8 Calculate the elongation percentage for a ductile material if the extension at fracture is 10 mm on a gauge length of 28 mm.

11.9 Discuss the relative advantages and limitations of the following methods of determining ductility:
   (a) tensile test
   (b) bend test
   (c) Erichson cupping test

11.10 As well as providing a measure of toughness, discuss the other characteristics of a material that can be determined from an impact test.

12 Materials testing (non-destructive)

> The topic areas covered in this chapter are:
>
> - The need for non-destructive testing.
> - Visual examination.
> - Crack detection using dye penetrants.
> - Ultrasonic testing.
> - Eddy-current testing.
> - Magnetic testing.
> - Radiography.

12.1 The need for non-destructive testing

In Chapter 11 we saw how material properties could be determined by *destructive testing*. Although such tests are invaluable for the quality control of materials, they cannot control the quality of components, fabrications and assemblies made from those materials.

For instance, a tensile test on a cast test piece may show the correct yield point, strength and elongation percentage. It cannot show whether a casting made from the material contains blow holes, porosity and cold shuts. Further, the grain structure of the test piece will be unlike the grain structure of a large casting which will have cooled more slowly, therefore the properties of the test piece will not accurately reflect the properties of the casting. In the case of a hollow casting, destructive testing cannot show the effects of a misplaced or distorted core.

For safety, welds on pipe runs and pressure vessels must be free from discontinuities such as porosity and slag inclusions. Such welds must also have correct fusion and penetration. They should show no reduction in cross-sectional area due to lack of filler material.

Again, for safety, the forged light-alloy hinges for the control surfaces of aircraft must be free from cracks due to forging at too low a temperature, or grain growth due to forging at too high a temperature.

A full programme of *non-destructive testing* can be very expensive, requiring each component to be handled individually through each series of tests. The tests themselves are often complex (e.g. radiography), requiring expensive equipment and highly skilled personnel in their execution.

For small components produced in batches and whose performance is not critical it is sufficient to take random samples from each batch, inspect them visually and, if hollow, section them to see that the cored holes are to specification. If the occasional blow hole turns up during machining, then the component can be rejected and melted down again. Sample forgings can be sectioned, etched and given a macro-examination.

However, such sampling followed by destructive examination is not appropriate for large castings produced singly or in small batches. A fault lying undiscovered in a turbine housing, or discovered only after much expensive machining has taken place, would be disastrous not only in terms of human safety but also financially. A faulty weld in the legs of an offshore oil-rig or an under-sea pipeline could have consequences out of all proportion to the cost of carrying out a full examination of each weld.

Similarly, each critical component of an aircraft must be fully and individually inspected. Obviously such tests and examinations must be *non-destructive* and in no way affect the properties and performance of the component or assembly being examined. The non-destructive testing techniques to be considered in this chapter are:

- Visual examination.
- Visual examination assisted by dye-penetrants.
- Ultrasonic testing.
- Eddy-current testing.
- Magnetic testing.
- Radiography (X-rays and gamma-rays).

12.2 Visual examination

This is the simplest and cheapest possible non-destructive examination. A visual examination of a casting can identify such defects as:

- Surface cracks.
- Scabs.
- Surface porosity.
- Warping and twisting.
- Inadequate filling of the mould.

Preliminary dimensional checks and marking-out will indicate whether or not sufficient machining allowance is present so that machined surfaces will 'clean-up'. Such checks can also determine whether cored holes are correctly positioned and of correct size, also whether such holes are correctly centred in their bosses.

In the case of welds, visual inspection can show whether or not there are any large surface cracks, inadequate fusion, lack of penetration, insufficient filler material, poor joint shape, surface porosity and undercuts. Figure 12.1 shows the appearance of two faulty welds which are easily identified by visual inspection.

Fig. 12.1 *Visual inspection of weld faults: (a) a manual metal-arc single 'V' butt weld, showing a lack of penetration and misalignment of the plate edges; (b) a manual metal-arc single 'V' butt weld, showing undercut and misalignment of the plate edges*

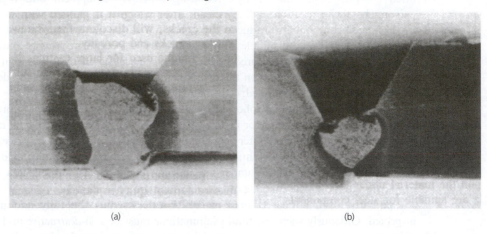

(a)   (b)

12.3 Use of dye penetrants

The following techniques are used to make surface cracks visible, particularly where they are so fine that they normally escape visual inspection. A very old technique, which is still widely used, is to immerse the casting in a bath of hot paraffin. Heating the paraffin reduces its viscosity so that, combined with its already high surface tension, it is easily drawn into the finest cracks and porosity by capillary attraction. The casting is removed and wiped thoroughly clean, after which it is painted with whitewash. Paraffin, seeping out from the cracks, will discolour the whitewash and reveal the presence of surface cracks and porosity.

A more modern technique used for large castings is to paint them with special penetrants which are readily drawn into the surface cracks and porosity at room temperature. The penetrants contain brightly coloured dyes. After allowing time for the penetrant to be absorbed into any surface cracks which may be present, the casting is thoroughly cleaned off and dusted with a white powder. As with paraffin, the penetrant seeps out from the cracks and stains the white powder. The coloured marks on the powder indicates where the cracks are.

A more sophisticated technique is to use specially developed fluorescent penetrants. The penetrant is painted over the surface of the casting or applied by a pressurised aerosol spray in the case of small castings. After allowing time for penetration the surplus penetrant is cleaned off and the casting is viewed under *ultraviolet light*. The presence of any penetrant seeping from surface cracks and porosity will appear as glowing lines and spots against a dark background.

Care must be taken when using ultraviolet light. Constant exposure to it can cause skin cancers and it can also cause eye injuries if viewed directly. These eye injuries are similar to 'arc-eye', which is caused by an arc welding flash when suitable eye protection is not being worn. Protective goggles with the appropriate filter lenses must always be worn when working with ultraviolet light.

12.4 Ultrasonic testing

Although it varies widely from person to person, the highest frequency a normal person can be expected to hear is pitched at about 18 kHz. Sound waves of an even higher frequency can be heard by most animals, as their hearing is more acute than that of human beings. Sound waves whose frequency is too high to be heard by human beings are said to be ultrasonic. The higher the frequency (shorter the wavelength) of the ultrasonic sound waves, the easier it is to beam them and control them.

The frequencies used for ultrasonic testing (ultra sound) lie between 0.5 and 15 MHz depending upon the material under test. Frequencies between 1 and 3 MHz are suitable for steel components. Pulses of high-frequency oscillations are generated electronically, amplified, and fed into suitable transducers for conversion into sound waves. Such transducers exploit either magnetostriction or a piezoelectric effect.

- *Magnetostriction* This relies upon the fact that soft iron changes dimensionally in the presence of an alternating magnetic field. Thus a rod of soft iron placed in a coil, through which an alternating current is being passed, will increase and decrease in length in step with the alternations in the magnetic field and, therefore, in step with the alternating current producing that field. The oscillations in length drives a diaphragm which produces ultrasonic waves in the air.
- *Piezoelectric effect* This relies upon the fact that certain crystals such as quartz change dimensionally when small electric currents pass through them. Thus, when an alternating current passes through the crystal, the crystal will change in size in step with the alternations. The 'bleep' signal in computers, electronic watches and alarm clocks are usually generated using piezoelectric devices. An advantage of the piezoelectric effect is that it is reversible. Not only can an electric current change the size of the crystal, straining the crystal mechanically generates a corresponding electric current. This latter effect is exploited in record player pick-ups and in microphones. Thus a transducer using the piezoelectric effect can be used both as an ultrasonic transmitter and as a receiver.

The basic principle of ultrasonic testing is shown in Fig. 12.2. The pulses of high-frequency oscillations are driven into the component by the transducer. When the sound waves meet any discontinuity, such as a crack, the waves are reflected back into the transducer where they are converted into electrical pulses which can be displayed on the screen of a computer. By using pulses of short duration, the same transducer can transmit and receive alternately, as shown in Fig. 12.2(a). The appearance of the display on the VDU of a computer is shown in Fig. 12.2(b).

Since the distance between the incident pulse and the reflection from the back surface of the component is proportional to the component thickness, the position of the reflection from the fault clearly indicates where it lies in the component relative to the front and back faces. Figure 12.3 shows a block diagram of the complete equipment.

The use of a single, common transducer for transmission and reception will detect most randomly oriented faults and is quick and easy to use and interpret. However, it may miss a long, narrow fault whose axis is parallel to the path of the ultrasonic waves, as shown in

Fig. 12.2 *Principles of ultrasonic testing: (a) ultrasonic testing; (b) typical VDU displays*

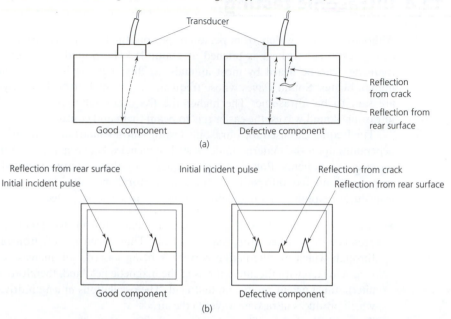

(a)

(b)

Fig. 12.3 *Ultrasonic testing equipment: (a) dual transducer equipment; (b) single transducer equipment*

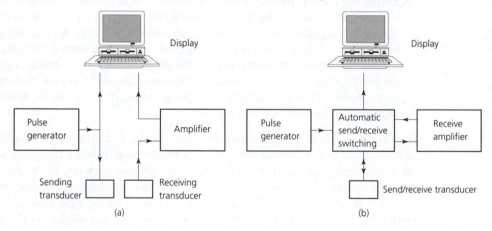

Fig. 12.4(a). To overcome this problem the more difficult technique of using separate transducers for transmission and reception may be employed, as shown in Fig. 12.4(b). Obviously, it is much more difficult to position the fault by this technique.

The transducers must be in intimate contact with the surface of the component or false echoes will occur. This contact is achieved by placing a film of oil between the transducer(s) and the component surface, so that no air gap can exist. One great advantage of ultrasonic testing is that it can equally well be used on magnetic and non-magnetic metals, and on non-metals.

Fig. 12.4 *Multiple transducer techniques: (a) a single transducer can miss a narrow defect parallel to the pulse; (b) defect detected by using dual transducers*

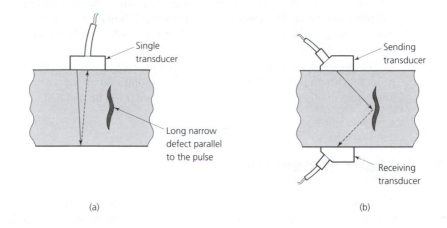

Single transducer

Long narrow defect parallel to the pulse

Sending transducer

Receiving transducer

(a)

(b)

12.5 Eddy-current testing

This technique depends upon the fact that if an alternating magnetic field is linked with any metallic object, eddy currents will be induced in the metallic object by electromagnetic induction. This applies to all metals whether they are ferromagnetic or non-magnetic. Figure 12.5 shows the principle of the test. A high-frequency alternating current is made to flow through a small coil and, as a result, a high frequency alternating magnetic field is set up around the coil. This field induces small electric currents, called eddy currents, which circulate in the component and produce magnetic fields in their own right.

Fig. 12.5 *Principle of eddy-current testing*

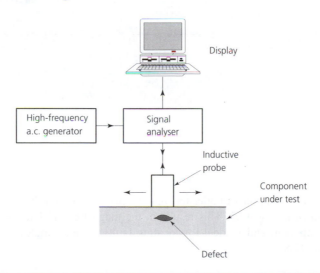

Display

High-frequency a.c. generator

Signal analyser

Inductive probe

Component under test

Defect

These secondary fields react with the field of the induction coil and affect the flow of current through that coil. Providing there is no change in section or composition as the induction coil is moved across the surface of the component the resultant current in the coil will remain constant. If, however, the induction coil moves over a crack or other fault in the component, the eddy-current system will be temporarily disrupted and the resultant current in the coil will change.

Therefore, if the induction coil is moved over the surface of the component and the resultant current through the coil is constantly monitored on the screen of a computer, then any cracks or other discontinuities in the component will be revealed.

12.6 Magnetic testing

This method of crack detection, although simple and reliable, can only be applied to ferromagnetic materials. Further, it is only appropriate for surface cracks and discontinuities not more than 10 mm below the surface of the component. This technique is based upon the fact that the magnetic susceptibility in the region of a discontinuity is inferior to that of the surrounding metal and that this distorts the magnetic flux distribution. The resulting distortion of the flux field is usually detected by means of magnetic powder (magnetic iron oxide) in suspension in light machine oil or paraffin. The suspension is spread thinly over the surface of the component and the magnetic powder 'bunches' in the vicinity of the fault, as shown in Fig. 12.6.

Fig. 12.6 *Principle of magnetic crack detection*

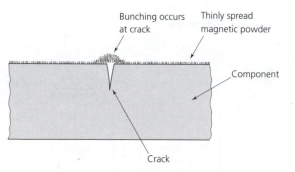

The component under test may be placed between the poles of a powerful magnet, as shown in Fig. 12.7. This results in the magnetic field lying parallel to the surface of the component and is ideal for locating faults at right angles to the flux field, as shown in Fig. 12.8.

Fig. 12.7 *Orientation of the magnetic field*

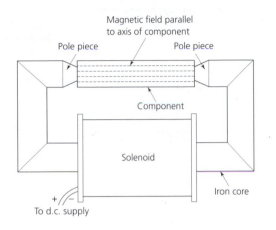

Fig. 12.8 *Detecting faults perpendicular to the component surface*

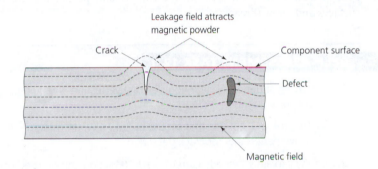

However, for long bars the faults usually lie parallel to the axis of the bar and would not be shown up by magnetisation in the plane of the axis, as shown in Fig. 12.9(a). For long thin bars it is better to pass a heavy, direct electric current through the bar from end to end as this sets up magnetic flux fields at right angles to the current flow, as shown in Fig. 12.9(b). Thus the flux field will again be perpendicular to the plane of the anticipated faults. Bars up to 4 metres long and 75 mm diameter can be tested in this manner. In this instance, the bars are immersed in the suspension of magnetic particles and slowly rotated so that the bars can be examined all over.

Magnetic crack detection, like other non-destructive tests, can be applied to finished components. It is effective in showing up surface cracks invisible to the unaided eye. Such faults are not uncommon in the wire or rod from which large coil springs are made. Frequently these cracks do not develop until after the springs have been wound and heat treated. Consequently, coil springs for critical applications are generally inspected magnetically after manufacture.

Fig. 12.9 *Detecting faults parallel to the component axis: (a) defects parallel to the magnetic field are not detected; (b) defects parallel to the component axis are perpendicular to the magnetic field when a direct current is passed through a component*

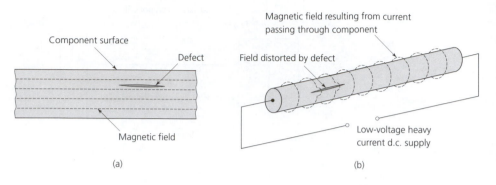

(a)

(b)

Magnetic crack detection is also used to show up hair-line surface cracks due to local overheating during grinding processes and which might lead to early fatigue failure in such devices as ball and roller bearings. After testing, care must be taken to clean any residual iron or iron oxide particles from the component, particularly in the vicinity of bearing surfaces and screw threads.

SELF-ASSESSMENT TASK 12.1

1. List **three** situations where non-destructive testing is required, giving reasons for your choice.

2. Discuss the precautions that should be taken when using an ultraviolet light source with dye penetrants for crack detection.

3. Briefly explain why eddy-current crack detection can be used on any metal, whilst magnetic crack detection can only be used on ferrous metals.

12.7 Radiography

This is a photographic process in which the 'illumination' of the component is by X-rays or the even more penetrating gamma-rays. These are electromagnetic radiations exactly the same as radio waves and light waves, except that they have a very much shorter wavelength (higher frequency). This enables X-rays and gamma-rays to penetrate solid objects. When photographic film is exposed to X-rays or gamma-rays and then developed, the film becomes dark. If a solid object is placed between the source of radiation and the film so that it casts a shadow on the film, the level of radiation reaching the film in the shadow area will be reduced and the shadow will appear on the film, after development, as a less dark area, as shown in Fig. 12.10.

Fig. 12.10 *Principles of radiography: (a) general arrangement; (b) appearance of film after development*

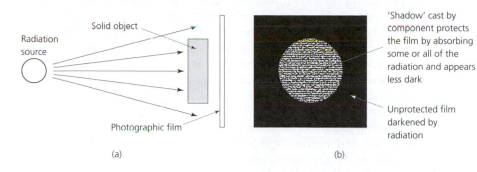

'Shadow' cast by component protects the film by absorbing some or all of the radiation and appears less dark

Unprotected film darkened by radiation

(a)

(b)

The thicker or more dense the material, the greater will be the level of absorption. Therefore, if there is a void in the material, the level of absorption will be lower at this point and the void will show up as a darker area on the film, as shown in Fig. 12.11.

Fig. 12.11 *Use of radiography: (a) general arrangement; (b) appearance of film after development*

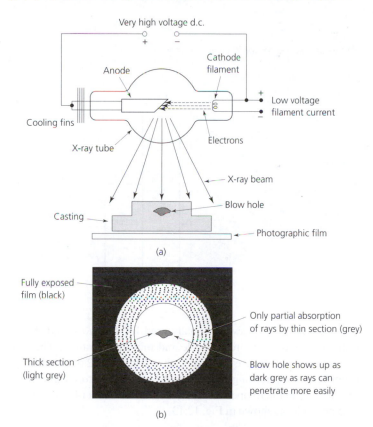

The smallest defect that can readily be detected by X-ray radiography is approximately 2 per cent of the section thickness. Below this the film cannot resolve the difference in radiation level. Thus, for a component 25 mm thick any flaw under 0.5 mm in size will remain undetected. Thus the process becomes less reliable as the section thickness increases. In this respect it is at a disadvantage compared with ultrasonic testing which can detect small flaws in steel up to 15 metres thick with much less costly equipment and no potential radiation hazard. However, radiography has the advantage of providing a permanent visual record of the defect, its precise form and its precise location.

X-rays are generated electrically in high-voltage discharge tubes. This method of generation provides a constant source of radiation, and the exposure time is easy to determine and control. Although widely used under laboratory or test department conditions, the size and weight of the equipment and the need for an electrical power supply makes it unsuitable for site work. It must be remembered that X-ray machines for examining metal components have to be much more powerful than those used for examining living tissues.

Figure 12.12 compares the *half-value thicknesses* for mild steel using X-ray and gamma-ray sources. Half-value thickness is the thickness of the material under examination which will reduce the intensity of the radiation to one half its incident value. The range of material thicknesses which can be examined by any given radiation source extends from one half-value thickness to about eight half-value thicknesses.

Fig. 12.12   *Half-value thicknesses for low-carbon steel*

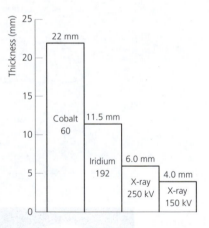

Gamma-ray sources are usually used for site work and thicker materials because the radiation source is more portable and more penetrating. The radiation source is a small amount of a radioactive material such as iridium 192 or cobalt 60. The radioactive material is continuously emitting radiation and has to be stored in a heavy, lead-lined container. Provision is made to open a 'window' in the container by remote control when an exposure is to be made, as shown in Fig. 12.13.

Fig. 12.13 *Typical gamma-ray source: (a) gamma-rays 'off'; (b) gamma-rays 'on'*

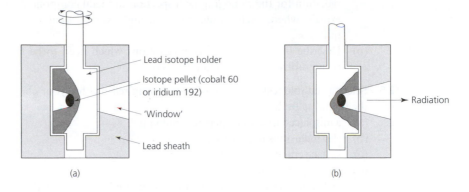

Lead isotope holder

Isotope pellet (cobalt 60 or iridium 192)

'Window'

Lead sheath

Radiation

(a)   (b)

Radioactive (gamma-ray) sources do not give a constant output like an X-ray tube, but steadily decay. The useful life of iridium 192 is about 40 days, and that of cobalt 60 is about 5 years. The rate of decay is exponential, rapid at first and then progressively slowing down. They remain too dangerous to handle outside their containers almost indefinitely.

Stringent safety precautions and codes of practice must be observed whenever X-ray or gamma radiation is used. Operators of radiography equipment must be protected by radiation-resistant screens and other persons must be kept well away from the radiation zone. Good ventilation is essential and the operator should change his or her protective overalls regularly. Normally the operator wears a radiation-sensitive tab which changes colour if exposed to excessive radiation, but below the danger level to the operator. Further, radiographers should be subjected to regular medical checks.

Gamma-radiation sources must be stored in containers which are not only radiation resistant, but are strong enough to resist mechanical damage if involved in an accident so that there is no leakage hazard.

EXERCISES

12.1 Explain why non-destructive testing is an essential element of a programme of quality control, and give examples of where it would be used.

12.2 Find out, and write a brief account, about a simple audible crack detection test that was regularly used on the wheels of railway rolling stock.

12.3 Compare the advantages and limitations of the following inspection techniques for finding surface and internal defects in components:
(a) dye penetrants
(b) ultrasonic testing
(c) eddy-current testing
(d) magnetic testing

12.4 (a) Compare the advantages and limitations of gamma-ray and X-ray sources of radiation for the radiographic inspection of metal components and assemblies, explain where each would be used, and explain what is meant by half-life thickness.

(b) Describe in detail the safety precautions which need to be taken when using radiography as an inspection technique.

12.5 Select a suitable test or group of tests, giving reasons for your choice, for detecting the following faults:

(a) a misplaced insert in a plastic moulding

(b) slag inclusions in a casting

(c) surface cracks in a forging

(d) porosity in a pipe weld

(e) inclusions and/or discontinuities orientated parallel to the axis of a bright drawn steel bar

13 Materials in service

The topic areas covered in this chapter are:

- Allowable working stress.
- Creep in metals and in polymeric materials.
- Fatigue and factors affecting fatigue.
- The corrosion of metals.
- Metals that resist corrosion.
- Chemical inhibition.
- Cathodic protection.
- Protective coatings.
- Plastic degradation.

13.1 Allowable working stress

Despite the tests described in Chapters 11 and 12, materials may still fail in service, sometimes with disastrous results (e.g. when the failure occurs in aircraft, bridges, ships, etc.). To try to avoid such disasters occurring, the designer avoids using materials continuously at their maximum allowable stress. This is done by employing a *factor of safety*.

Unfortunately, increasing the strength of a component in the interests of safety not only increases the initial material costs, but also the operating costs. For example, the stronger and heavier the structural members of an aircraft, the fewer passengers it can carry and the more fuel it consumes. Therefore a balance has to be maintained between safety, initial cost and operating costs. The designer is constantly striving to improve the former whilst reducing the latter.

Allowable working stress is taken as a proportion of the yield or proof stress; that is, the component is only stressed within its elastic range when in service. For example, consider the screwed fastening shown in Fig. 13.1. When the nut is tightened normally the bolt is stretched slightly and, providing it is stressed within its elastic range, it will behave like a very powerful spring and will pull the joint faces together very firmly.

Fig. 13.1 *The effect of stressing a bolt and nut*

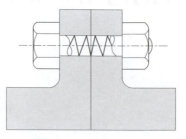

The stress in the bolt is made up of two elements. Firstly, the stress imparted by the initial tightening of the bolt; secondly, the stress imparted by the load on the fastening in service. The sum of these stresses must not be allowed to approach, let alone exceed, the yield stress for the material or it will cease to act in an elastic manner. Therefore the designer proportions the fastening (and other components) so that there is a factor of safety. Usually the designer assumes an allowable working stress of only half the yield stress for the material.

If the fastening shown in Fig. 13.1 is overstressed by applying excess torque to the nut (for example, by extending the length of the spanner with a tube), the bolt will be stressed beyond its elastic range. Once the yield stress for the bolt has been exceeded it exhibits plastic properties and takes a *permanent set* – that is, it becomes permanently lengthened, the 'spring back' is seriously reduced and the joint faces are no longer held firmly together. Thus for critical assemblies, the designer will seek to control the stress in the fastening and associated components by specifying the *torque* to be applied to the nut as it is tightened by the use of a 'torque spanner' set to a specified value.

13.2 Creep

Creep is defined as the gradual extension of a material under a constant applied load. It is a phenomenon which must be considered in the case of metals when they are required to work continuously at high temperatures. For example, the blades of jet engines and gas turbines. Figure 13.2 shows a typical creep curve for a metal at high temperature. A constant tensile load is applied to a test piece in a tensile testing machine whilst the test piece is maintained at a constant elevated temperature.

The creep curve obtained from this test shows three distinct periods of creep.

- *Primary creep* This commences at a fairly rapid rate but slows down as work hardening (strain hardening) sets in and the strain rate decreases. It can be seen from Fig. 13.2 that the extension due to creep is additional to the instantaneous elongation of material to be expected when any tensile load is applied (see Section 10.3). For calculating creep as a percentage elongation, the initial elongation is ignored and creep is considered to commence at point A on the curve.

Fig. 13.2 *Creep*

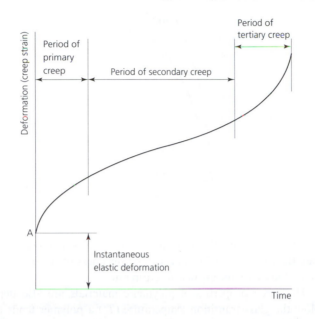

- *Secondary creep* During this period of creep the increase in strain is approximately proportional to time. That is, the strain rate is constant and at its lowest value.
- *Tertiary creep* During this period of creep the strain rate increases rapidly, necking occurs and the test piece fails. Thus the initial stress, which was within the elastic range and did not produce early failure, did eventually result in failure after some period of time.

13.3 Creep in polymeric materials

Polymeric materials, however, are subject to creep effects even under normal ambient conditions. When used for engineering components, rigid polymers must be capable of withstanding reasonable tensile and compressive forces over long periods of time without dimensional change. Figure 13.3 shows some typical creep curves for a cellulose acetate material at 25 °C. As is to be expected, the creep rate is highest for large applied loads and almost negligible for small applied loads, after the initial deformation has taken place.

Creep, for all materials, is calculated in the same manner as elongation when associated with a tensile test. That is:

$$\text{creep } \% = \frac{\text{elongation}}{\text{original length}} \times 100$$

The difference between the elongation determined from a tensile test and the creep for the same material is that the former reflects the immediate response of the material to the applied load, whereas the latter reflects the response of the material after the load has been applied for a very long period of time. (This can be several thousand hours.)

Fig. 13.3 *Typical creep values (cellulose acetate)*

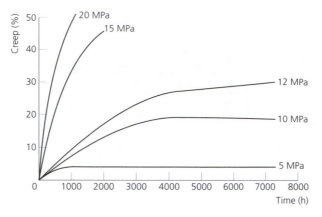

The creep rate reflects the amount of creep which occurs in unit time. This is greatest when the load is initially applied. The steepness of the curve indicates the creep rate. The steeper the curve, the greater the creep rate.

The creep properties of polymeric materials are also dependent upon temperature. Below the glass transition temperature (T_g) a polymer tends to be rigid and, for a given stress, have a low creep rate. As the temperature rises towards the T_g for the material, the creep rate and the elongation increase. At the T_g a much greater elongation is achieved by a given stress than at lower temperatures, although there will be no appreciable increase in the creep rate. Beyond the T_g even greater elongation will occur, but the creep rate will actually decrease.

13.4 Fatigue

Since more than 75 per cent of failures in engineering components are attributed to *fatigue failure*, and as the performance from engineering products is continually increased, the need to understand the failure of materials from fatigue becomes increasingly important.

In service, many engineering components undergo between thousands and millions of changes of stress within their working life. A material which is subjected to a stress which is alternately applied and removed a very large number of times, or which varies between two limiting values, will fracture at a very much lower value of stress than in a normal tensile test. This phenomenon is referred to as fatigue failure.

The source of these alternating stresses can be due to the service conditions of the component – for example, the flexing of the valve springs in a car engine, or the vibration of the axles of a vehicle caused by irregularities in road surfaces. The fatigue crack which ultimately causes fatigue failure usually starts at a point of stress concentration such as a sharp corner (incipient crack), a tooling mark due to machining with too coarse a feed, or a surface crack due to faulty heat treatment.

Most fractures which have occurred due to fatigue have a distinctive appearance, as shown in Fig. 13.4. The point of origin of the failure can be seen as a smooth flat elliptical area. Surrounding this is a burnished zone with ribbed markings. This is caused by the rubbing together of the surfaces of the spreading crack due to stress reversals. When the cross-section of the component has been sufficiently reduced by the spread of the fatigue crack, the component will no longer be able to carry its designed load and will fail suddenly, leaving a crystalline area visible as shown.

Fig. 13.4 *Appearance of fatigue failure*

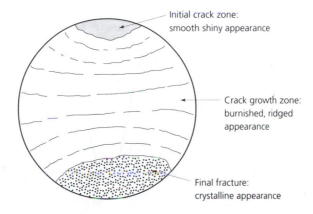

Three conditions of loading may occur, and these will now be described.

- *Fluctuating load* This varies between two positive limits, as shown in Fig. 13.5(a).
- *Pulsating load* This is also known as a 'repeated load' and applies to a load which fluctuates between zero and a positive value, as shown in Fig. 13.5(b).
- *Alternating load* This is also known as 'reversed loading'. The load fluctuates between a positive maximum load and a negative maximum load repeatedly, as shown in Fig. 13.5(c).

Let's now look at how the test load can be applied. One of the most popular tests is to apply an alternating load in the bending mode, as shown in Fig. 13.6(a). The test piece is supported in a rotating chuck at one end and loaded as a cantilever by a dead weight at the other. As the specimen rotates the stress at any point on the circumference of the specimen alternates between tension and compression once during each revolution. To reduce the variation in applied stress from the centre (neutral axis of bending) to the outside of the specimen, it is sometimes made hollow.

Alternatively, 'four-point loading' can be used as shown in Fig. 13.6(b). This loading gives a pure bending moment on the centre portion of the specimen and eliminates the shearing stress that exists when the cantilever specimen is used. The testing machine is fitted with a revolution counter to count the number of revolutions which equals the number of stress reversals. There is a trip to stop the machine when failure of the specimen occurs.

Fig. 13.5 *Conditions for fatigue loading: (a) fluctuating load; (b) pulsating or repeated load; (c) alternating load*

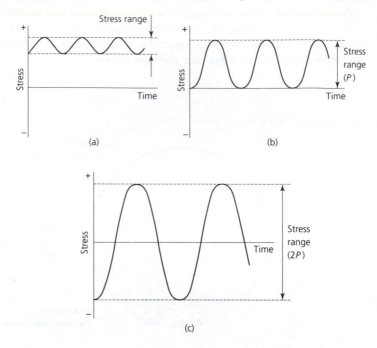

Fig. 13.6 *Fatigue testing: (a) cantilever loading – alternating stress; (b) beam loading – alternating stress*

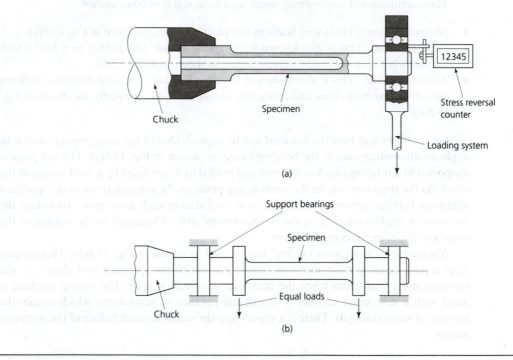

If a large number of tests are made on a correspondingly large number of specimens of the same material, and if the stress range (2P) is gradually reduced, the number of reversals necessary to produce failure can be plotted against the range of stress. The graph produced will be similar to that shown in Fig. 13.7(a). This is called an *S–N diagram*. It implies that failure will never occur provided the range of stress is kept within the value of S_D, which is called the *fatigue limit*.

Fig. 13.7 *Stress reversal curves: (a) S–N diagram for a typical steel; (b) S–N diagram for a typical non-ferrous alloy*

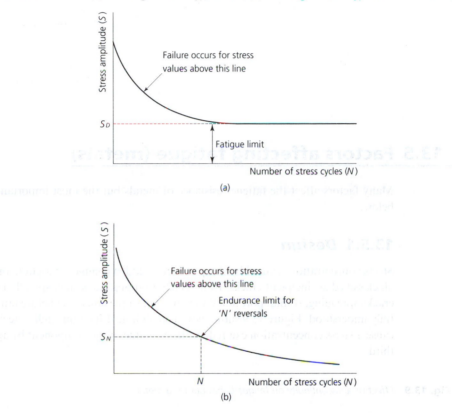

From the nature of the test it is not possible to prove this and, in practice, the limiting range of stress which will not produce failure in 10×10^6 reversals is generally taken as the *fatigue strength* of the material. For some materials 'fatigue limit' values are not applicable, as shown in Fig. 13.7(b). For such materials an endurance limit of S_N is quoted. This defines the maximum stress amplitude which can be sustained for N reversals.

Non-ferrous alloys usually have an *S–N* diagram similar to 13.7(b) and, since non-ferrous alloys are used in the manufacture of aircraft, this accounts for the reason why critical components have to be replaced after a specified number of flying hours (and therefore stress reversals) even when they appear – visually – to be in good condition. The fatigue test results depend upon the type of machine used and the type of test used, but similar test conditions give similar results. A typical *S–N* curve for a wrought aluminium alloy of the 'duralumin' type is shown in Fig. 13.8.

Fig. 13.8 *S–N diagram for a wrought aluminium alloy*

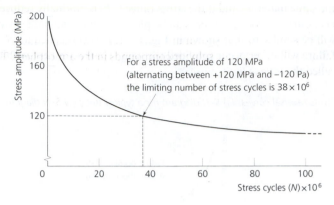

For a stress amplitude of 120 MPa
(alternating between +120 MPa and −120 Pa)
the limiting number of stress cycles is 38×10^6

13.5 Factors affecting fatigue (metals)

Many factors affect the fatigue resistance of metals but the most important are discussed below.

13.5.1 *Design*

Stress concentrations caused by sharp corners, sudden changes of section, or undercuts are all classified as 'incipient cracks' from which a fatigue crack may spread. To prevent such cracks spreading, round 'port holes' were used in ships' hulls long before fatigue failure was fully understood. Figure 13.9 shows how a small hole drilled through a steel component to cause a stress concentration can reduce the S_N value of the component by approximately a third.

Fig. 13.9 *Effect of a discontinuity on fatigue behaviour for a metal*

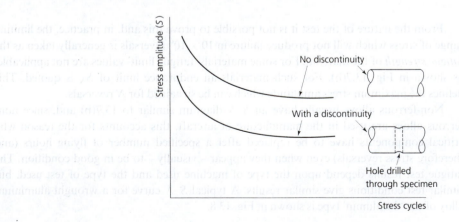

No discontinuity

With a discontinuity

Hole drilled
through specimen

Stress cycles

13.5.2 *Surface finish*

Scratches in highly finished surfaces or tooling marks left when machining also represent stress concentrations which can lead to fatigue failure in highly stressed components. Surface discontinuities left by heat-treatment processes, hot working and cold working can also cause fatigue failure.

13.5.3 *Temperature*

Changes of temperature at which the test is carried out, and at which the material is subsequently used in service, can have a significant effect upon the fatigue resistance of that material. This is related to the effect of creep at elevated temperatures (see Section 13.2).

13.5.4 *Residual stresses*

Such stresses left by processing can also substantially affect the fatigue resistance of the work piece. Processes which leave compressive stresses in the surface of the material improve its fatigue resistance, whilst processes which leave tensile stresses in the surface of the material reduce its fatigue resistance. In the latter case a stress relief treatment such as normalising can remove the internal stresses.

13.5.5 *Corrosion*

This may be atmospheric corrosion, oxidation during heat treatment or saline attack due to marine environments. Figure 13.10 compares the S–N curves for a plain carbon steel before and after exposure to sea-water. It can be seen that prior to exposure to sea-water the curve showed a clear fatigue limit (S_D) below which fatigue would not occur, whilst after exposure the steel showed no clearly defined fatigue limit and an 'endurance limit' would need to be specified. Surface treatment by galvanising and painting prior to saline exposure would prevent corrosion and would result in normal fatigue characteristics for the material.

Fig. 13.10 *Effect of corrosion on fatigue resistance of a metal*

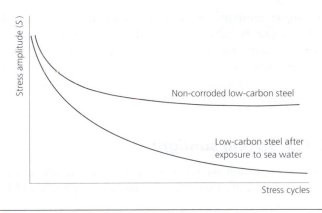

13.6 Factors affecting fatigue (polymers)

13.6.1 *Temperature*

This has a much greater effect on polymeric materials than it has on metals (see Section 13.2) and substantially reduces the fatigue resistance of the material. Therefore temperature must be taken into account not only during the fatigue test but also during the service conditions in which the material must operate. Further, polymeric materials heat up internally when subjected to rapidly alternating stresses. The greater the stress and the more rapidly it alternates, the greater will be the temperature rise. Since this temperature rise results in a lowering of the fatigue resistance of the material, an imitation of the frequency of alternation must be placed on the test providing it is not reduced below the frequency of any stress alternation the material would be subjected to in service. Figure 13.11 shows a typical *S–N* diagram for unplasticised PVC.

Fig. 13.11 *S–N diagram for an unplasticised PVC*

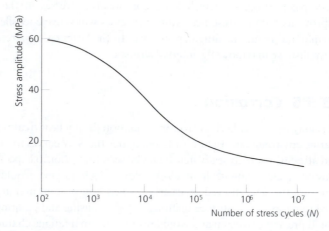

13.6.2 *Environment*

The fatigue resistance of polymeric materials along with their general strength and toughness can be adversely affected by the presence of organic substances such as detergents, soaps and alcohols. 'Solvent cracking' also affects the fatigue resistance of polymers when they are exposed to the fumes of such substances as toluene, acetone, benzene, etc. It is thought that the solvent is absorbed into the polymer chains under stress causing them to straighten and slip more easily.

13.6.3 *Strong sunlight*

Ultraviolet radiation and the presence of ozone can also cause some polymers to degrade and fail in fatigue at a lower than normal value (see Section 13.14).

SELF-ASSESSMENT TASK 13.1

1. (a) Explain why the allowable working stress is taken as a percentage of the yield stress and **not** as a percentage of the ultimate tensile stress.
   (b) If the yield stress for a material is 300 MPa, state what would be a generally acceptable allowable working stress and explain how you obtained this result.

2. Explain what is meant by the term 'creep' and state the conditions under which it occurs.

3. The life of an aircraft engine crankshaft was extended by removing the machining marks in order to improve the surface finish. Discuss the probable cause of the original failures.

4. Explain why a 'fatigue limit' (S_D) is not appropriate for some materials and an 'endurance limit' (S_N) is used instead.

13.7 The corrosion of metals

Corrosion is the slow but continuous eating away of metallic components by chemical or electrochemical attack. That this is costly and destructive will be vouched for by any motorist who has seen his valuable 'pride and joy' eaten away before his or her eyes as body rot sets in. Three factors govern corrosion.

- The metal from which the component is made.
- The protective treatment the component surface receives.
- The environment in which the component is kept.

All metals corrode to a greater or lesser degree; even precious metals like gold and silver tarnish in time, and this is a form of corrosion. Prevention processes are unable to prevent the inevitable failure of the component by corrosion; they only slow down the process to a point where the component will have worn out or been discarded for other reasons before failing due to corrosion. Let's now look at the three ways in which metals corrode.

- *Dry corrosion*   This is the direct oxidation of metals which occurs when a freshly cut surface reacts with the oxygen of the atmosphere. Most of the corrosion-resistant metals such as lead, zinc and aluminium form a dry oxide film which protects the metal from further atmospheric attack.
- *Wet corrosion*   This occurs in two ways:
  (a) The oxidation of metals in the presence of air and moisture, as in the rusting of ferrous metals.
  (b) The corrosion of metals by reaction with the dilute acids in rain due to the burning of fossil fuels (acid rain) – for example, the formation of the carbonate 'patina' on copper. This is the characteristic green film seen on the copper clad roofs of some public buildings.
- *Galvanic corrosion*   This occurs when two dissimilar metals, such as iron and tin or iron and zinc, are in intimate contact. They form a simple electrical cell in which rain,

polluted with dilute atmospheric acids, acts as an electrolyte. An electric current is generated and circulates within the system. Corrosion occurs with one of the metals (depending upon its position in the electrochemical series) being eaten away.

Fortunately the reactivity of a metal and the rate at which it corrodes is not related. For example, although aluminium is chemically more reactive than iron, as soon as it is exposed to the atmosphere it forms an oxide film which seals the surface and prevents further corrosion from taking place. On the other hand, iron is less reactive and forms its oxide film more slowly. Unfortunately, the iron hydroxide film (rust) is porous and the process continues unabated until the metal is destroyed. Table 13.1 indicates the average rates of corrosion for unprotected steelwork in various environments.

Table 13.1 *Rate of corrosion*	
Type of environment	Typical rate of rusting for low-carbon steel in temperate climates (mm/year)
Rural	0.025–0.050
Urban	0.050–0.100
Industrial	0.100–0.200
Chemical	0.200–0.375
Marine	0.025–0.150

13.8 Atmospheric corrosion

The cost of corrosion and its prevention is largely related to *atmospheric corrosion*. This is the form of corrosion that will be considered in this chapter. The corrosion prevention problems associated with chemical engineering, marine engineering, food processing, etc., are more specialised and the remedies that are resorted to, after a careful study of the environmental and service conditions, are correspondingly more specific and more complex.

Any metal exposed to normal atmospheric conditions becomes covered with an invisible, thin film of moisture. This moisture film is invariably contaminated with dissolved solids and gases which increase the rate of corrosion. The most common example of corrosion due to dissolved oxygen from the atmosphere is the rapid surface formation of 'red rust' on unprotected ferrous metals. This 'red rust' is a hydroxide of iron and should not be confused with the blue-black oxide of iron called 'mill-scale' which is formed by heating iron in dry air. This is the scale you find on hot-rolled steel bars and girders.

Figure 13.12 shows that air (oxygen) and moisture must both be present for rusting to occur. If either the air or the moisture is removed, rusting cannot commence. Once 'rusting' commences the action is self-generating – that is, it will continue even after the initial supply of moisture and air is removed. This is why all traces of rust must be removed or neutralised before painting, otherwise rusting will continue under the paint, causing it to blister and flake off. The rate of rusting slows down as the thickness of the rust layer increases, but as rain washes off the surface film of rust the process speeds up again. The cycle is continuous and, once started, is difficult to control.

Fig. 13.12 *Corrosion of steel: polished nails in (a) dry air or (b) air-free (boiled) water will not rust; but polished nails in (c) water containing air will rust*

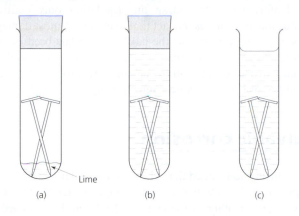

Lime

(a)   (b)   (c)

Atmospheric pollution rapidly increases the rate of rusting of iron and steel. It also attacks, but more slowly, corrosion-resistant metals such as copper and zinc. Lead is virtually unaffected and so is pure aluminium providing it has been correctly pretreated and is regularly washed clean. How the gases given off by burning fossil fuels and by industrial processes are converted into corrosive substances is shown in Fig. 13.13. Gases which become acids when dissolved in water are called *anhydrides*. In coastal areas the problem is aggravated by the presence of salt spray in the atmosphere.

Fig. 13.13 *Corrosive pollution: the waste gases combine with rain water to form sulphurous acid, sulphuric acid, carbonic acid, nitrous acid and nitric acid – which are highly corrosive in the dilute state and will attack unprotected metals*

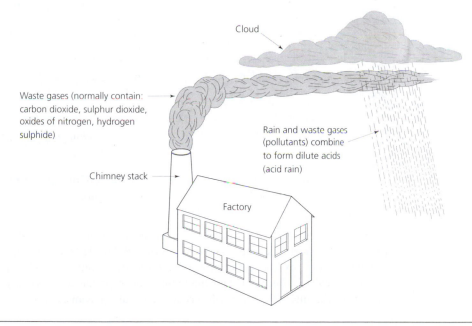

Cloud

Waste gases (normally contain: carbon dioxide, sulphur dioxide, oxides of nitrogen, hydrogen sulphide)

Rain and waste gases (pollutants) combine to form dilute acids (acid rain)

Chimney stack

Factory

The most important gaseous pollutant as far as corrosion is concerned is sulphur dioxide. Many 'clean' or 'smokeless' fuels are guilty of producing large quantities of this gas. Unfortunately the 'clean air' legislation only controls the visible products of combustion and ignores the fact that many of the most damaging pollutants are colourless gases which are invisible to the naked eye. It has been estimated that approximately 10 million tonnes of sulphurous and sulphuric acids are produced in the United Kingdom annually by the combustion of coal and coal products alone. This is about four times the amount of such acids produced intentionally.

13.9 Galvanic corrosion

It has already been stated that when two dissimilar metals come into intimate association in the presence of an electrolyte that a simple electrical cell is formed resulting in the eating away of one or other of the metals. Metals can be arranged in a special order called the electrochemical series. This series is listed in Table 13.2 and it should be noted that, in this context, hydrogen gas behaves like a metal.

Table 13.2 *Electrochemical series*		
Metal	*Electrode potential (volts)*	
Sodium	−2.71	*Corroded (anodic)*
Magnesium	−2.40	
Aluminium	−1.70	
Zinc	−0.76	
Chromium	−0.56	
Iron	−0.44	
Cadmium	−0.40	
Nickel	−0.23	
Tin	−0.14	
Lead	−0.12	
Hydrogen (reference potential)	0.00	
Copper	+0.35	
Silver	+0.80	
Platinum	+1.20	
Gold	+1.50	*Protected (cathodic)*

If any two metals come into contact in the presence of a dilute acid, the more negative metal will corrode more rapidly and will be eaten away. Figure 13.14 shows what happens during electrolytic corrosion. For example:

- In the case of galvanised iron (zinc-coated mild steel) the zinc corrodes away whilst protecting the steel. The zinc is said to be *sacrificial*. The destruction of the zinc coating can be retarded by the application of a protective paint film. The zinc forms an excellent 'key' for the paint, and the paint film keeps the wet corrosive atmosphere away from the zinc and prevents the occurrence of any galvanic action.

- In the case of tin plate, the mild steel base is corroded away if the coating is broken at any point. Hence the cut edges of tin plate should always be sealed by soft soldering or painting with a suitable lacquer. Marking out bend lines should be done with a soft lead pencil and not a scriber so that the tin-plate film is not cut.

Fig. 13.14 *Electrolytic corrosion: (a) protection by a sacrificial coating – coating is eaten away whilst protecting the base; (b) protection by a purely mechanical coating – coating only protects the base if intact, and if coating is damaged base is eaten away more quickly than if coating were not present*

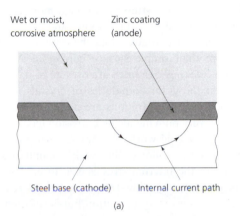

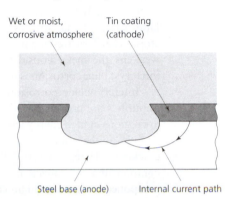

(a) (b)

13.10 Factors affecting corrosion

13.10.1 *Structural design*

The following factors should be observed during the design stage of a component or assembly to reduce corrosion to a minimum.

- The design should avoid crevices and corners where moisture may become trapped, and adequate ventilation and drainage should be provided.
- The design should allow for easy washing down and cleaning.
- Joints which are not continuously welded should be sealed, for example, by the use of mastic compounds or impregnated tapes.
- Where dissimilar metals have to be joined, high-strength epoxy adhesives should be considered since they insulate the metals from each other and prevent galvanic corrosion.
- Materials which are inherently corrosion resistant should be chosen or, if this is not possible, an anti-corrosive treatment should be specified.

13.10.2 *Environment*

The environment in which the component or assembly is to spend its service life must be carefully studied so that the materials chosen, or the anti-corrosion treatment specified, will provide an adequate service life at a reasonable cost. It would be unnecessary and uneconomical to provide a piece of office equipment which will be used indoors with a protective finish suitable for heavy-duty contractors plant which is going to work on construction sites in all kinds of weather conditions.

13.10.3 *Applied or internal stresses*

Chemical and electrochemical corrosion is intensified when a metal is under stress. This applies equally to externally applied and internal stresses, although more common in the latter case. Internal stresses are usually caused by cold working and, if not removed by stress-relief heat treatment, results in corrosive attack along the crystal boundaries. This weakens the metal considerably more than simple surface corrosion. An example of intercrystalline corrosion is the 'season cracking' of a brass after severe cold working.

Intercrystalline corrosion occurs at the grain boundaries of crystals not only when impurities are present but also when stress concentrations are present. Grain boundaries are regions of high energy levels, even in very pure metals, so corrosion tends to occur more quickly at the grain boundaries. Severely cold-worked α brasses are prone to 'season cracking'. Here, intercrystalline corrosion follows the grain boundaries until the component is no longer able to sustain the internal stresses due to cold working. The component then cracks. This can be prevented by a low-temperature stress relief annealing process. This low temperature does not cause recrystallisation but is sufficient to remove the locked up stresses by allowing the atoms to move small distances nearer to their equilibrium positions.

13.10.4 *Composition and structure*

The presence of impurities in non-ferrous metals reduces their corrosion resistance. Hence the high level of corrosion resistance exhibited by high-purity copper, aluminium and zinc. The importance of grain structure has also been mentioned above, and a fine-grain structure is generally less susceptible to corrosion than a coarse-grain structure. The inclusion of certain alloying elements such as nickel and chromium can also improve corrosion resistance – for example, the stainless steels and cupro-nickel alloys.

13.10.5 *Temperature*

For all chemical reactions there is a critical temperature below which they will not take place. Since corrosion is the result of chemical or electrochemical reactions, corrosion is retarded or stopped altogether at low temperatures. On the other hand, corrosion is at its worst in the hot, humid atmosphere of the tropical rain forests, and equipment for use in such environments has to be 'tropicalised' if it is to have a reasonable service life. High temperatures alone do not increase the rate of corrosion, and corrosion is virtually non-existent in arid desert areas of the world. Failure of mechanical devices in desert environments is due generally to the abrasive effect of the all-pervasive sand.

Let's now consider the methods by which corrosive attack can be reduced or prevented.

13.11 Metals which resist corrosion

It has already been made clear that metals combine with atmospheric oxygen and/or atmospheric pollutants to a greater or lesser extent. The following metals, which resist corrosion, react to form impervious, homogeneous coatings on their surfaces which prevent further corrosion from taking place, providing these coatings remain undisturbed.

13.11.1 *Copper*

When copper is exposed to atmospheric pollution for a long time, as for example when used as a roofing material, the surface develops a green coating or 'patina'. This 'patina' is caused by the action of the acids in the atmosphere attacking the basic oxide coating of the copper to form a protective layer of sulphate and carbonate salts, as shown in Fig. 13.15. Atmospheric acids do not attack the copper directly.

Fig. 13.15 *Formation of 'patina' on copper: (a) bare copper is attacked by atmospheric oxygen; (b) atmospheric acids plus basic oxide coating forms salts and water*

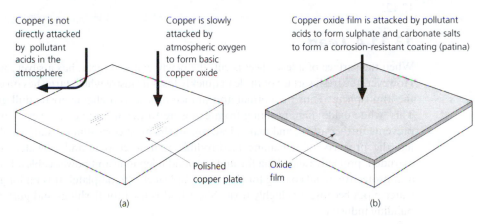

The 'patina' which forms on copper, giving it a protective skin, must not be confused with the darker green compound 'verdigris' which forms on copper by reaction with organic matter. Verdigris corrosion will, in time, completely destroy copper. Thus copper vessels used in the food-processing industry, and for cooking, are heavily coated with tin to prevent the formation of verdigris.

13.11.2 *Zinc*

The results of atmospheric attack upon zinc is very similar to copper, especially for outdoor purposes. In this instance it forms a protective carbonate coating which strengthens with time. This coating is grey in colour, not unlike the parent metal itself, and does not crack or peel off with any expansion or contraction of the metal due to temperature changes.

For this reason zinc is an excellent exterior building material and is widely used for flashings and roof cladding. Similarly it gives excellent protection when coated onto steel (see galvanising, Section 13.12). Its useful life tends to be foreshortened, particularly as a galvanising material, in areas of industrial pollution where there is a high concentration of sulphur compounds in the atmosphere. Therefore, galvanised hardware has a shorter life in an urban environment than it has in a rural environment, unless protected by sealing the galvanised surface with a paint film.

13.11.3 *Aluminium*

The reaction of aluminium and its alloys with the atmosphere is somewhat different to that of copper and zinc, but the effect of the initial reaction is, as in the previous cases, to retard further corrosion. Aluminium has a great affinity for oxygen and even highly polished aluminium surfaces quickly develop a thin, transparent film of aluminium oxide or 'alumina' which, if the metal is kept indoors, prevents further oxidation and retains the bright, polished appearance. However, exterior use of aluminium results in the oxide film thickening. When this happens the film becomes grey in colour but, when sufficiently developed, protects the parent metal from further attack. The oxide film on aluminium can be artificially thickened and coloured by a process called *anodising*, as described in Section 13.12.

13.11.4 *Lead*

When the surface of a lead sheet is cut, the newly exposed metal has a silvery appearance. However, the fresh surface of the lead quickly loses its lustre which tarnishes on exposure to the atmosphere. Thus the normal appearance of lead is a characteristic dull grey colour. This 'white oxide' film, resulting from exposure to the atmosphere, is very tenacious and prevents further attack and makes lead one of the most corrosion resistant of all metals. It has the property of remaining incorrodible year after year and is widely used as an extruded, protective sheathing for underground telephone and power cables. It is also used as a protective metal coating for sheet iron and steel – 'terne-plate'. It is no longer used for water pipes because it is highly toxic. Sheet lead is used for flashings and gutterings in the building industry.

13.11.5 *Stainless steel*

Stainless steels are outstanding corrosion-resistant metals. Unlike the metals considered so far in this section, stainless steels have high structural strength and high corrosion resistance even at the elevated temperatures that would melt other corrosion-resistant metals. One familiar stainless steel is known as '18/8' because the alloy contains 18 per cent chromium and 8 per cent nickel, the remainder being iron.

As with aluminium and lead, it is the formation of a complex oxide film which protects the surface from attack. The grades containing the additional alloying element molybdenum are recommended for architectural applications in heavily polluted areas as they will retain their polish and remain unchanged in appearance indefinitely. The presence of molybdenum and/or titanium also reduces the likelihood of 'weld-decay' in fabricated stainless steels. Such alloys are said to have been 'proofed'.

Stainless steels are also widely used in chemical plant and in food-processing and handling equipment where, to a wide range of substances at elevated temperatures, they combine corrosion resistance, high strength and non-toxic properties. Some grades can also be hardened and sharpened to a cutting edge for cutlery. No finishing treatment is required by stainless steels other than polishing or brushing. Although expensive in initial cost, it is economical in the long term as maintenance is minimal.

13.11.6 *Nickel*

This metal has already been mentioned as a constituent of stainless steels. It is also used for 'electroplating' onto other metals to provide a protective coating. As an alloying element, it forms corrosion-resistant alloys with copper, and these were discussed in Section 7.11.

13.11.7 *Chromium*

This metal has already been mentioned as a constituent of stainless steel. It is also used for 'electroplating' but, unlike nickel which can be deposited directly onto a variety of base metals, chromium is used as a finishing film over a nickel-plated substrata. It has a more pleasing colour than nickel and does not tarnish.

SELF-ASSESSMENT TASK 13.2

1. Briefly explain the difference between the terms 'corrosion' and 'erosion' as applied to metals.

2. Describe the main differences between 'wet' and 'dry' corrosion.

3. Briefly explain why:

    (a) the tin film coating of tin plate must not be broken during forming processes
    (b) the zinc coating of galvanised wire netting is eaten away before the wire it is protecting

4. Explain why sheet copper is resilient to urban pollution when used as a roofing material.

13.12 Chemical inhibition

When unpacking a new set of slip gauges or similar highly finished equipment, a sheet of VPI paper is often found in contact with the metal. These letters stand for *Vapour Phase Inhibitor* and the paper is impregnated with chemical substances which give off a vapour neutralising any corrosive atmosphere which may remain within the packing or which may penetrate the packing. Any residual moisture, which could form an electrolyte and cause electrochemical corrosion, is rendered more alkaline so that the metal becomes passive.

13.13 Cathodic protection

The protection of iron by galvanising has already been introduced. Zinc is more corrosion resistant than iron and is less likely to be corroded. In the event of corrosion occurring, the zinc is eaten away rather than the iron or steel. This is because the iron is electropositive compared with zinc – that is, the iron forms the cathode of the electrolytic cell which exists when iron and zinc are in contact in a corrosive environment. Thus the use of zinc to protect ferrous metals is called cathodic protection. Microporous zinc electrodes are often fitted to the steel hulls of ships adjacent to the bronze propellers. This prevents electrolytic attack of the hull (the highly saline sea-water being the electrolyte) and the easily replaced zinc block is eaten away instead. This is *cathodic protection* and the zinc is said to be *sacrificial*.

13.14 Protective coatings

Where the natural protective film which builds up on the surface of a metal is inadequate, or where (as in the case of most ferrous metals) that film is porous, additional protective coatings have to be applied. For short-term protection, oiling or greasing up a component may be adequate. However, in the long term this is not satisfactory for the following reasons:

- The protective film will dry up and no longer seal the surface from the corrosive environment.
- Oils and greases tend to absorb water from the atmosphere and corrosion can take place under the oil or grease.
- Cheap oils and greases often contain active sulphur and acids that will themselves attack and corrode the very metal surfaces they are supposed to be protecting.

Therefore more reliable and permanent protective coatings are required and some of the more common will now be considered briefly. For a more detailed study of the causes of corrosion and its prevention, see *Engineering Materials*, Volume 2.

13.14.1 *Hot dipping*

The finished components are cleaned, fluxed and immersed in molten metal (usually zinc). This is the principle of 'hot-dip galvanising'. It has the advantage over electrolytic galvanising of providing a thicker coating and, since it is applied to the finished work, it seals all the raw edges. In the case of agricultural hardware such as buckets, feeding troughs, etc., it has the added advantage of sealing the joints and making them watertight.

13.14.2 *Electroplating*

This process is used to coat components with a thin layer of metal which may be protective, decorative or both. The components to be plated are immersed in a solution called an electrolyte since it allows the passage of an electric current. The component(s) to be plated are made the *cathode* of the cell – that is, they are connected to the *negative* terminal of a low-voltage, heavy current, d.c. supply as shown in Fig. 13.16. To complete the circuit, *anodes* are immersed in the electrolyte and connected to the *positive* terminal of the supply.

Fig. 13.16 *Electroplating*

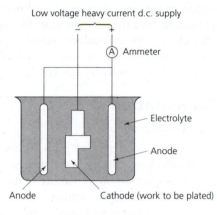

Low voltage heavy current d.c. supply

(A) Ammeter

Electrolyte

Anode

Anode   Cathode (work to be plated)

Upon the passage of an electric current metal ions are transferred from the electrolyte onto the surface of the cathode (work) to build up a protective metallic coating. The strength of the electrolyte is usually maintained by dissolving metal ions from the anode. The greater the current and the longer it flows, the greater will be the mass of metal electroplated onto the work.

13.14.3 *Cladding*

Originally this process was developed for the production of 'Sheffield Plate'. Prior to the development of electroplating a thick coating of silver was applied to a sheet copper base by pressure welding a 'sandwich' of the metals between rolls. Nowadays, the same principle is used for the production of such metal composites as 'Alclad' where a corrosion-resistant pure aluminium facing is pressure welded to a high-strength aluminium alloy base, as shown in Fig. 13.17.

Fig. 13.17 *Cladding: (a) section through clad metal composite; (b) proportional reduction of clad billet*

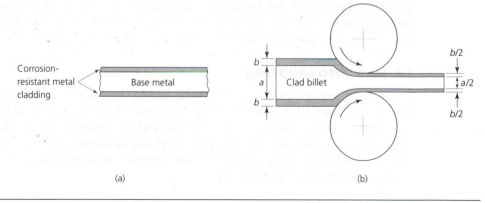

Corrosion-resistant metal cladding

Base metal

b

a

b

Clad billet

$b/2$

$a/2$

$b/2$

(a)   (b)

13.14.4 *Spraying*

Corrosion-resistant or wear-resistant metals are melted in an oxyacetylene flame or by electric arc and sprayed onto the component surface by compressed air blast. This process is generally used to build up worn bearing journals on shafts, although it is sometimes used on new components where a surface with special properties is required. The surface is rough and uneven and has to be finished by grinding.

13.14.5 *Cementation*

There are three process in common use:

- *Sherardising* Cleaned and pickled plain carbon steel components are placed in a rotating steel barrel along with zinc metal powder and heated to 370 °C for up to twelve hours. A zinc film is formed on the surface of the components and this, in turn, forms an excellent 'key' for subsequent painting.
- *Calorising* As above, except that aluminium powder is used at a higher temperature. This provides protection against high-temperature oxidation rather than against ambient temperature corrosion (e.g. car exhaust systems).
- *Chromising* (Not to be confused with *chromating* – see below.) The work is baked in a mixture of aluminium oxide and metallic chromium powder at temperatures in excess of 1200 °C in a hydrogen atmosphere to prevent oxidation. This is a very expensive process giving extreme protection against corrosion. It is used for chemical plant.

13.14.6 *Anodising*

Aluminium and aluminium alloy components are cleaned and degreased, after which they are etched, wire-brushed or polished depending upon the surface texture required. The work is then made the anode of an electrolytic cell (see electroplating, where the work is the cathode) and a direct current is passed through the cell. The electrolyte is a dilute acid and varies with the finish and protection required. Colours may be integral or applied subsequently by dyeing. The purpose of the treatment is to increase the thickness of the natural, protective oxide film and improve the corrosion resistance of the metal.

13.14.7 *Chromating*

(Not to be confused with *chromising* – see above.) This process is applied to magnesium alloy components which are dipped into a hot solution of *potassium dichromate* to form a hard, protective oxide film. The surface is then sealed by the use of *zinc chromate paint*, after which it is finished with a decorative paint.

13.14.8 *Phosphating*

Traditionally, phosphating processes were known by such names as Parkerising, Bonderising, Granodising and Walterising. More recently the whole range of phosphating processes have been standardised under BS 3189. The components to be treated are pickled and degreased so that they are chemically clean and immersed in the hot, phosphating solution. The metal surface is converted into complex metal phosphates, chromates or oxides depending upon the process used.

Finally they are sealed by oiling, waxing or painting. The process is widely used as a pretreatment for painting as it provides a good 'key' for the paint and prevents corrosion occurring under the paint film. Since phosphates have a high lubricity, phosphate conversion is often applied to bearing surfaces for gears, tappets, pistons, etc., and as a pretreatment for metals which are subjected to severe cold-drawing processes.

13.14.9 *Plastic coating*

Plastic coatings can be functional, corrosion resistant and decorative. It provides the designer with a means of achieving:

- Corrosion resistance.
- Cushion coating (up to 6 mm thick).
- Electrical and thermal insulation.
- Flexibility over a wide range of temperatures.
- Non-stick properties.
- Permanent protection against weathering and atmospheric pollution.
- Reduction in maintenance costs.
- Resistance to corrosion by a wide range of chemicals.
- The sealing of welds and porous castings.
- Decorative finishes.

The method of application depends upon the type of plastic to be used, its form of supply and the type of component to be coated. It is essential, in all cases, that the surface of the work piece is free from grease, dirt, oxidation or any other form of contamination. This process is described more fully in *Engineering Materials*, Volume 2.

13.14.10 *Painting*

A paint system consists of:

- *Primer* This is designed to 'key' firmly to the component being treated and usually contains a corrosion inhibitor to prevent corrosion taking place under the paint film as this would cause the paint film to blister and flake off. As stated above, galvanising, sherardising or phosphating is often used as a pretreatment to provide an improved 'keying surface' for the primer and to prevent corrosion taking place under the paint film.
- *Undercoat* This builds up the paint film and gives 'body' to the final colour. The undercoat must be built up to the required thickness by a number of thin coats rather than one thick coat. It is finally 'flatted down' to provide a suitably smooth surface for the finishing coat.
- *Gloss coat* This is the finishing coat. It provides a hard, tough protective film which has a high gloss and the final colour. It seals the previous films against moisture and pollutants and is the most important element in the system for protecting the surface of the work.

Paint may be applied by brushing, spraying and dipping. It may be dried naturally or by stoving. The paint system must be chosen to suit the application process as well as the degree of protection required and the material and surface to be treated. Like all finishing processes, it is a highly complex technology if satisfactory protection is to be provided at a reasonable cost. Also, like all finishing processes, success depends upon the preparation of the surface to be treated to ensure it is clean, free from corrosion and has a suitable texture to receive the finish. These finishes and their application are considered in greater detail in *Engineering Materials*, Volume 2.

13.15 Plastic degradation

Polymers degrade when exposed to heat, sunlight and weathering. They are also susceptible to 'solvent cracking' when exposed to the fumes of such substances as toluene, acetone and benzene (see also Section 13.5).

Degradation is usually accompanied by colour change, deterioration in mechanical properties, cracking and surface crazing. These changes usually occur as the result of chain scission and cross-linking. Polymers based on olefins are particularly susceptible to such degradation. Overheating during the moulding process can also cause degradation.

Stabilisers can be added to reduce the rate of degradation, and these include:

● *Anti-oxidants* to prevent atmospheric oxidation.
● *Anti-ozonants* to prevent ozone attack.
● *Ultraviolet filters* to absorb the ultraviolet radiation of sunlight.

The ultraviolet radiation of sunlight has sufficient energy to break down most of the chemical bonds in polymers. To prevent this happening, 'carbon black' is a pigment used as a radiation absorber in polythene film and sheet, but unfortunately it absorbs visible light as well as ultraviolet light.

Where the black colour is unacceptable, benzophenones and benzotriazoles are used as ultraviolet filters. However, although they are much more expensive, their spectrum transmission curves, which are shown in Fig. 13.18(a), compare very favourably with the ideal curve shown in Fig. 13.18(b).

Fig. 13.18 *Ultraviolet light absorption additives: (a) transmission curves for actual ultraviolet light absorption additives; (b) transmission curve for an ideal ultraviolet light absorption additive*

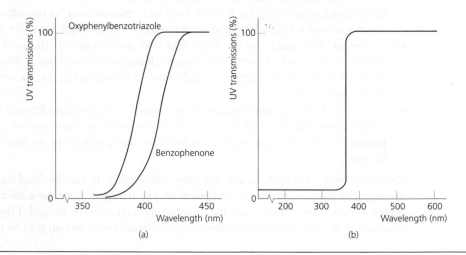

The weathering properties of polymers must be carefully considered when selecting a material for outdoor use. Materials which give excellent results indoors may be quite unacceptable out of doors. For example, nylon is highly water absorbent and this reduces its suitability for use out of doors. A material which has a low cost may require more frequent replacement than one only a little more expensive. Although the former may be quite acceptable for indoor purposes, for outdoor applications it may be more economical in the long term to use the rather more expensive but more weather-resistant and longer lasting material.

SELF-ASSESSMENT TASK 13.3

1. Describe the essential difference between electroplating and anodising.

2. Describe the essential difference between electroplating and the cementation processes.

3. Differentiate between the processes of chromating and chromising.

4. Give **three** examples where you might expect to find plastic-coated metal components in your home.

5. Explain the importance to a paint system of:

   (a) the primer
   (b) the gloss coat

EXERCISES

13.1 With the aid of a diagram distinguish between primary, secondary and tertiary creep and discuss the importance of creep in highly stressed engineering components operating at elevated temperatures.

13.2 With the aid of sketches describe how a fatigue test may be performed and the appearance of a typical S–N curve derived from such a test. Explain how the S–N curve is interpreted.

13.3 With the aid of sketches explain what is meant by the terms *fluctuating load*, *pulsating load* and *alternating load* when applied to fatigue testing.

13.4 Discuss **four** factors which affect the fatigue resistance of a component and give examples of where these factors may apply in practice.

13.5 Describe what is meant by the terms:
   (a) atmospheric corrosion
   (b) galvanic corrosion

13.6 Discuss **four** factors which affect the corrosion of metallic components and assemblies.

13.7 (a) Describe in detail how metals such as aluminium, copper and zinc resist corrosion.

(b) Discuss the main factors responsible for the degradation of polymeric materials in service.

13.8 (a) Explain how low-carbon steel components can be protected by the process of galvanising and explain why the coating is said to be 'sacrificial'.

(b) Explain why the surface of a metal component must be carefully prepared prior to painting and how this may be performed.

(c) Explain how paint systems are classified according to the method by which the paint dries and hardens, and describe the various coats which are used to build up a complete paint system.

13.9 State briefly what is meant by the terms:

(a) dry corrosion

(b) wet corrosion

(c) galvanic corrosion

13.10 Explain why sherardised steel work in an industrial, urban environment should also be painted.

14 Select materials for engineered products

The topic areas covered in this chapter are:

- Criteria for the selection of materials for engineered products.
- Property requirements.
- Processing requirements.
- Economic requirements.
- Failure of materials in service.
- British Standard Specifications.
- Case studies and lessons from failures.

14.1 Criteria for the selection of materials for engineered products

At the start of this book we looked at a machine tool casting, a connecting rod forging and an electric cable connector. We established that the choice of materials for these products was based on the properties of the materials and upon economic considerations. In the succeeding chapters we looked at the composition and properties of a wide range of materials. We looked at how these properties can be modified by heat treatment and by processing, and we looked at how material properties can be tested.

To round off this book let's complete the circle of events and look once more at material selection – this time, however, in greater detail and with the benefit of the knowledge gained from the preceding chapters.

The main criteria to influence the selection of materials for any particular engineered product are summarised in Fig. 14.1. It show that these criteria can be grouped under three main headings, namely:

- Property requirements.
- Processing requirements.
- Economic requirements.

Ultimately our final choice will involve a compromise. There is rarely, if ever, an ideal solution.

Fig. 14.1 *Selection of materials*

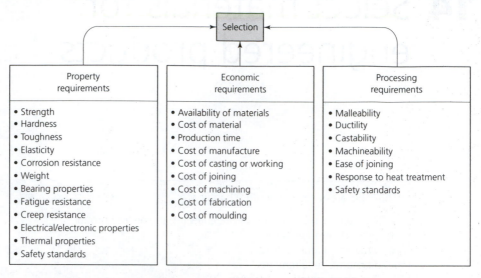

14.1.1 *Property requirements*

In order for a material to be successful in the service for which it has been chosen it must have the most suitable properties. It would be no good trying to build an aircraft entirely from a nickel–chromium steel alloy just because this material is very strong. It would be far too heavy and, even if the aircraft could get off the ground, it could not carry enough passengers to be an economic proposition. So, as a compromise, aluminium alloy would be used because of its lower density. The clever part lies in making the airframe strong and safe despite the relatively lower strength of this material compared with alloy steel. A *compromise* between weight and strength has to be made.

Again, choosing a material for, say, the back axle of a motor vehicle just because it has a high tensile or compressive strength would be foolish if the toughness of the material was not also considered. Such an axle would require toughness in order to withstand the impact loading on the axle resulting from normal driving conditions over a variety of surfaces. A *compromise* between strength and toughness is required.

There would be little point in manufacturing a kitchen sink unit from dead mild steel despite the fact that this would be the cheapest and easiest material to process. It would soon become rusty and prove to be unsuitable for this application. Although more expensive and more difficult to process, stainless steel is a far better material for making sink units. It is corrosion resistant, pleasing in appearance, easy to keep clean and hygienic. A *compromise* between cost, corrosion resistance and ease of processing has to be made.

14.1.2 *Processing requirements*

Materials, particularly metals, can be classified as either casting materials or wrought materials and the process suitability and ease with which each can be cast or worked can be very important. It is equally important to consider how easily a cast or wrought material can be machined, and the quality of surface finish that can be obtained, when using standard machine shop equipment and tooling.

The ease of joining materials must also be considered, depending upon whether it is to be welded, brazed, soldered or subjected to adhesive bonding. It is also necessary to consider the effect of the high temperatures involved on the structure and properties of the material when such joining methods are used. Aluminium cables would be lighter and cheaper than copper cables and aluminium is more readily available. However, aluminium has a slightly lower conductivity so the cables would have to be thicker for the same current-carrying capacity and this would offset some of the weight advantage. Further, aluminium is difficult to solder, so crimped cable lugs would have to be used, as shown in Fig. 14.2. Since these are mechanical joints, there is always the possibility of, corrosion setting in, resulting in high resistance joints causing failure through overheating.

Fig. 14.2 *Crimping: (a) crimped terminal; (b) hand-operated crimping tool*

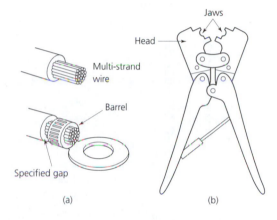

If the material is to be heat treated, consideration must be given to its ability to respond to the treatment prescribed and the point in the processing at which heat treatment should be carried out.

In addition, it is necessary to consider both the duty and the shape of the component in order to select a material with the required processing characteristics, whether it be metallic or non-metallic. For instance, consider the grass box of a mowing machine. Traditionally these boxes were made from sheet steel and painted to improve their appearance and provide a level of corrosion resistance. In time, they became dented, shabby and the paint became chipped allowing rusting to set in. However, the duty did not demand the strength of steel and, when impact-resistant plastics such as ABS became available, mowing machine grass boxes were made as one-piece plastic mouldings. Such mouldings are tough and rubbery, and can withstand minor collisions with pieces of rockery and other obstructions without damage and without losing shape. They are light in weight, easily processed in the large quantities required for the domestic market, and are corrosion resistant. Large, heavy-duty commercial mowing machines for sports grounds still have mainly fabricated sheet steel grass boxes as the production volume does not warrant the tooling for plastic grass boxes.

14.1.3 *Economic requirements*

Economic considerations are important and should take into account the total cost, which comprises the cost of raw material plus the cost of its manipulation, machining, joining and finishing. Unless a specific material is required by a customer, the overall cost of a number of alternative materials, both metallic and non-metallic, should always be compared, together with their availability, by every manufacturing engineer. A compromise has to be met between cost, availability, ease of processing and fitness for purpose.

The availability of a particular material is important as this will influence its cost. Figure 14.3 shows the availability of some common metals. However, a material is not necessarily inexpensive just because the mineral ores from which it is derived are plentiful. The extraction and processing costs must also be taken into account. Adverse geographical conditions and the political stability (or lack of it) in the country where such mineral deposits are found can also influence material prices.

Fig. 14.3 *Metallic constituents of the earth's crust by weight*

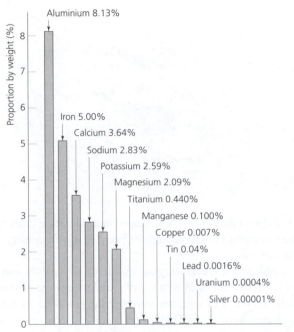

Environmental issues are also becoming increasingly important. One often reads of attempts to use battery electric vehicles instead of petrol or diesel vehicles to reduce pollution. However, until we have a satisfactory substitute for the lead–acid accumulator or the nickel–cadmium accumulator this is not a realistic solution. For instance, Fig. 14.3 shows that lead is one of the metals that are least available to us, and cadmium doesn't even register on the figure. There just would not be enough of these metals to go around if we all had battery electric vehicles. Further, the extraction of these metals requires large amounts of energy and produces large amounts of toxic waste. Environmentally, the usage of such metals should be discouraged rather than encouraged.

14.2 Failure of materials in service

It is essential that engineered products are designed in such a way that any stresses that are encountered in service are insufficient to cause failure. We have already looked at the need to introduce a *factor of safety* (see Section 13.1) where the design stress does not exceed 50 per cent of the yield stress.

However, despite such allowances, components still fail in service and designers now recognise that operating conditions produce brittleness, fatigue, creep and/or environmental attack which, if ignored, will ultimately lead to failure. For instance, during the Second World War, cargo ships were being built quickly and cheaply using welded construction in place of the traditional riveting. A disturbing number of these ships broke up under storm conditions in the North Atlantic despite being correctly stressed. It wasn't known at that time that the cold – near arctic – conditions of the Atlantic winter caused embrittlement and failure of welded joints in the materials then being used.

14.2.1 *Fracture*

Fracture can be classified either as brittle fracture or ductile fracture, and depends upon the stress at which it occurs in relation to the elastic/plastic properties of the material. This is shown in Fig. 14.4.

Fig. 14.4 *Types of fracture: (a) brittle; (b) ductile*

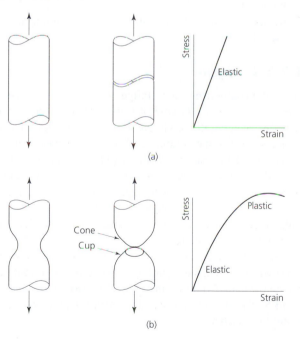

In *brittle fracture*, as shown in Fig. 14.4(a), failure occurs suddenly and before any appreciable plastic deformation takes place. The fracture follows the paths between adjacent crystal planes. Brittle fracture generally occurs in materials with the weakest atomic bonding, such as cast iron, glass and concrete. Metals with body-centred-cubic and close-packed-hexagonal structures are also susceptible to brittle fracture. Brittle fractures usually start at some discontinuity in the material or component which results in a point of stress concentration. This is not always the case and the 'pitting' caused by corrosion, for example, can also act as 'stress raiser' in the metal.

Ductile fracture takes place at some stress figure above the yield point of the material so that some plastic flow precedes failure. The resulting fracture is of the 'cup and cone' variety shown in Fig. 14.4(b). It is associated with face-centred-cubic crystals such as those found in low-carbon steels. This type of fracture occurs in destructive tensile tests, as discussed in Section 11.2.

Since the fracture occurs in the plastic zone of the tensile test curve, it is obviously the result of a design fault or severe overload. As previously stated, the maximum working stress should not exceed 50 per cent of the yield stress for the material, let alone reach the stress levels required to produce plastic deformation.

The type of fracture produced is dependent mainly on the nature of the material and its particular lattice structure. However, other factors can also influence the onset of fracture, for example:

- The rate of application of stress (see Section 11.5).
- Environmental and temperature conditions (our previous example of ships breaking up in the North Atlantic).
- The amount of cold working the material has received during any previous processing.
- The shape of the component and whether or not it has sharp corners and sudden changes of section.
- The surface finish of the components (machining marks are incipient cracks (stress raisers)).

14.2.2 *Fatigue*

Failure of a component due to fatigue has already been considered in Section 13.4. It can be seen to be a particular case of brittle fracture were failure occurs under conditions of repeatedly changing stress. The change can be in magnitude, direction or both.

Aircraft wings and rotating shafts are examples of members subjected to alternating stresses, and where similar stresses are to be encountered careful consideration at the design stage and during subsequent maintenance is extremely important if fatigue failure is to be avoided. Failure occurs below the normal levels of stress for static loading and is difficult to predict. Materials used for components subjected to stress variations should be subjected to Wohler type tests, as described in Section 13.3, to determine the susceptibility of such materials to fatigue failure. The design criteria for such components should be based on case history as well as mathematical modelling. Prototypes of critical components, such as aircraft wings and engine mountings, should be tested to destruction in test rigs designed to simulate the load conditions met with in service. The stages of fatigue failure are shown in Fig. 14.5. The fracture always starts from a point of stress concentration caused by a discontinuity such as:

- Poor surface finish.
- A crack.
- A machining mark.
- An inclusion (foreign particle) in the metal.

Fig. 14.5 *The stages of fatigue failure*

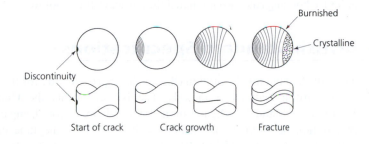

14.2.3 *Creep*

This has been introduced in Section 13.2. Unlike fatigue failure, creep occurs under constant stress conditions. It usually occurs at elevated temperatures since slip in the lattice structure is easier at such temperatures. Since creep leads to dimensional change, it becomes an important design factor in steam and gas turbines. The materials selected for the rotor and stator turbine blades must be carefully chosen to minimise this effect. It would be catastrophic if the rapidly rotating blades of the rotor touched the stator blades due to dimensional change through creep.

This is why 'nimonic alloys' with their high nickel content are widely used for turbine blades. As well as being heat resistant, nimonic alloys are resistant to creep. Like all materials which offer improved creep resistance, nickel has a close-packed crystal structure. All materials with close-packed-hexagonal or face-centred-cubic crystals have good creep resistance.

At the other end of the scale, lead, with its long plastic range at normal ambient temperatures and its low melting point, has a very low creep resistance. Sheet lead was widely used for roofing churches and other ancient public buildings. During restoration it was normal to find that the solid lead had flowed down the roof so that it was thick at the bottom by the eaves but paper thin at the top by the ridge. This was entirely due to creep over a time span of several hundred years. Thermoplastic type polymer materials are also susceptible to creep.

14.2.4 *Corrosion*

Corrosion and its prevention has been introduced in Chapter 13. The selection of materials that are corrosion resistant is an important design consideration. The conditions encountered by materials used on site or in the hostile environment of a chemical plant are quite different to those encountered in a domestic or office situation. It would be uneconomic to use the same level of corrosion resistance and anti-corrosion coatings in all these situations. Each application must be considered on its own merits.

Many of the problems associated with 'body-rot' in cars immediately after the Second World War have been eliminated by better design to allow drainage and ventilation of hollow box section members. Also anti-corrosion treatment of the body panels before painting, and the use of improved paint systems, have all added to the improved life of car bodies. Changes in materials have also helped, with the use of high-impact polymers in those areas most susceptible to spray and salt during winter motoring. Continuous adhesive bonding has also eliminated the crevices left by spot welding, bolting and riveting.

14.3 British Standard Specifications

The British Standards Institution (BSI) is the approved body for the preparation and issue of the British National Standards. The BSI represents the United Kingdom in the International Standards Organisation (ISO) and, apart from being actively engaged in the preparation of international standards, it is at the same time ensuring compatibility between international and British Standards wherever this is possible.

The British Standards Institution aims to coordinate the efforts of producers and users in the improvement, standardisation and simplification of engineering and industrial components and materials. It does this by:

- Setting standards of quality.
- Setting standards of dimension (size).
- Setting standards of performance.
- Promoting the general adoption of British and/or international standards.

British Standards have been referred to from time to time throughout this text and, although every endeavour has been made to ensure that they are currently valid at the time of writing, they should only be considered as examples. The standards are continually being reviewed and updated and, for definitive information and data, all product designers and users of engineering materials should regularly check that they are using the latest editions of the appropriate standards. So let's see how we can ensure that we are using an up-to-date Standard.

First, if a Standard has a BSEN number instead of just a BS number it shows that the Standard is compatible with European Common Market requirements. You should also refer to the latest edition of the British Standards Catalogue.

The revisions to British Standards range from relatively minor amendments to the complete withdrawal of earlier standards and their replacement by new standards reflecting technological changes and harmonisation with ISO requirements. The currency and validity of any Standard can be identified as set out in the following notes, which are based on the preface to the British Standards Catalogue.

14.3.1 *Basic details of entries*

The list of BSI publications in their catalogue is arranged in numerical order, within each series. The series can be identified from the alphabetical characters which precede the number of the Standard. For example: BS AU = automobile series, or BS EN = European Standards adopted by the BSI. There is also an alphabetical index of titles at the end of the catalogue.

The current use of EN to signify compatibility with European standards must not be confused with the original edition of BS 970 for wrought steels, which is now obsolete. All the steel specification codes were prefaced EN. Unfortunately, although this original Standard has been obsolete for many years, old habits die hard, and the original EN specifications still appear on drawings and in the catalogues and stock sheets of many major steel stockholders.

Current publications can be identified by the use of **bold type** for the number of the publication and its title. The revision of any publication automatically supersedes all previous editions of the publication. Only current editions are listed. Withdrawn publications can be identified by the use of *light type* for the number and title of the publication and the word 'withdrawn' in parentheses (brackets).

14.3.2 *Definition and interpretation of entries*

An entry in the catalogue or as part of the heading of a Standard has various *elements*. In the example shown in Fig. 14.6, the various elements are labelled A–J. The key explains each element.

Fig. 14.6 *Interpretation of the validity of a British Standard Specification (source: BSI)*

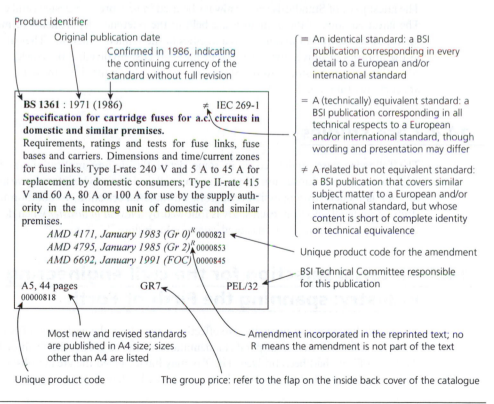

14.3.3 *Amendments*

All separate amendments to date of despatch are included with any main publication ordered. Prices are available on application. With the next reprint of the Standard the amendment is incorporated into the text which then carries a statement drawing attention to this and includes an indication in the margin at the appropriate places on the amended pages.

14.3.4 *Review*

The policy of the BSI is for every Standard to be reviewed by the technical committee responsible not more than five years after publication, to establish whether it is still current and, if not, to set in hand appropriate action. Circumstances may lead to an earlier review. When reviewing a Standard, a committee has *four* options available:

- *Withdrawal*   This indicates that the Standard is no longer current.
- *Declaration of obsolescence*   This indicates by amendment, that the Standard is not recommended for use in new equipment, but needs to be retained for the servicing of existing equipment that is expected to have a long working life.
- *Revision*   This involves the procedure for new products.
- *Confirmation*   This indicates the continuing currency of the Standard without full revision. Following confirmation of a publication, stock copies are overstamped with the month and year of confirmation.

The latest issue of Standards should always be used in new product designs and equipment. The latest editions of the Standards are held in the reference departments of most of the main public libraries of major towns and cities in the United Kingdom. They are also held in the libraries of universities. Because of the high cost involved, most colleges of further education limit their stocks to the Standards specifically required for the courses that they are currently running.

14.4 Case studies

The following case studies cover in some detail the design implications and material selection criteria of this section. Hopefully, you will find them to be informative and helpful to you, as models, during assignment work on material selection. In addition to the case study of material selection, each study is followed by a *lesson in disaster* when design faults and incorrect material selection caused catastrophic failure.

14.5 Material selection for the civil engineering industry: spanning the Firth of Forth

The Forth Bridge was a necessity. No effective route along the east coast of Scotland was possible until those two great parallel indentations of the coastline, the Firth of Forth and the Firth of Tay, had been bridged. The Tay may have seemed the greater obstacle because of its greater width, but this was not the case since the Firth of Tay is comparatively shallow.

It was possible to sink foundations at regular intervals across the whole width of the Firth of Tay, erect supporting columns and build an ordinary girder bridge. This bridge was of exceptional length but presented no particular difficulty from the point of view of design or construction. Bridging the Forth was an entirely different matter. Immediately north of Edinburgh the Forth has a depth of 50 metres and more. Whatever type of bridge was to be used, it was impossible to avoid spans of unprecedented length. The bridge was built slightly to the north-east of Edinburgh where the Firth of Forth has a width of 1600 metres (over 1 mile). At this point the mainstream islet of Inchgarvie served as a foundation for the central and most massive cantilever of the bridge. No intermediate supporting piers could be sunk at the time it was built because of the great depth of the water. The technology was not then available. A schematic diagram of the structure is shown in Fig. 14.7. In addition to the bridge proper, fifteen high approach viaducts also had to be built and a number of smaller masonry arches at the extreme ends. The total overall length was 2500 metres (1.5 miles approx.).

Fig. 14.7 *Schematic layout of the Forth Bridge*

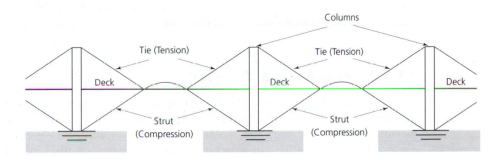

The choice of candidate materials was not difficult. At that time the only relatively high-strength metals that were available in the quantities required were low-carbon (mild) steel (tension and compression members) and cast iron (compression members). As you will read later, the cast iron columns used on the Tay bridge collapsed with serious loss of life and were not considered for the Forth bridge. This narrowed the candidate metals to steel and wrought iron. Wrought iron had superior corrosion-resistance properties, but steel was chosen because of its superior strength over wrought iron and its availability in the quantities required. Corrosion-resistant steel alloys containing chromium were not at that time available. Copper alloys would not have had a suitable strength to weight ratio, neither would they have been available in sufficient quantity. Their cost would also have been prohibitive. Corrosion-resistant light alloys were non-existent at that time.

Masonry (stone-work) was used for the lower arches of the approach roads and for the foundations. This was an established technology at the time and had a proven record of strength, service life and durability. Obviously corrosion was not a problem. Reinforced concrete was not an established technology at that time.

A major design problem for such a large structure was the effect of thermal expansion and contraction due to atmospheric temperature changes. The overall expansion in the entire length of the three cantilevers was 2.5 metres and this had to be allowed for by various means. Certain joints in the immense lattice work were designed as hinges. The bed plates that transmitted the weight of the cantilevers to the stone piers were provided with oval holes for the holding-down bolts. The bed plates are maintained in a highly polished and greased condition to allow for ease of movement. The rail tracks also had to have corresponding expansion joints to allow movement on this scale, yet provide a smooth running surface.

The matter of corrosion has already been mentioned. This was further aggravated by the salt spray in the air. The only solution at that time (and still is today for this structure) was regular and scrupulous cleaning and painting. The area of steelwork to be painted is 135 acres and it takes a gang of 45 men three years to apply one coat of paint weighing 54 tonnes. By the time they have finished it is ready for cleaning and painting again. At the same time that painting is taking place the maintenance team also inspect the structure for defects which they report back to the engineers so that repairs can be carried out before catastrophic failure occurs.

Thousands of cubic metres of masonry and cement were used, together with 54,000 tonnes of steel. The members were fabricated and joined using 6.5 million rivets. Suitable welding technology was not available at that time.

14.4.2 *Lessons of a disaster*

Although wider than the Forth, the Tay was spanned first because its shallowness made the design and task easier. Unfortunately lightweight cast iron columns were used after the fifteenth masonry pier, and the superstructure was strangely slender for such a massive bridge. The effects of gale force winds funnelling up the Firth of Tay were ignored. Also the cast iron columns were filled with Portland cement concrete to increase their compressive strength. The concrete expanded as it set and cracked the cast iron outer skin causing considerable weakness despite the use of wrought iron bands to repair the damage.

In December 1879, in gale force conditions, disaster struck. The wind resistance of the high girder spans and a train passing over the bridge caused major lateral forces on thesupporting columns. Such bending forces could have been resisted by steel but not by the cast iron piers. Remember that cast iron and concrete, although very strong in compression, are very weak in tension. This weakness, aggravated by the cracking of the cast iron (already mentioned) caused the bridge to collapse. Seventy-three passengers were hurled to their death as the bridge and the train were flung into the waters of the Tay.

This terrible accident produced a number of valuable lessons. The official inquiry found that the bridge was badly designed, *material selection was incorrect*, and that the construction and subsequent maintenance was inadequate and of poor workmanship. Insufficient allowance had been made for the greatest strain any bridge has to bear – namely, the lateral forces caused by wind pressure. Needless to say, cast iron columns filled with Portland cement were *never* again used for the piers supporting a giant viaduct.

14.6 Material selection in the aviation industry: airframes

The design and manufacture of aircraft involves a series of compromises. The need for great size, which may involve heavily loaded structures requiring great strength, may conflict with the equally pressing need for lightness in the interests of economical operation. Further, the need for great range, which will require allocating a lot of space and weight-carrying capacity for the accommodation of fuel. This will inevitably reduce the capacity to carry passengers, the *pay load*. Therefore, material selection for aircraft airframes has always been (and still is) a search for materials with improved strength to weight ratios.

The earliest airframes were being produced before light alloys based on aluminium were as highly developed as they are today. Fabric-covered wooden frames were used, liberally braced with steel wires. During the First World War, the Junkers aircraft company used corrugated aluminium alloy skins over wooden frames with considerable success to give a more durable aircraft. Anthony Fokker, the Dutch aircraft designer, used conventional fabric-covered wooden frames for the fuselages (bodies) of his aircraft but used stressed plywood skin wings. It was not until 1926 that William B. Stout – an American – designed the first all-metal aircraft. This was the Ford Tri-motor based on the Junker's corrugated aluminium alloy skin philosophy. Nowadays all aircraft are of all-metal construction.

We have established that an aircraft has to be as light as possible, yet strong enough to withstand the effects of changing air pressures associated with altitude changes and cabin pressurisation, and the constant buffeting from turbulence in the airflow over the wings, body, engine housings and control surfaces. In addition, the airframe has to withstand the impact loadings associated with taking off and landing. The airframe must also be rigid to ensure the alignment of the control surfaces under varying load conditions. No one material can satisfy all the requirements and the candidate materials must be compatible so as not to set up the conditions that could lead to electrolytic corrosion where the materials come together. With changing altitudes comes changing temperatures and pressures, which can lead to the formation of condensation moisture in the airframe. This will be exacerbated by overseas flights to the humid conditions of tropical airports. Corrosion then becomes an important design factor. The material is required in large sheets in order to provide a skin with as few joints as possible in the interests of strength and an uninterrupted surface airflow. The sheet material must be easy to flow form in the cold condition to the shape of the aircraft. We can summarise our requirements as:

- Strength and stiffness.
- Low density.
- Availability and cost.
- Ease of production.
- Corrosion resistance.

Let's now look at what materials meet our requirements. Our choice will lie between several materials or combinations of materials which we will refer to as the 'candidate materials'. The material or materials selected must have good *structural efficiency*. They must satisfy all the requirements listed above. We will select materials with the *highest yield stress*, the *highest value of elastic modulus*, yet have the *lowest density*.

Having selected a short list of candidate materials to satisfy these requirements we must consider their toughness – that is, their resistance to failure under dynamic (impact) loading and their resistance to crack propagation. This is important because if cracks start to grow and are left undetected and unrepaired they can easily grow to the *critical crack length*. Once they have reached the critical crack length, catastrophic failure can occur without any further increase in stress. Tough materials are not only more resistant to cracking in the first place but they can tolerate a greater critical crack length before failure occurs.

The cyclical stressing of the airframe and its skin due to the cabin pressurisation system, and the vibration caused by the engines and air turbulence, can result in premature failure by *fatigue* at stress levels that would leave the same material unaffected under static load conditions. To satisfy the requirements of its 'Certificate of Airworthiness', the critical components of an aircraft have to be changed at a specified number of flying hours depending upon the duty imposed on the aircraft. The greatest stressing occurs during take off and landing, so 'short-haul' aircraft which take off and land frequently have to have critical components changed after fewer flying hours than are required for 'long-haul' aircraft.

Corrosion resistance is important since the eating away of metal due to corrosion *reduces the cross-sectional area* of the components being corroded. Since stress is load per unit area, any reduction in cross-sectional area leads to a corresponding increase in stress for any given load. When we look at the electrochemical series, we see that metals such as gold, silver and copper are less likely to corrode than metals at the other end of the series such as magnesium. The exception to this rule are the *passive* metals which form a protective oxide layer. Such metals are aluminium, titanium and stainless steel. Gold, silver and platinum will not be considered because they have poor structural efficiency which more than offsets their good corrosion resistance properties.

The physical properties also have to be considered. These include the density which will affect the strength to weight ratio, and the electrical conductivity since the frame of the aircraft is usually used as an earth return path for the electrical installation to save the weight of using earth return conductors. The thermal conductivity and the coefficient of thermal expansion are also factors of importance because of the wide range of temperatures an aircraft encounters during every flight. Modern supersonic aircraft such as Concord and many military aircraft become very hot due to air friction, so for these special aircraft strength at high temperatures has to be considered.

Finally we have to consider the ease with which the material(s) can be formed and fabricated to the rounded and complex shapes of a modern aircraft. We must also consider the availability and cost of the material(s) chosen if our aircraft is to be a commercial proposition.

Limiting ourselves to the airframe of a passenger airliner, let's first consider materials with good structural efficiency – that is, high strength and low density. Here we might consider such candidate materials as:

- Aluminium alloys.
- Titanium.
- Magnesium.
- Carbon fibre.

High-strength materials such as steels and nickel alloys such as the 'nimonic' alloys have too high a density and their use is limited to high-strength engine components and hinge and pivot pins for the control surfaces. They are also used in the more highly stressed components of the landing gear such as axles, pivots pins, hydraulic and brake components. As well as being too heavy, steels (with the exception of stainless steel) are not *passive* metals and corrode easily.

Corrosion-resistant materials such as stainless steel, monel metal and copper and its alloys have too high a density. However, carbon fibre and glass fibre-reinforced plastic materials are increasingly finding their way into stressed members. They are not as yet widely used in passenger-carrying airliners except for cabin trim, but they are used widely in the manufacture of small aircraft and competition gliders. However, synthetic adhesives are increasingly used even in the most highly stressed members since there are no holes needed (as when riveting) and the stresses are transmitted uniformly along the joint and not at the points where the rivets are sited. Rivets and rivet holes can be considered as stress raisers and potential sources of failure. Also the use of adhesives *insulates* the various materials from each other electrically and removes a cause of electrolytic corrosion.

So let's examine the materials in the bulleted list above.

Aluminium alloys

These have a relatively high strength to weight ratio and are an established material for aircraft manufacture. There is much documented information on their performance in service. They are relatively good electrical conductors and are relatively corrosion resistant, particularly when anodised. They are reasonably tough and fatigue resistant and have been successfully used over a number of years. They are easy to cast, forge and cold form from sheets. They are difficult to joint by welding and, traditionally, riveting is the most widely used method of jointing. However, synthetic adhesive bonding is now being increasingly widely used with all the advantages listed above. Unfortunately the structural efficiency of aluminium alloys falls off as the service temperature increases. Therefore, for supersonic aircraft, such as Concord, alternative materials have to be found that can withstand the high temperatures generated by the friction of the airflow over the skin of the aircraft.

Titanium alloys

These are excellent materials with a high strength to weight ratio, good corrosion resistance, high structural efficiency that is maintained at elevated temperature, and good electrical conductivity. They are also tough and resistant to fatigue. Unfortunately they are difficult to fabricate and are very costly. Their prime use is mainly in high-performance military aircraft which are smaller than passenger airliners and the initial cost of the materials is only a small part of the overall cost. The only passenger aircraft containing large amounts of this metal in its construction is Concord because of its need to maintain its structural efficiency at the temperatures associated with supersonic speeds.

Magnesium alloys

These are lighter than aluminium alloys and have a good structural efficiency. Unfortunately they have a poor resistance to corrosion, their structural efficiency falls off rapidly as their temperature increases and they are more flammable than the aluminium alloys. Therefore they will be rejected.

Glass fibre-reinforced plastics and carbon fibre-reinforced plastics

These are commonly known as GFRPs and CFRPs, respectively, and have been discussed previously. They have a good strength to weight ratio and good structural efficiency. They are easily fabricated to the most complex shapes and do not corrode. In fact they are totally unaffected by all atmospheric conditions and most solvents and pollutants. Their main disadvantage is their lack of electrical conductivity which results in the build up of static electrical charges. If these charges are not dissipated on landing, the ground crews could be at risk from electric shocks, and the possibility of sparks during refuelling represents a serious fire hazard. The inherent electrical conductivity of metal aircraft enables these charges to be easily dissipated on landing. Where large amounts of reinforced plastics are used, it is usual to incorporate a percentage of metal fibres in the reinforcement to improve their electrical conductivity.

Summary

Summing up, our candidate materials for this application would be aluminium alloys, GFRPs and CFRPs for large civilian aircraft and the larger and slower flying military aircraft. Titanium alloys would be the most suitable for high-performance, supersonic military aircraft.

14.6.1 *Lessons of a disaster*

The De Havilland Comet was the first significant jet-powered passenger airliner when it flew in 1949. Unfortunately, Britain's lead in this field was dashed when two of these aircraft exploded in mid-air in 1953 and 1954 with total loss of life. This was found, in one of the disasters, to have been due to the explosive disintegration of the pressurised passenger cabin. This was caused by the cyclical pressure changes associated with cabin pressurisation causing fatigue failure. The site of this failure was a small and poorly designed aperture in the structure. A sharp corner in this aperture behaved as a stress raiser which caused a fatigue crack to spread rapidly and catastrophically throughout the fuselage structure.

The Comet disasters led to some concentrated research into metal fatigue and the findings taught aircraft designers some important lessons.

- Certain aluminium alloys are no longer used in load-bearing structures.
- The permitted overall stress levels have been reduced.
- Much greater attention is given to design details to avoid sharp corners that could lead to stress concentrations.
- Fatigue cannot be entirely eliminated, so all aircraft structures are now designed to one of two fatigue categories: 'fail safe' and 'safe-life'.
- A fail-safe structure will not completely fail even if a crack develops in one member of the structure.
- A safe-life structure has a specified fixed life and then has to be automatically replaced as part of a routine maintenance programme, even though it may show little or no evidence of wear or fatigue.

14.7 Selection and application of materials

The use of the foregoing information in the selection and application of engineering materials for various components is an essential exercise for all engineers. The following exercises are based upon the selection of materials for the various components that make up the typical lathe tailstock shown in Fig. 14.8. Components requiring materials not included in this text will be disregarded. In all the following exercises discuss the reasons for your choice of materials in the same way as the examples introduced in the case studies.

Fig. 14.8 *Tailstock*

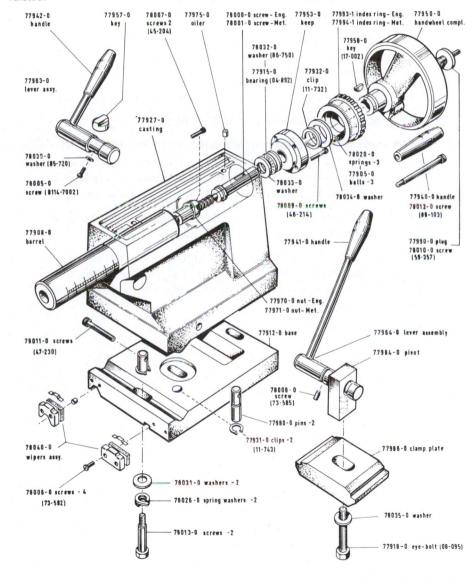

77942-0 handle
77957-0 key
78007-0 screws 2 (45-204)
77975-0 oiler
78000-0 screw – Eng.
78001-0 screw – Met.
77953-0 keep
77993-1 index ring – Eng.
77994-1 index ring – Met.
77950-0 handwheel compl.
78032-0 washer (86-750)
77958-0 key (17-002)
77915-0 bearing (04-892)
77932-0 clip (11-732)
77963-0 lever assy.
77927-0 casting
78039-0 washer (85-720)
78005-0 screw (8114-7002)
78033-0 washer
78020-0 springs -3
77905-0 balls -3
78034-0 washer
77940-0 handle
78012-0 screw (88-103)
78009-0 screws (46-214)
77908-0 barrel
77941-0 handle
77990-0 plug
78010-0 screw (59-357)
77970-0 nut – Eng.
77971-0 nut – Met.
78011-0 screws (47-230)
77912-0 base
77964-0 lever assembly
77984-0 pivot
78008-0 screw (73-585)
77980-0 pins -2
77931-0 clips -2 (11-743)
77986-0 clamp plate
78040-0 wipers assy.
78006-0 screws - 4 (73-582)
78031-0 washers -2
78026-0 spring washers -2
78035-0 washer
78013-0 screws -2
77918-0 eye-bolt (08-095)

14.1 Select a suitable material for component 77927-0 (Casting). Describe the precautions which must be taken:
(a) to ensure uniform cooling and minimum internal stresses
(b) to ensure dimensional stability in service

14.2 Select a suitable material for component 77942-0 (Handle).

14.3 Select a suitable material for component 77970-0 (Nut), paying particular attention to its anti-friction properties and the relative wear between the screw and the nut.

14.4 Select a suitable material for component 77908-0 (Barrel) and describe any heat treatment which may be necessary to ensure toughness and wear resistance.

14.5 Select a suitable material for component 77993-1 (Index ring) and describe any surface finish treatment which may be required to ensure that the engraved scale is easy to read and wear resistant.

14.6 Select a suitable material for component 77950-0 (Handwheel).

14.7 Select suitable materials for 78040-0 (Wiper assembly).

14.8 Select a suitable material for component 77953-0 (Keep) paying particular attention to cost and ease of production since this component is not subject to wear. From the drawing, this component would appear to have its bore bushed with an anti-friction metal to support component 78000. Specify a suitable bearing metal for this bush.

14.9 Select a suitable material for the component 77986-0 (Clamp plate). (*Note*: This component is subject to bending forces.)

14.10 Select a suitable material for component 78000 (Screw) and specify any heat treatment which may be necessary. Pay particular attention to any problems of cutting the screw, wear and cost of replacement relative to the nut with which it is associated.

Index